A ZOO LOVER'S COMPANION

BEASTLY BEHAVIORS

Pandas Turn Somersaults, and Crocodiles Roar

A WATCHER'S GUIDE TO HOW ANIMALS ACT AND WHY

JANINE M. BENYUS

Illustrations by Juan Carlos Barberis

▲ Addison-Wesley Publishing Company

Reading, Massachusetts   Menlo Park, California   New York   Don Mills, Ontario
Wokingham, England   Amsterdam   Bonn   Sydney   Singapore   Tokyo   Madrid
San Juan   Paris   Seoul   Milan   Mexico City   Taipei

OTHER BOOKS BY JANINE M. BENYUS

The Field Guide to Wildlife Habitats of the Eastern United States
The Field Guide to Wildlife Habitats of the Western United States
Northwoods Wildlife: A Watcher's Guide to Habitats

Many of the designations used by manufacturers and sellers to distinguish their products are claimed as trademarks. Where those designations appear in this book and Addison-Wesley was aware of a trademark claim, the designations have been printed in initial capital letters.

Library of Congress Cataloging-in-Publication Data
Benyus, Janine M.
    Beastly behaviors : a zoo lover's companion : what makes whales
  whistle, cranes dance, pandas turn somersaults, and crocodiles roar
  : a watcher's guide to how animals act and why / Janine M. Benyus ;
  illustrations by Juan Carlos Barberis.
        p.   cm.
    Includes bibliographical references and index.
    ISBN 0-201-57008-4
    1. Animal behavior.   2. Wildlife watching.   3. Zoo animals–
  -Behavior.   I. Barberis, Juan Carlos.   II. Title.
    QL751.B368   1992
    591.51—dc20   92-5381
                                                             CIP

Jacket design by Diana Coe
Jacket illustration by Pedro Gonzalez
Text design by Joel Avirom
Set in 11-point ITC Berkeley by TSI Graphics

1 2 3 4 5 6 7 8 9–MU–95949392
First printing, August 1992

THIS BOOK IS DEDICATED TO THOSE
WHO STRUGGLE TO SAVE WILD LANDS

CONTENTS

ACKNOWLEDGMENTS

To gain perspective on a subject this vast, a writer must stand on the shoulders of giants. Hoisting me aloft were the thousands of researchers who braved white-outs, monsoons, bone-chilling nights, and scorching noons to bring home news of animal behavior. I am indebted to them, as well as to the zoo professionals who shared with me their enthusiasm and fierce devotion to wildlife conservation.

The technical reviewers were: Dr. George Archibald, Director, International Crane Foundation, Wisconsin; Dr. Cheryl Asa, Reproductive Biologist, St. Louis Zoo; Dr. John Behler, Curator of Herpetology, Bronx Zoo; Dr. Daryl Boness, Research Zoologist, National Zoo; Dr. Donald Bruning, Curator of Ornithology, Bronx Zoo; Teresa DeLorenzo, Research Affiliate, Metro Washington Park Zoo, Portland; Dr. Sue Ellis-Joseph, Conservation Education Specialist, Minnesota Zoo; Dr. Susan Evarts, Post-doctoral Associate, Bell Museum of Natural History, University of Minnesota; Dr. Martha Hiatt-Saif, Senior Trainer, New York Aquarium; Dr. Brian Joseph, Veterinarian, Minnesota Zoo; Dr. Jeffrey Lang, Professor, University of North Dakota; Dr. Frank McKinney, Curator of Ethology, Bell Museum of Natural History, University of Minnesota; Dr. Jill Mellen, Conservation Research Coordinator, Metro Washington Park Zoo, Portland; Dr. Jackie Ogden, Research Fellow, Center for Reproduction of Endangered Species, San Diego Zoo; Lisa Rappaport, Behavioral Researcher, Riverbank Zoo, Georgia; Bruce Read, Curator of Mammals, St. Louis Zoo; Dr. Michael Robinson, Director, National Zoo; Conrad Schmitt, Associate Curator, Cheyenne Mountain Zoo; Peter Shannon, Curator of Birds, Audubon Park, Louisiana; Dr. Steve Sherrod, Director, Sutton Avian Research Center, Oklahoma; Dr. Beth Stephens, Research Biologist, Zoo Atlanta; Dr. Lisa Stevens, Mammal Collection Manager, National Zoo; and Scott Swengel, Assistant Curator of Birds, International Crane Foundation, Wisconsin. Their guidance was invaluable, and I thank them.

Beastly Behaviors was lovingly birthed by talented midwives. I was fortunate to work with visionary agent Jeanne Hanson, editor extraodinaire Nancy Miller, the illustrious Juan Carlos Barberis, the conscientious Melissa Turk, design greats Diana Coe and Joel Avirom, and the ever-calm production maestro John Fuller. The fact that these people cared about the book made the whole process a pleasure.

Although writers are not a particularly social species, I have been blessed with a warm circle of friends and family. My thanks to Cynthia Robinson for lending a generous ear during the early stages. For typing gigabytes of notes, editing with a fine eye, and buoying me with her company, I am deeply thankful for Mary Ann Hatton. Laura Merrill, who was as constant as the northern star, gave me the immeasurable gift of hope and laughter, for which I cannot begin to thank her enough. Laura also pored over the galleys, as did Daniel Benyus and La Rue Moorhouse. Even Eight Ball, the enigmatic tomcat who crowded me at the keyboard, played a part. Any errors or pearls of wisdom found on these pages were probably typed in by his paws.

Finally, I want to thank the moonlit mountains of Montana. Living here, I've learned that there really is no place like home and that every last, wild acre is worth protecting.

BEASTLY
BEHAVIORS

A koala sleeps soundly in its eucalyptus home.

THERE'S NO PLACE LIKE HOME

It was the wee hours of the morning when Neil Armstrong took his first small step on the moon. My parents and sister, after gallantly waiting up for hours, had yawned their last yawn and fallen asleep. I was only 11, but I was wide awake in front of the television, anxious not so much for Armstrong to set foot on the moon as for him to turn around and tell us what the earth looked like from there.

I know I wasn't the only one. Twenty-three years later, I've noticed that we don't hang pictures of the lunar landscape on our office walls. It's not the Sea of Tranquillity that distracts us from our work, but that homeward-bound shot—that misty, turquoise globe sailing all by itself in a sea of black. Like any good symbol, the earth's portrait speaks straight to the heart.

Look, it says, we're all in one boat here, and it's the only one we've got.

Knowing that hasn't stopped us. In a few reckless decades, we've punched a hole in the ozone above Antarctica and sprung thousands of other environmental leaks that are just as hard to fix. Reaching the moon didn't put our problems any farther behind us. All it made us realize was how unusual the earth is, how it alone contains the right levels of light, temperature, gases, and water to keep us alive. This is where we evolved, and every cell in our body is designed to cope with and take advantage of this place. Even if we could find a similar planet to beam ourselves up to, wouldn't we miss the plants and animals and even the insects we grew up with?

After all, we're not the only ones who evolved in this petri dish of soil, sun, air, and water. Millions of species live here, and all of them, from amoebas to zebras, were shaped by the peculiar demands of their home habitat. The earth turned out to be a wildly diverse place. Rain puddles, jungles, ice caps, and searing deserts were mothers to very different sorts of biological inventions.

The wonderful thing about a zoo is that you can see an entire globeful of nature's inventiveness in a few accessible acres. You can see, smell, hear, and be splashed by the earth's best and brightest, the products of a cornucopia of habitats. In the new zoos, exhibit designers are working hard to re-create these habitats, both for our education and for the animals' well-being. They've found that the best way to call up the full range of natural behaviors is to put animals in the environments they evolved in. What a tall order! An arctic icescape in San Diego, a coral reef in Minnesota, a steamy tropical jungle in the Bronx. These exhibits are amazing, to be sure, and the animals seem more at home here, which of course delights the designers and zoo-keepers. But, like space travel, it also humbles them.

Planning for life on another planet is a lot like exhibit design; you quickly learn how hard it is to re-create a habitat. You can choose the right plants and animals, but how do you stage the interplay of soil and water, predators and prey, allies and enemies? How do you re-create the perfect, perplexing *balance* of it all? In the end, it's not how habitats look, but what they do that is magical. Wild, healthy habitats do more than support life; they challenge it, inspiring new life forms and fine-tuning the ones that already exist. No matter how sophisticated our exhibits become, we can't touch that.

Maybe that's why so many people in the zoo world are committed to saving the genuine article. They know that an animal is part and parcel of the forest, grassland, desert, or ocean that it evolved in. Without its native habitat, even the fastest, strongest, most clever organism is a sorry fish out of water. By the same token, if we stop and think what will happen if we finish off the rain forests, eat up more ozone, and keep on reproducing like there's no tomorrow, something dawns on us. We realize that we're just as vulnerable as the disappearing sea turtles, spotted owls, elephants, and manatees; we're as bound to this irreplaceable place as they are. If we suddenly found ourselves out of home waters, we'd be gasping for breath too.

Ask an astronaut.

When I look back, I think maybe the real Giant Leap for humankind was for a few people to travel far enough away from earth to feel homesick for the first time. The planet they flew back to is not as serene as it looked from a distance. It is warping under the G-force of too many people, and among those feeling the effects most keenly are the incredible animals you will meet in these pages.

As you read their stories, I hope you'll be entertained and enamored and feel like staying up late to learn about them. But even more than that, I hope you'll feel a little homesick for the paradise that these animals evolved in. I'd like to think it's not too late to patch our holes, bail like crazy, and make this turquoise vessel seaworthy again. In the meantime, we need to honor and protect all the species that still sail with us, and dedicate our next Giant Leap to them.

Housed in habitats instead of cages, today's zoo animals are displaying more natural behaviors than ever before.

WHAT'S NEW WITH ZOOS?

If you haven't been to a zoo since your eighth-grade field trip, you're in for a wild surprise. Zoos have weathered a tough soul searching over the last 15 years, and the good ones have re-created themselves from the inside out. They've sprung the cages and turned the animals loose in startling simulations of their home habitats, some so lifelike that you'll swear you're being stalked by that leopard in the leaves or that wolverine on the hill. The authenticity seems to agree with the animals as well. They get to vine-swing in tropical forests, dive in living coral reefs, dabble in creeks, and burrow to their heart's content in prairies. In a wonderful turnabout, it's the zoo visitors who are now hemmed in by railings, not the residents.

What a welcome change! Instead of feeling your heart sink at the sight of a despondent gorilla, you'll feel it race as a barrel-chested silverback explodes into view, then disappears in the greenery. You may have to search for him there, but that's a good sign; it means gorillas and other animals are blending with the landscape in a way they never could when marooned on tile floors and manicured lawns. For the first time, zoo animals have the space and privacy to prowl, howl, court, build nests, and defend their territories. Besides being more at home, the animals are also in better company. No longer the lone representative of their species, they now romp in herds and pods, troops and bevies. Some have even decided to put down roots, and if you search closely, you'll see kits, calves, joeys, and cubs, some of which are the first of their kind born in captivity.

While these new exhibits make zoos all the more entertaining, their real agenda is to educate. By immersing us in the animal's world, they show us how the animal evolved in tandem with its habitat, and how, from crest to claw, it is adapted to live where it lives. The exhibits also bring out natural behaviors in the animals, prompting them to act more like themselves than they ever have in captivity. In fact, scientists who once scorned zoos are now bringing binoculars and clipboards to study animal behavior close up. The findings are helping zoos fulfill what has become their foremost mission: to successfully breed the endangered species that have landed in their lifeboat.

Just how did zoos move from their days of bars-and-shackles to the new look and outlook of today? To answer that, we have to rewind a few million years, to the very beginnings of our ancient relationship with animals.

Our Love Affair with Animals

It was no more than an evolutionary eyeblink ago that our apelike ancestors were crawling on their hands and knees in pursuit of big, dangerous, delicious animals. For 5 million years (99% of our time on earth), our survival hinged on finding animals that we could eat. Stalking an animal meant knowing everything about its habits: where it slept, where it drank its daily water, how fast it would run once it caught wind of us. By necessity, we became astute observers of our fellow creatures and learned a respect that comes with intense study of a subject. Although we've stepped further and further from the wild in our last 5,000 years of agriculture and industry, an ancient awe still stirs deep within us. It's that to-the-bone shiver we feel when we see a wolf trot lightly from the woods or hear the thin song of a rising whale.

What is this uncanny ability animals have to amaze, delight, and at times, frighten or repulse us? Part attraction, part fear, and part admiration still make us curious about wild animals even though we no longer need them to fill our stomachs. In this country alone, 115 million people pour through the turnstiles of zoos each year. That's more people than go to all the professional sporting events combined! Worldwide, 357 million people attended the 757 zoos listed in the *1984 International Zoo Yearbook*—the equivalent of all the people in the United States, Canada, France, and the United Kingdom. And it's not just school groups either. For every wide-eyed child, there are three astonished adults rushing to the rail to see the dolphins leap.

Thousands of generations and evolutionary steps later, humans are still fascinated by wild animals.

These days, animals in zoos represent more than just a shadow of our ancestral past. They are the last ambassadors of a world that is rapidly becoming less and less wild. The red-alert sirens are screaming, and, for the first time in history, we are admitting to our destructiveness and grappling to right the wrongs. Zoos that were once primarily amusement parks are now on the front lines of that fight, working to brighten the future for rare animals in their keeping. To track this sea change in the zoo world, we have to go back thousands of years, to the creation of the very first zoos.

Outgrowing the Old Zoos

Zoos were originally menageries kept by royalty as a show of power and wealth. After all, if you could arrange to ship a giraffe 1,500 miles down the Nile (as Queen Hatshepsut of Egypt did 3,500 years ago), you could probably get anything you wanted. Thousands of years later, when zoos finally opened their doors to the public, they became sources of pride for local communities. The goal was to have one example of as many different species as possible, like a stamp collection of animals. Various zoos guarded their collections jealously, and if one zoo knew the secret of, say, keeping a rare turtle alive, it wouldn't dare tell another zoo. With each collecting trip to Africa or Asia, the zoo's acquisitions grew. It didn't matter that the animals were confined to cramped, barren cages or that they often died of disease or stress; it was easy enough to get replacements from the wild. Although the majority of the public stayed silent about the bad conditions at zoos, not everyone was satisfied.

A small revolution began at the turn of this century when Carl Hagenbeck of Germany did away with the cages and put his animals on wide green lawns surrounded by hidden moats. Visitors applauded the change, but it took some time for the exhibits to catch on in other zoos. One of the problems was that zookeepers found it hard to control diseases that spread in the soil and grass of outdoor exhibits. A cement box was easier to hose down, and advocates claimed it was better in the long run for the animal's health, if not for its spirits. After World War II, medical technology improved and so did animal health care. Many zoos began to feature Hagenbeck-like exhibits, at least for their hoofed animals and cats of the African plains. Still, the majority of zoos put animals behind bars and replaced their losses with captures from the wild. These zoos enabled people to see animals, but instead of engendering respect, the pacing jaguars and psychotic polar bears often provoked pity and a feeling of guilt among visitors.

In the 1960s, this collective guilt became sharper, focused through the lenses of writers, activists, and opinion-makers. People were becoming aware that we share this planet with other creatures that have every right to be here. As our consciousness of animal rights matured, true animal champions suddenly had a dilemma. They wanted to see wild animals and reconnect with their roots, but the deplorable conditions in many zoos kept them away. Some took issue with the restricted quarters and with the practice of capturing animals from the wild. Others protested the heavy emphasis on vaudeville-style entertainment such as chimp tea parties, penguin parades, and walrus ballets. These rumblings of discontent came not only from outside, but also from within the zoo staffs themselves. It became clear that zoos were losing credibility and would soon be forced to evolve or to shut their gates for good.

Big Shoes to Fill

Meanwhile, as the debate about captive animals raged, sobering news about their wild counterparts started to trickle in. Scientists reported that many of the species thriving in zoos were just barely hanging on in the wild. For some species, they predicted that the numbers in captivity would soon be greater than those in the wild. Overnight, it seemed, zookeepers were caring for some of the most precious cargo on earth. Realizing this, zoo directors and conservationists from around the world started to draft a whole new role for zoos.

The days of collecting wild specimens were over. Zoos began to experiment with breeding, and by the mid-1980s, more than 90% of zoo mammals were born in captivity. Although this meant that zoos could resupply themselves rather than deplete wild populations, it was not enough. As populations on the outside teetered, zoos began to realize that they might someday host the only examples of a species left on the planet. When that day came, the responsibility, not only for individuals, but for an entire species—its physical and genetic health and its possible reintroduction to the wild—would rest squarely on the zoo world's shoulders.

That day has come. Extinctions are occurring at a rate unprecedented in the planet's history, rising from a loss of one species every 5 years in 1850 to a current estimated rate of one species an hour, or nearly 44,000 species lost every 5 years. You may notice the proliferation of Vanishing Species signs in front of the animal exhibits at your zoo. The prognosis is not good for many of these faltering species; experts predict that, within the decade, the last of the last will perish from the wild forever.

Preventing extinction isn't just a matter of protecting animals from poachers. The real problem is that wild animals are running out of space, and, where they do have space, their habitats are being degraded. It's hard to find a place on earth that hasn't felt the grip of human greed and its alibi, the so-called license to "subdue and dominate." Even the woolliest of wildernesses are now surrounded by a lasso of development that cinches tighter each day. Tropical deforestation (at 100 acres a minute) combined with pollution (possibly warming the world by as much as 3 to 8 degrees Fahrenheit) and the prolific human population machine (producing a quarter of a million new people a day) threaten to push many species over the edge. In fact, conservative estimates say that 25% of all species will be in trouble within the next few years. Faced with statistics like that, zoos have become modern-day arks, called as Noah was to protect the future of entire evolutionary lines.

Protecting and breeding these animals, as important as that is, is just the first part of the solution. The ark is only as good as the promise that someday the flood will subside and we can release the passengers to suitable habitats. Unfortunately, suitable habitats are getting harder and harder to find. If the ark is to be anything more than a pipe dream, we must devote ourselves wholeheartedly to the other part of the solution: *we must stop the flood of habitat destruction at its source.* The fact is, after all is said and done, a species of wild animal doesn't really belong in a zoo. It belongs in the wild, living by its wits and evolving in the face of natural challenges.

Zoos are in a unique position to tell this story to millions of people who are smiling, enjoying themselves, and primed to learn and listen. Zoos can tune us in and turn us on. They can make us angry about the assault on wild habitats and show us how to harness our energy in votes, dollars, lifestyle changes, and volunteered talent. In the meantime, while we struggle to salvage what's left of the wild, the new zoos can act as safety nets, centers for research, and places where we can enjoy the exquisite pleasure of spending time in the company of animals.

ZOOS AS SAFETY NETS

Biologists agree that the only way to really save animals is to save their habitats. Although bits and pieces of wild land are being protected throughout the world, the current trend is a net loss of habitats. Even the 1% of the world that is set aside in national parks and refuges is not free of encroachment by poachers and people who are desperate for food and fiber. As writer Colin Tudge reminds us, hungry people have no time for rare animals.

The human population passed a landmark of 5 billion in 1987. Demographic experts say that if we manage to avoid epidemics, wars, and ecological collapse, there will be 10 billion of us on earth by the middle of the twenty-first century. At that point, they predict, our population will level off for 5 centuries and then begin to decline, relieving the pressure on what is left of the land. Here's where zoo people prove to be true optimists. They hope to maintain a reservoir of animals and a bank of frozen sperm and eggs so that habitats, if they recover, can be repopulated. On the fragmented habitat "islands" where animals have survived, sperm and eggs from zoo animals could breathe fresh genetic air into stale populations.

There are plenty of pessimists who have no hope that any of this will be possible. They point to high mortality rates in previous attempts to reintroduce species to the wild. After living in captivity, they say, predators will forget how to hunt, and prey species will lose their wariness. Imprinting on humans may cause animals to have no fear of human hunters, and perhaps leave them unable to have healthy sexual relationships with members of their own species. Captive breeders counter all these claims, pointing to examples of feral dogs that have returned to the wild after being house pets for several generations. To ease this reentry for zoo animals, breeders are experimenting with reintroduction training programs and trying to raise young without imprinting them. While they admit that many captive-bred animals will die when released into the wild, they feel that even a small survival rate is better than none.

Though captive breeding is a number one priority at many zoos, it is not always obvious to the casual visitor. Much of the work is done behind the scenes at off-site facilities like the National Zoo's Front Royal Breeding Center in Virginia. Here animals are raised outdoors in proper social groups and bred according to a sophisticated genetic plan that involves not only the Center's animals, but animals at zoos throughout the world. You see, unlike Noah, we can't simply herd the animals two by two onto the ark and pray for sun. If we hope to release these animals into the wild someday, we have to keep them genetically sharp enough to be able to react to changes in their environment. A species is at its best when enough individuals are

Thanks to captive breeding efforts, rare creatures such as mountain tapirs are giving birth to a new generation.

breeding to keep the gene pool diverse. Having a diverse gene pool is like having a large toolbox filled with tools that you don't use daily, but that someday you might need. If the population of breeding animals is too small or genetically limited, it leaves fewer tools to solve problems with, should disaster strike. Besides having less genetic ingenuity, small populations may start inbreeding, which can perpetuate rare genetic weaknesses.

To avoid these pitfalls, zoo managers are cooperating in a worldwide strategy called the Species Survival Plan. For the purposes of breeding, all captive rhinoceroses, for instance, no matter where they reside, are considered part of the same population. Stud records, showing who was born to whom, are stored in a database called ISIS (International Species Information System), housed in Minnesota. When a female is ready to go into heat, ISIS is used to find a male with "fresh" genes, even if he must be flown from Moscow to Chicago for the occasion. The goal of this family planning is to enlarge the breeding population in captivity and then subdivide the population among zoos (a precaution against epidemics). If the matchmakers at ISIS have their way, the genetic structure of zoo populations will resemble that of wild populations, so that one day, we'll have a relatively intact species, instead of just individuals, to release into wild areas.

Knowing what we must do is not the same as being able to do it, however. Dr. Michael Soulé, founder of the Society for Conservation Biology, estimates that nearly 2,000 species of vertebrates will need a place on the ark in the next several decades. The number of at-risk invertebrates, which are even more critical to ecosystem health, could soar into the millions. Keeping viable populations of even a portion of these species would be a formidable task given the small size and budget of most zoos. Consider, for instance, that all the zoos in the world could fit into the borough of Brooklyn, New York, a mere 20,000 acres. Space considerations aside, yawning gaps remain in our knowledge of rare animals. We probably know as much about breeding these soon-to-be-extinct creatures, for example, as people knew about raising livestock 4,000 years ago. Even the technology of sexing animals (determining their gender) is just now allowing us to find out whether we have males or females in our bird and reptile collections.

On the horizon is the hope that we will one day keep "frozen zoos" of sperm and embryos on hand, with obvious benefits in terms of shipping ease and reduced risk to individuals. Before we can fill the test tubes, however, we have to decipher the basic breeding cycles of rare creatures, a job that calls for coordinated record keeping among the world's zoos. In years past, record keeping was often a hit-or-miss operation, and sharing information was unheard of in the competitive atmosphere of zoos. Today, using ISIS, we have been able to assemble family trees for nearly 100 endangered species and have hopes of adding new ones as our knowledge and store of records increase. In the meantime, there's a lot to learn about how animals reproduce, and not much time left for many species. In the shadow of the ticking clock, the new zoos have inherited yet another important mission: to be centers of animal behavior research.

WHAT ZOO ANIMALS CAN TEACH US

As the zoo world repairs its reputation, a wonderful resource is being taken out of its shroud. For years, scientists studied the remains of dead animals at museums, scrutinized lab animals, but simply overlooked the opportunities at their local zoos. Here they had a chance to study live animals at close range—being born, growing, learning, resolving conflicts, building a home, winning a mate, parenting, and aging. Those who did take advantage of zoo studies laid legendary groundwork in the field of animal behavior. It was in a zoo setting, for instance, that the facial expressions of wolves were first studied in detail. Given the skittish nature of wolves, this subtle "language" would have been nearly impossible to decode in the wild. The panda was equally difficult to study in the wild. Its solitary habits and remote, forested haunts kept researchers from learning about reproduction—until the first panda cubs were born in zoos.

Today, as immersion exhibits become more lifelike, animals are rounding out their repertoire of natural behaviors, making zoo study all the more revealing. Behaviorists get an eyeful, since zoo animals generally have more time to devote to social interactions than wild animals do. Courtship, mating, and parenting are especially fertile subjects, given that animals are not quite as secretive as they would be if predators were afoot. Close-up looks at development and anatomy are also naturals for zoo research, along with studies on nutrition and on animals' reactions to captivity.

There are, of course, limitations to what we can study at zoos. Because zoo animals are fed, inoculated, and shielded from harm, some of the most important ecological facts of life are missing from the equation. We can't, for instance, study predator-prey relationships, migration, or animals' reactions to seasonal shortages of food. This book describes the behaviors we *can* see in zoos: the everyday routines of feeding, body care, and movement, plus social behaviors such as bonding, aggression, courtship, and parenting. Naturally, some animals perform these behaviors differently in their exhibit than they would in the wild, depending on how sensitive they are to human presence. Even when you factor in the human influence, however, zoo-based studies can still provide valuable baseline data for field studies. These results can also be used to manage semiwild populations in the outdoor megazoos called national parks.

Far from their home turf, zoos have spearheaded and funded some of the most famous studies of wild populations. In turn, data from these studies, such as Jane

Goodall's work on chimpanzees, are often used to design zoo exhibits, closing the circle and uniting field researcher and zoo professional in a spirit of cooperation. This spirit is absolutely essential if we are to scale the walls that face us. As part of this effort, the last and most pressing mission of zoos is to recruit the public's help in stemming the tide of habitat destruction.

RECRUITING EVERYONE'S HELP

Besides being places where animals can breed and scientists can study, zoos are the only places where most people can watch, hear, smell, and *meet* rare animals in living color. All the television shows, museum dioramas, and encyclopedias in the world can't match the chemistry that occurs when animals and people look into each other's eyes. This chemistry works its own magic, touching people's hearts in a way that lasts. For many children, a trip to the zoo is their first real encounter with the animals they have read about, sung about, and drawn since their earliest years. The fact that animals figure so prominently in children's fables as well as in mythology, art, and language says volumes about our connection to wild creatures. At the zoo, myth becomes reality, and the connection is reaffirmed in a new way.

Spending time with an animal in a naturalistic exhibit puts us in its world, kindling an affection and, eventually, a concern for the animal's well-being. Like our ancestors, we want to know more about the animals we watch: where they live, how they live, and whether they will survive the crushing assaults on their wild habitats. Happily, learning happens naturally at a zoo; the animals pique our interest, and the exhibits, if they are well done, quench our thirst for information.

Zoo people know that the crucial step in the making of a conservationist, after awareness and education, is a commitment to change the status quo. It is at the precise moment when people are falling in love with animals that they are best able to hear the cry for help and respond in a personal way. As the entrance sign at the Bronx Zoo says, "In the end, we will conserve only what we love. We will love only what we understand. And we will understand only what we are taught."

Meeting an animal face-to-face can awaken a lifelong love of nature.

After years of writing exposés about the sad and shocking decline of zoos, the media is now trumpeting the zoo renaissance. Indeed, zoos around the world have spent hundreds of millions of dollars the last few years to improve living conditions for animals and to educate the new generation of zoogoers. People who were once uneasy at zoos are breathing a sigh of relief, and zoo attendance is higher than it has been in years.

What will be even more fantastic are the zoos of tomorrow, now on the drawing boards of a few visionary individuals such as Michael Robinson, the director of the National Zoo. Robinson's dream is to create bioparks, which will incorporate the contents of aquariums, botanical gardens, archaeological digs, natural history museums, and more. Juxtaposing these windows to the natural world will allow visitors to experience the underlying unity of life. Robinson's holistic touch has already found its way into the National Zoo in the new Invertebrate Exhibit. People who come to see the giant pandas are surprised to find themselves enjoying Madagascar hissing cockroaches and other members of the invertebrate clan, a maligned but ecologically crucial group that makes up 99% of all species on earth. Chicago and Cincinnati zoos are also celebrating invertebrates in their new insect exhibits.

Making the ordinary extraordinary will be one of the greatest challenges of the future zoo, says Robinson. Minnesota has already designed an award-winning exhibit around one of its relatively commonplace residents, the beaver. Though the beaver is not itself endangered, some of the species that use its habitat are. Only when people can see that water beetles, ferns, and spotted frogs are as fascinating as Siberian tigers can we begin to save not just a handful of species, but the entire biome, the wellspring of life.

A day at a habitat-oriented zoo is a journey through unforgettable neighborhoods like this bustling prairie-dog town.

Wandering through bioclimatic zones of plants and animals will be a far cry from shuffling past row upon row of caged specimens as we did in the past. After a day of absorbing wild creatures in their lush natural contexts, we'll feel as if we've traveled from mountaintop to ocean floor and back again. That night, as we drift off to sleep in our own home habitats, it will be the *connections* between living things that we remember, and the fact that our place in the puzzle is not on top, or apart, but somewhere in the middle.

This book is designed to be your companion not just in today's best zoos, but also in the bioparks of tomorrow. In these pages, you'll find animals "exhibited" in their bioclimatic zones, and you'll read about what they do in the wild as well as in captivity. Zoos are changing, and I'm hopeful that even if you can't see all these wild behaviors today, you'll at least see glimpses of them. More importantly, you'll know what to look for in the days ahead when, as exhibits get wilder and wilder, what you see at the zoo will too.

INSECT-EATERS *Chameleons use gripping feet, a long sticky tongue, and phenomenal balance to garner a bit of protein.*

HOW ANIMALS BEHAVE: A PRIMER

WHAT IS BEHAVIOR?

Behavior is a survival maneuver. It's everything and anything an animal does to keep itself alive today and its genes alive tomorrow. Building a home, courting a mate, finding food, stalking prey, threatening an enemy, and soaking up the sun's warmth are common survival moves, but, as you will see in this book, each species performs them with its own flair. These differences distinguish a mongoose from a marmot, and tell us volumes about where an animal lives and what it has to face each day.

When we visit animals in zoos, it's sometimes hard to remember that they evolved in wild habitats where they were not fed by zookeepers, assured of shelter, fenced from enemies, or checked regularly by the zoo vet. Yet it was the harsh reality of limited food, weather, predators, and diseases that shaped their bodies and behaviors in the first place. Though the struggles in a desert are different from those in a tropical jungle, the principle behind them is exactly the same. So before we talk about what makes animals different from one another, let's celebrate what unites them.

A CRACK AT IMMORTALITY

If there's one thing that all of us—from slime molds to slide guitar players—have in common, it's DNA. Units of DNA called genes contain blueprints that could be used to re-create our bodies from scratch if need be. In fact, when an embryo is starting to grow, it follows the instructions on those genes to a tee. In asexual reproduction, or cloning, the parent simply creates an identical copy of its genetic blueprint; barring a rare mutation, the young is a carbon copy with no new features. In sexual reproduction, however, two individuals conspire to create a new blueprint made up of one-half of the mother's genes and one-half of the father's genes. This conspiracy allows new genetic combinations, new instructions, to emerge, paving the way for innovative new features.

The animals we share the earth with today are here because the combinations that cropped up were winners. They gave rise to new physical and behavioral features that gave animals an edge over other contestants in their habitat. This edge enabled the recipients to live longer and have more offspring than their competitors, making them, in the parlance of natural selection, the fittest of the fit. Because their

WHY ANIMALS BEHAVE THE WAY THEY DO

winning designs were blueprinted on genes, animals could pass these advantages on to their offspring, and they in turn passed them on to the next generation.

Because genes live on in the offspring when a parent perishes, genes have a crack at immortality, or, at least, a longer tenure than the bodies that carry them. This gene immortality idea gives us one way of understanding why animals behave the way they do. Richard Dawkins, in a remarkable book called *The Selfish Gene,* says animals do what they do not because the animals themselves want to survive, but because the genes within the animals "want" to be replicated and thus increase their frequency in the larger gene pool. Our bodies are survival machines—carrying, protecting, and promoting our genes. As Samuel Butler once said, a hen is the egg's way of producing more eggs. And we, says Dawkins, are the hens for our genes.

This concept makes a lot more sense when you imagine what would happen to genes that built a survival machine *without* the behavioral and physical tools for survival. The faulty machine would die out, perhaps without leaving offspring. Before too long, those genes would disappear from the gene pool, while at the same time, the genes that were building better and better carriers would be getting more numerous.

It's an interesting idea, but how can it help us understand what zoo animals are doing? Simple. Whenever you see a behavior, think to yourself, is this behavior adaptive or nonadaptive? If it's adaptive, it must somehow help the animal's genes increase their frequency in the gene pool. Does the behavior make the animals more attractive to the opposite sex? Does it help them deal with food shortages in winter? Does it allow them to sneak by dominant animals without a scratch? Does it enable them to eat and avoid being eaten? If the behavior jeopardizes the animal's chances of passing on its genes, it may be what's called nonadaptive. Eventually, if a nonadaptive behavior is detrimental enough, the animals that carry it will do poorly, and the genes that control it will be edited out of the gene pool.

On the surface, some perfectly adaptive behaviors look as if they would be detrimental for the animal's genes. Let's take an example. Say you are watching the chimpanzee exhibit and you notice that a mother gives up her baby for a while to another chimp, who baby-sits and even nurses the infant. This helpfulness seems to contradict everything we've said about the selfish gene. What do these nannies have to gain by helping young that are not their own? The answer goes back to the common denominator of life, the genes.

Because chimps travel in troops of closely related individuals, the baby-sitter is likely to have some genes in common with the mother. If she is a sister or a daughter, 50% of their genes are the same. Since some of the helper's genes are bound to be present in the infant, the helper does have a valid genetic stake in its survival. She is also getting good practice for the day when she will raise her own infant, perhaps more successfully because of her early experiences. Once this helping behavior proves beneficial, the gene that controls or influences it will continue to be passed on, making the chimp an even better survival machine.

By knowing the genetic realities that drive animals, we can avoid slipping into anthropomorphism, or attributing human traits to animals. Think, for instance, how easy it would be for us to say, "Look, that female chimp is doing that new mother a favor. Chimps must have big hearts." This would be denying the fact that wild animals, unlike most modern-day humans, have to fight for their survival. *The genes that support survival will live on, while the genes that detract from it will eventually be edited out.* And, as genes go, so goes animal behavior. In the end, genes pull the strings, dictating how and when an animal behaves the way it does.

Some people may say that thinking about animal behavior in terms of genes somehow takes away its mystery, but I think it makes the riddle more enticing. As Marion Stamp Dawkins, author of *Unravelling Animal Behavior,* says: "The fight for a place in the gene-pool has given rise to some of the most beautiful and intricate phenomena on this earth. Animals . . . have developed the power to swim and to fly, to care for their young, to stalk their prey, to play, to sing, and to be curious about the world around them. To know all this comes from such simple beginnings can enhance and deepen the wonder, not diminish it."

THE ROLE OF HABITAT

While the selfish gene idea helps us explain the common motives behind behavior, we need something else to illuminate all the differences that make animal watching so absorbing. If the ultimate goal is the same throughout the world, why is there such marvelous variety in the way animals go about surviving? The simplest answer is this: though the game is the same, the playing fields—or habitats—are different throughout the world.

ADAPTATION TO HABITAT *An animal's body echoes its habitat. This tree-dwelling barred leaf frog has suction-cup toes and light-gathering eyes to help it maneuver in its leafy, shaded habitat.*

A habitat is the place where an animal finds what it needs to survive. More than a geographical place, a habitat is all the opportunities and challenges an animal faces, including competition, climate, food availability, predators, and a host of other conditions that can't be described on a map. Through the wonders of natural selection,

HOW DOES HABITAT SHAPE ANIMAL BEHAVIOR?
Questions to Keep in Mind While Watching Zoo Animals
Keeping in mind that each behavior is an adaptation to habitat, much like keen eyesight or an opposable thumb, see if you can spot the evolutionary logic in what zoo animals are doing. If you're stumped, the following questions might shake out a clue:
1. What was the animal's wild habitat like? ■ What opportunities were available to the animal in its wild home (food, water, shelter, nesting opportunities, hiding places, and so on)? ■ What challenges did the animal have to overcome to meet its needs (weather, competitors, predators, diseases, human intervention, and so on)?
2. How would the behavior you are watching help the animal survive and promote its genes in its wild habitat?
3. Why is this behavior *better than any other* at helping the animal excel in its habitat?
Note: This is assuming that the behavior you are watching is natural. If it makes no sense at all, it may be a stress-related, captivity-induced behavior.

each survival machine is custom crafted to excel in its particular constellation of living conditions. Its body and behavior echo the habitat it evolved in.

Figuring out where an animal evolved takes a bit of detective work. We do it all the time when we try to guess which part of the world someone is from. We look at a person's mannerisms, dress, and accent to figure out where he or she grew up. In the same way, an animal's mannerisms are a clue to its homeland. An animal that climbs up the exhibit when frightened, for instance, probably lived its natural life in trees. Those that feed as a group with one sentinel keeping watch probably evolved on an open plain, where constant surveillance for predators was a necessity.

AN INSIDER'S PERSPECTIVE

Though every species on earth behaves uniquely, all behaviors are just variations on a theme. The following sections explain the universal themes or categories of behavior, from basic behaviors such as feeding and sleeping to sophisticated social behaviors such as mating and parenting.

Having an overall framework for animal behavior is like understanding the culture of a country you are visiting. Once you appreciate the everyday needs and compelling forces that drive animals, you'll be able to make an educated guess about what any animal is doing, be it at the zoo, in the wild, on safari, or even on TV. This insider's perspective will change the way you look at the animal world, transforming you from a tourist who takes pictures to a traveler who makes connections.

BASIC BEHAVIORS

Every animal, even one with a social lifestyle, spends some time alone, keeping itself comfortable, out of harm's way, well-fed, rested, and even entertained. Your chances of seeing these basic behaviors in the zoo are good, because they occur even if an animal is exhibited alone.

Basic behaviors aren't designed to communicate a message, but rather to help the animal get on with the business of living. Think of the things that a fly on the wall might witness you doing: running to catch the bus, getting a snack from the refrigerator, drying yourself after a bath, putting new sheets on your bed, napping, or locking your door at night. Now watch the clouded leopard at your zoo: leaping up to a branch, tearing apart its food, licking its fur, arranging its sleeping nest, dozing, or freezing when a loud noise sounds. You can arrange your behaviors as well as the leopard's into categories such as: locomotion, feeding, body maintenance, shelter building, sleeping, and avoiding enemies. In this section, we'll take a look at these routine behaviors and see how they help animals stay alive in the wild.

LOCOMOTION

Animals have solved the problem of moving through every conceivable medium: air, water, soil, snow, even volcanic ash. You can guess how an animal moves by looking at its extremities. Are its arms longer than its legs, giving the animal an advantage when swinging through trees? Do the claws curl automatically when it lands on a branch, or do they spread and grip tight on the surface of a tree trunk? Are the feet webbed to push water efficiently, or are they horny and tough to allow the animal to

LOCOMOTION   *The way animals move says something about their habitat, the predators they face, and how they make their living. Springboks pronk on the plains, gibbons brachiate through the branches, flying squirrels glide from tree to tree, and basilisk lizards (also called Jesus Christo lizards) scutter across water on large, spidery feet.*

run over hard, sun-baked surfaces? Are the legs more muscular than the wings? Are the flippers the real powerhouse, or is it the tail? Are the feet used for steering, as they are in the sea-swimming penguin, or for springing, as in the red kangaroo? Are the wings narrow for swift, darting flight or wide and flat for steady soaring?

Locomotion can tell you a lot about how an animal makes its living, how it escapes from predators, and even where in the world it lives. When a sandhill crane ascends, for instance, it needs to run along the ground for a while, whereas a grouse can burst immediately out of a thicket. The grouse's tight takeoff tells you it evolved in a densely vegetated habitat, while the crane probably had the generous runways of a sedge meadow habitat. Similarly, you can tell that loons evolved in a watery habitat, since a loon trying to walk on land is a clumsy sight. Meanwhile, in the tree canopy, nothing moves quite as slowly as the three-toed sloth. This tells you that sloths probably had few predators to worry about and not much energy to waste, perhaps because of their leafy, low-nutrient diet. Matching the movement to the habitat is a fascinating exercise and can easily keep you busy the whole time you're at the zoo.

FEEDING

We are what we eat in more ways than one. All animals use food to fuel their engines and ultimately build and rebuild body tissues, organs, and countless other spare parts. What an animal eats, therefore, winds up in each and every cell. Its diet also dictates how it will find its food and, ultimately, how it will live its life—whether it will be sedentary, roving, social, or solitary.

Meat-Eaters

Meat-eaters can find food in one of two ways. Predators kill other animals, called prey. Although you can't see the actual chase and capture at the zoo, you can imagine which role an animal plays in the wild by studying its behavior and its body. The hunter and the hunted are locked in a biological cat-and-mouse game that's been going on for millions of years. Each time the predator evolves a faster, surer, more clever way of finding and overcoming its prey, the prey evolves a countermeasure; just as the stalking lion's movements become quieter, the zebra's ears become more attuned to slight noises. This spiral dance of adaptations is a beautiful example of how natural selection constantly tinkers with its original designs.

Finding, subduing, or preparing prey to be eaten costs an animal both time and energy. Animals that specialize in prey many times smaller than themselves must contend with the problem of collecting enough to make a meal. Filter-feeders solve this problem by sucking in water that is clouded with small organisms and then straining out the good stuff. Baleen whales use their giant filtering mouth to trap shrimplike krill, while flamingos use their upside-down smile to dredge organisms from the mud. Insect-eaters, such as anteaters, shrews, and bats, are also handily equipped to gather enough of their tiny prey to satisfy their hunger.

Some meat-eaters, such as vultures, eagles, and hyenas, let other animals do the work. These scavengers eat the leftovers from other animals' hunts, or simply take advantage of animals that die of old age, injury, or illness. Their keen sense of smell allows them to ferret out decaying flesh no matter where it has fallen. Although they don't worry about getting attacked (corpses don't fight back), scavengers do risk a certain amount of disease. To compensate, natural selection has provided them with a tolerance to foods that would make other animals sick. Many scavengers, such as the bald eagle, can also hunt live prey if need be.

The way an animal gathers and eats its food in the wild can influence how it will be exhibited at the zoo. For instance, if an animal is used to defending a feeding territory, it may not take too well to having other animals in its enclosure. The urge to guard its food and the space around it is just too great. Species that hunt alone, such as forest-dwelling leopards, are also more comfortable on their own or, at most, in pairs. Lions, on the other hand, crave company in captivity, perhaps because in the wild they depend on other lions to help them raise young and kill large, fleet-footed prey.

EVOLUTION *Predators like this horseshoe bat are in a biological cat-and-mouse game with their prey. As bats hone their echolocation skills, moths develop better ways to escape. Some moths have evolved listening membranes to detect bat calls, while others produce calls of their own that jam the bat's sonar system.*

How specialized an animal is may also affect its comfort level in captivity. For instance, some animals, such as eagles, snakes, and large cats, are extremely adept hunters. Their accuracy rate is so good and their prey is so filling that they need to spend only a fraction of their time hunting, allowing them to laze away the rest. In the zoo, these are the animals you'll see sleeping or just gazing from a branch or hiding spot. This behavior is completely normal, true to what they would do in the wild.

Animals that don't have a feeding specialty are called opportunistic feeders, and they have quite a different lifestyle in the wild. Instead of zeroing in on one type of food, they spend much of their time searching for anything edible, devouring it as fast as they find it. In sterile quarters, these opportunists tend to go stir-crazy because their natural tendency is to search and sniff, explore and check out their world. To keep these animals healthy, exhibit designers must build environments that continually challenge them.

Plant-Eaters

Even plant-eaters can be specialists, adapted for a particular feeding niche. You can recognize grazers, such as zebras, white rhinos, and buffaloes, because they spend most of their time with their head down, cropping grasses like giant lawn mowers. Since the grass they eat is not very nutritious, they must eat large quantities and pass it through their digestive tract quickly (look for lots of manure). This makes grazers the dryland counterparts of filter feeders; their digestive tract filters nutrients from the grasses, then flushes the remains through to make room for more.

Browsers look for hardier fare; they eat the twigs, buds, and leaves of woody plants. Koalas, sloths, and leaf-eating monkeys spend their lives in the branches, literally surrounded by food. Giraffes, elephants, and gerenuk antelopes use their anatomical stepladders (necks, trunks, and long hind legs) to transport them to the canopy layer. Shrubs and trees in their zoo exhibits are usually pruned to the highest spot they can reach.

Fruit-eaters concentrate on the juicy seed containers that plants produce. These are typically rich in water, carbohydrates, vitamin C, and sometimes oil. The more animals that fruit-bearing plants can attract, the better their chances of having their seeds sprout next spring. The seeds eaten with the fruit travel through the animal's digestive tract, often losing their hard shell so they are ready to germinate when deposited (conveniently) in a pile of fertilizing manure.

Flower-feeders have adapted ingenious ways to exploit the sweet, energy-rich nectar hidden deep in blossoms. As the long-billed hummingbird hovers at the mouth of a flower, its wings beat 50 to 80 times a second, up to a million times without pause. To fuel this stunt, the tiny birds must visit as many as 60 blossoms a day, consuming half their body weight in nectar. In a complementary adaptation, the flowers use the birds as a kind of pollinating service. They attract them with bright red petals and then bribe them with sugary nectar. Each hummingbird that dips its bill to drink brushes the flower with pollen from the last stop or picks up new pollen to bring to the next flower.

Seed-eaters, such as mice, chipmunks, and finches, have homed in on a truly nutritious and abundant source of food. Seeds are packed with fat, carbohydrates, and

FOOD HOARDERS *Acorn woodpeckers bore hundreds of small holes in a dead tree to store acorns for the winter.*

protein, and they usually keep well when stashed away in larders. As a result, seed-eating rodents and birds have prospered, becoming two of the world's most numerous and widely distributed classes.

Food Storage

Whenever there is a surplus of food, some zoo animals look as if they are playing with their dinner. A wolf that gets a bone for a treat, for instance, may eat some now, then bury the rest for later. Your own dog may try the same trick, digging fruitlessly into your living room carpet, trying to store the bone for a rainy day. In the wild, jackals, coyotes, foxes, bears, wolverines, mink, martens, and weasels have all been known to stash away what they can't eat right now (a wild version of doggy bags).

Other animals hoard food by scattering it throughout their territory or piling it in one safe spot. Burrowing rodents cache their food in underground chambers so they can eat in peace, away from predators' prying eyes. Animals that must contend with winter, such as Arctic foxes, squirrels, pikas, beavers, and acorn woodpeckers, diligently stockpile food to get them through inclement times. To find out if a zoo animal is a food hoarder, watch for it tucking bits of food in the nooks and crannies of its enclosure.

DRINKING

A zoo animal's drinking habits can also reveal something about its life in the wild. Some animals won't take a drop of water from conventional sources such as pools or rivers. For slow-moving canopy-dwellers such as the koala, an overland trek would be more arduous or dangerous than it's worth. Instead, they stay in the branches and derive moisture from leaves and drops of dew.

In habitats where open water is scarce, animals have adapted ways to wring moisture from the most surprising places. Desert-dwelling spadefoot toads, for instance, mop up the water from dew-soaked soil through their belly skin. Kangaroo rats place bone-dry food in a cool burrow and wait for it to sponge up humid air vapor before eating it. Desert sand grouse wade in puddles that form after a rain and then bring their soaked breast feathers to the nest so the young can drink. Young sand grouse are so accustomed to sucking water from feathers that zookeepers acting as surrogate parents must give them soaked cotton.

Another way to deal with a dry environment is to "camel-up" (drink as much as possible) when water is available. A dromedary camel can down 15 gallons at a time, then go for 17 days without another drop. Once a dryland dweller is lucky enough to find water, the real trick is to retain it at all odds. Most desert animals produce dry feces, excrete almost no urine, and cool themselves in ways other than sweating or panting, both of which waste water.

Marine mammals, birds, and reptiles have their own problems when it comes to thirst: water is everywhere, but there's not a fresh drop to drink. Luckily, these animals are adapted to drink saltwater, then purge the salt by means of special kidneys, nasal passages, or tear ducts. If your zoo keeps its seabirds in ocean water, watch for signs of what looks like a constant head cold as the salt drips from their nostrils. Turtles "cry" their salt out in big crocodile tears.

Animals also differ in their drinking techniques. Felines curl their tongues into a spoon shape and throw a little water into their mouth with each lap, swallowing the accumulation every fourth or fifth lap. Elephants go for the gusto; they fill their trunk with 2 to 3 gallons of water at a time and then squirt it into their mouth. Pigeons and their relatives use a pumping action to suck up water, but this is unusual in the bird world. Most birds drink chicken-style, dipping their bill, filling it, and then lifting up to let the water run down their throat. Though the lift-and-drain method looks laborious, it may work to their advantage because it gives them a chance to lift their head and look around for predators.

Alertness at water sources is an almost universal animal trait. After all, predators soon learn where and when the local animals drink each day, and they make it a point to be there too. Keeping their head down too long while drinking can be an Achilles' heel for prey animals. Even the skyscraper giraffe, able to see a predator when it is still a speck on the horizon, is in danger when it dips its head to drink. Because of its unusual architecture, it must splay its front legs apart to lower its mouth down to water level. It takes a few heart-pounding seconds to hitch itself upright again, giving the predators-in-waiting an edge. Now that you know what a trial drinking can be, carefully watch the animals at your zoo when they imbibe. Though they need not worry about predators in their captive niche, their wild wariness often remains.

ELIMINATION

As with so many other behaviors, we humans rarely think about ridding ourselves of wastes. It's only when our systems go on the fritz that we appreciate how important this routine behavior is. We then realize it's part of the one endeavor we *do* spend a lot of time thinking about: eating.

When you see how wild animals eliminate wastes, you begin to understand why it's so easy to housetrain a cat and why it's almost impossible with a monkey. Small, wild cats return to a single spot for defecation, and often bury the smell and sight of their feces. Animals that swing through trees, however, practice the gravity method of waste disposal. They go whenever they please, not needing to be concerned with fouling their living space. Birds confined to nests can simply aim over the rim to keep the homesite clean. Young birds, not yet agile enough for a rim shot, commonly produce a membrane-covered sac of feces that the parent bird can neatly carry away and drop. Animals with ground nests, such as deer, have nowhere to drop the wastes of their young. Rather than tipping their hand to keen-nosed predators, many mothers simply eat the feces. Animals in burrows face a similar disposal problem. Wastes that build up underground invite disease, increase humidity, and send a fatal beacon to predators. The cautious prairie dog and other burrowers build their homes the way we do, with lavatory chambers separate from their bedrooms or nurseries.

We can't leave the subject of "wastology" without mentioning the one indelicate behavior that all animal watchers are bound to see eventually: coprophagy, or eating one's own wastes. When an animal's nutrition is not what it should be, it may eat its own feces to send nutrients through a second time. Animals such as rabbits and mice do this as a matter of course to get everything they can from their low-nutrient grass diet. Usually, apes and other primates reingest their feces in the wild only after famine has dried up their food sources. In the zoo, however, coprophagy can be caused by boredom or stress (see Captive Behavior, p. 63), indicating to keepers that the animal needs more stimuli or safety in its environment.

Coprophagy is not the only way to get double duty from feces and urine, however. Disturbed herons in a breeding rookery will defecate and even vomit over the side of a nest to discourage predators from climbing up (a good thing to know if you're walking beneath rookery trees). Stress-related diarrhea sends a clear "I give up" signal, which, in ritualized fights, can defuse a dominant animal's aggression. Urine and feces may also carry scents that identify an individual by gender, sexual readiness, and other characteristics that we can only guess at. Hippos defecate while swishing their tail in circles, and when the feces hit the fan, their scent is broadcast near and far. This same scent-identification system is why your dog insists on dribbling urine on every fireplug in the neighborhood. Using wastes in this way is really an act of communication, so we'll take it up again in the section on page 40.

CLEANING UP: SCRATCHING, RUBBING, PREENING, BATHING

Just because we're the only species with a cosmetics industry doesn't mean we're the only ones who care about cleanliness and good grooming. In fact, other animals are far more keenly interested in being clean than we are. To them, grooming is not a matter of manners: it's a matter of life and death.

Consider the grooming habits of birds. Without meticulous oiling and combing, feathers would grow dry and ragged, failing to function as airfoils, insulators, or water shields. To keep their feathers in peak shape, birds preen each and every day. If you watch closely, you'll see a preening bird reach its bill back to pick up some oil from a gland located at the base of its tail feathers. It draws one feather at a time through this anointed beak, laying down a layer of oil with each stroke. The oil not only helps repel water, but also breaks down into vitamin D when warmed by the sun. This vitamin is then absorbed into the bird's system through its skin and swallowed during preening. Each stroke of the bill also combs the feather, realigning the individual strands and zipping together their interlocking barbs.

PREENING   *All birds, including this southern carmine bee-eater, must comb and oil their hardworking feathers to keep them flightworthy.*

Many birds relish a good water bath as part of their preening ritual. Though they may seem to be splashing about haphazardly, they are performing the exact cleaning procedure performed by every bird of that species. Next time you watch a bird shaking its head, ruffling its feathers, and throwing water about, watch each movement closely. Other birds of the same species will go through the same choreographed routine.

Some birds, especially those in the pheasant family, bathe in dust as if it were a pool of water, perhaps using the grit to smother or dislodge parasites. Other birds may fumigate the bugs by lingering in a plume of smoke escaping from a chimney. Even more unusual is the bathing habit called "anting." To bathe in ants, a bird flops on top of an anthill and lets the angry residents swarm all over it. It may even take one of the ants in its bill and wipe the body onto its feathers. One explanation is that the bird is using the ant's formic acid to discourage parasites. Or, says another theory, given the bird's dazed look, the acid may simply be an addictive substance.

Birds that are bathing in the sun can also exude an air of sensual pleasure. They throw back their head and spread their wings and tail feathers, seemingly oblivious to the threat of predators. The sun is no doubt providing welcome heat (which may scare off parasites), while magically transforming their preening oil into vitamin D.

Mammals are as fastidious as their winged cousins when it comes to cleaning, and their regimes are every bit as predictable. Rodents usually begin by licking their paws and rubbing them over their lips, whiskers, and face. To reach up and over the ears, they get the whole foreleg involved. Next, they clean down each flank with both paws, tackling the hind legs and finally the tail. Primates use their facile fingers to part their fur and look for dirt, or, even better, something edible. After hours of combing and nibbling at their own skin, they may invite a partner to help them inspect places they can't see. A serious itch calls for a thorough rubbing on the ground, complete with flopping, writhing, and grunting. Mammals will also rub vigorously against vertical objects in their environment, such as rocks, termite mounds, trees, or companions that happen to be standing nearby.

Some mammals get clean in ways only a 4-year-old child could love. When it's bath time, hippos, rhinos, elephants, and other wallowers ooze eyeball-deep in a good sloppy pool of mud. Part of the attraction is the mud's cooling power, especially "après wallow" when the slow evaporation from the mud pack continues to cool them. A mud pack also keeps biting flies from reaching the skin, while smothering parasites that are already embedded. Once the mud dries and flakes, taking old skin with it, the wallowers vigorously rub their bodies against hard surfaces, unveiling to the world their new, improved hide.

Snakes and lizards renew themselves each year without mud. If you happen to be at the zoo at the right time of year (ask your zookeeper), you may be able to watch them shed their old skin and begin with a whole new layer of clean cells. Once the old skin loosens and begins to split around the head region, the snake rubs against stone or wood until its snout breaks through. It then wriggles through the split, and, like a finger pulling out of a tight glove, it leaves an inside-out skin behind it. If the job is not executed cleanly, the snake must rub against rocks for days until all the patches are scraped off. Amphibians, such as frogs and salamanders, don't have the convenience of a new skin. Besides immersing themselves in water, the most they do to clean themselves is to rub their eyes and mouth with their front feet.

DRYING OFF

Being wet is a burden for some animals and a downright liability for others. Birds, for instance, can't be "cleared for takeoff" if their wings are sopping wet. After a successful fishing dive, large water birds such as anhingas must perch for a while, holding their wings straight out from their sides to let them dry. Mammals have their own trick for shedding water and speeding the drying process. In fact, if you've ever been next to a big dog when it comes out of a lake, you know it well. The ever-popular shake-till-you-dry maneuver is practiced by almost every animal with a coat.

REGULATING BODY TEMPERATURE

All birds and mammals (including humans) are warm-blooded creatures that need to metabolize food in order to produce heat. They also use shade and sunlight and water to keep themselves comfortable, but these external aids are not as important to them as they are to cold-blooded species.

Amphibians and reptiles are called cold-blooded because they don't regulate their temperature internally; they simply take on the temperature of their surroundings. To keep temperatures constant and avoid the indignities of freezing or broiling, they reposition themselves in warm and cool places. Turtles, for instance, will stagger onto a sunny log after a cold night's torpor to warm themselves back into action. When they've had enough, their thermostat complains, and they plop back into the water to cool themselves down. A crocodile that gets a little hot under the collar will open its gigantic mouth and gape for a while.

The principle behind gaping is why panting and sweating work, not just for cold-blooded species, but for all animals. As water evaporates from the moist tissues of an animal's open mouth, its body temperature starts to drop. You'll know why if you think back to high school physics: when water evaporates, it turns into a gas, and that process releases heat. That's why you shiver when you come out of a pool on a windy day, and why sweating actually helps cool you off. Evaporation is also behind a dog's panting; as it breathes in and out, it circulates air over the moist tissues of its mouth and lungs, thus speeding evaporation and cooling. Vultures have a more exotic approach: they urinate on their legs and let the breeze do the work.

Animals can also cool down by retreating to the shade or by pressing their body against a patch of mud. Desert dwellers often have bare-skinned bellies so heat can quickly escape into the shade

KEEPING COOL *The veins in a jackrabbit's ears bring hot blood to the surface where it can cool a bit before returning to the rest of the body.*

created by their own bodies. Legs and ears may also be bare and crisscrossed with veins so that heat from the blood can come close to the surface and escape.

TAKING SHELTER

For many animals, the difference between surviving and succumbing is having a hideaway from weather and predators. Some animals use shelters only when over-wintering or raising young (see Parenting Behavior, page 57), while others crawl in daily for comfort and safety. Shelter may be as simple as a bowl scratched out of some gravel or as elaborate as an underground labyrinth.

Some innovative exhibits in zoos allow us to peek inside these underground shelters without disturbing the animals. One-way glass at the end of a weasel's nesting log or a video camera inside a beaver lodge give zoogoers something they could never experience in the wild: a privileged look at some very private lives.

Above ground, you can watch gorillas make new sleeping nests every night or see hammerhead birds make some of the largest tree-house nests you will ever see. Other unusual shelters to look for at the zoo are:

- Woven nests of weaverbirds
- Decorated bowers of bowerbirds
- Woodpeckers' holes in tree trunks
- Underground dens of wolves, coyotes, or foxes

SLEEPING

If you're the kind of person who needs at least 8 hours of sleep before you can operate the toaster, the following news might make you jealous. Giraffes, no matter how far they gallop in a day, need only 20 minutes of deep sleep a night. Reptiles and amphibians don't sleep deeply at all, but at least they rest. Hyperactive shrews are so consumed with finding their next earthworm meal that they don't even slow down!

Sleep or no sleep, many animals can still run, leap, lurk, flap, and lunge with precision, which is more than you can say for humans. When we lose too much sleep, we actually suffer from mental anguish. Are we victims of some evolutionary quirk when it comes to downtime? Not really. On the scale of slumbering, we are not nearly the most lethargic. Koalas sleep for 18 out of the 24 hours, squirrels for 14, and lions a lazy two-thirds of their lives compared with our one-third.

The extreme disparity in sleep requirements among animals has experts battling for a hypothesis. At present, there are two main camps of thought. One says that sleep recharges our batteries, but not necessarily our physical ones. If we wanted to refresh only our bodies, we could do that by lying down and relaxing all our muscles, say researchers. They claim the real reason for deep sleep (complete with dreaming) is to recharge our mental batteries, to sort through the sensations of the day and file them into place. As our memory banks tidy up the information, hard-to-categorize pieces come back as awkward dream sequences. Five of these dream sessions occur on a typical night, leaving us refreshed and able to find our car keys in the morning.

This recuperative theory, though it seems to work well for humans, bogs down when we begin looking at other animals. If recuperation is a must, wouldn't the high-metabolism, busy animals like shrews need it the most? And wouldn't slow-moving, lethargic animals like sloths have less to recuperate from? Then why is it that slothful sloths sleep 20 hours a day while harried shrews don't sleep at all? If recuperation is for restoring mental powers, wouldn't humans need more sleep than, say, armadillos?

Perhaps, claims the other camp of sleep theorists, animals are doing something else when they sleep. Maybe sleep is not the great recharger, but merely the great immobilizer. Maybe sleep knocks animals out during the times of day when their adaptations fail them: when it is too hot, too cold, too dark, or too bright to function safely. Asleep during these off-times, they may be less visible to predators than they would be if they were awake and fretting over every disturbance.

Of course, how long you sleep depends on whether you can afford to be immobile. If you must constantly eat to stay alive, a long sleep would be suicidal. Or if you are intensely vulnerable to predators both day and night, the way giraffes and other animals on the plains are, your sleep must also be brief. Some sleep habits may instead hark back to days when prehistoric predators made caution a survival imperative.

Today, the luxury of lethargy is often reserved for those at the top of the food chain, as well as animals that are large enough or heavily armored enough not to worry. As always, there are exceptions. Although elephants haven't worried about natural enemies for eons,

SLEEP REQUIREMENTS *To make the most of their nutrient-poor diet, three-toed sloths are careful not to burn too much energy. When they aren't sleeping (which they usually are), they operate in extra slow motion.*

they snooze for only 2 hours a night. For them, it's not the pressure of a predator that keeps them awake, but the pressure of their own enormous weight bearing down on their organs. To soften the blow, elephants often bunch up grasses beneath their head and flank before they turn in.

Other animals also prepare for a good night's (or day's) sleep, and it's one of the behaviors you can look for at the zoo. Gorillas use vegetation or straw provided by zookeepers to create a soft, springy bed. Peacocks ascend by the dozens to communal roosts in tree branches, a habit that may give them more than just a good night's sleep. Researchers think group roosts may double as information centers, where birds stand to learn about the good feeding spots their comrades found that day.

Flamingos also sleep in a group, but they stand up, tucking their beak behind one wing and keeping one eye open for predators. Dolphins are one-eye-open sleepers as well. By switching eyes every now and then, they can shut down one hemisphere of the brain at a time, resting both halves without ever being completely vulnerable.

One of the problems zoos face is that not all animals sleep at night the way most humans do. There are those that work during the day (diurnal), those that roam at night (nocturnal), and those that are active only during dawn and dusk (crepuscular). In the wild, this staggered sleep-shift allows different species to fully utilize a habitat without getting in one another's way. Zoo designers have come up with a way to keep night workers awake even when it's high noon. By keeping their exhibits dark during the day and bright at night, they induce nocturnal animals to switch their biological clocks and sleep during the same hours that we do. The red lighting (which animals see as black) in "nocturnal world" exhibits allows us to see them operating in what they think is the dead of night.

If you do happen to catch animals sleeping, however, don't be disappointed. Remember that sleeping is a very legitimate and necessary activity, one that we ourselves spend 20 to 30 years doing. Besides, it's fun to try to identify the sleep cycles of various animals. Higher mammals and birds experience periods of deep sleep alternating with periods of light sleep. You can look for evidence of rapid eye movement (REM) under the eyelids along with muscle twitches and occasional changes in posture.

YAWNING AND STRETCHING

Believe it or not, the sleepy subject of yawning happens to be on the cutting edge of animal behavior. There are very few behaviors that birds, mammals, reptiles, amphibians, and even fish share, but yawning is one of them. Since yawning represents a point where all vertebrate classes intersect, it excites researchers who are searching for crossroads in the evolutionary roadmap. Surprisingly, though, this excitement is relatively recent. Until a few years ago, most of the scientific community believed an early ornithologist's statement that birds do not yawn. Well, of course they do. The question is why.

A yawn can be many things. It can be a sign of trembling fear, intense aggression, or groggy fatigue. It can prepare an animal for battle or for sleep. It can resupply the bloodstream with fresh oxygen, stretch a tired jaw muscle, or simply give an animal a break from a stressful situation. Perhaps your own zoogoing experiences will help shed some light on this long overlooked link among the classes of animals.

So far, here's what we know. Stretching and yawning usually follow each other or occur in various combinations, usually just before or after a long rest period. The impulse to yawn may come from within the animal or it may come as a suggestion from outside (which is why I just yawned, and you may be thinking about it).

The internal trigger for yawning and stretching is part of the body's automatic "self-righting" system. When an animal rests, its blood flow eventually becomes sluggish. Oxygen, which rides in the bloodstream, may not be reaching the brain in sufficient quantities, and this shortage sets off the deep yawning response. Yawning vacuums in oxygen and speeds it to the brain. Certain amounts of carbon dioxide are also needed, however, to restimulate the part of the brain responsible for breathing. A few luxurious stretches cause the muscles to contract, pump-

YAWNING   *When the lion's brain needs more oxygen it calls for a yawn.*

ing carbon dioxide where it's needed. Yawning and stretching are like a jump-start for the body, prepping it for the rigors of a new day. Lions, which spend most of their time in drowsy slumber, yawn quite often to get their sluggish blood coursing again.

The external motivators of yawning and stretching are even more interesting, hinting at what might be a social function for these activities. Some species, such as baboons, yawn as a form of threat behavior, perhaps because it does such a good job of exposing their sharp teeth. Humans and many other animals, including fish, yawn when they are stressed, perhaps because the body senses danger and is automatically preparing itself for action by rushing oxygen to its muscles. Sometimes, however, we yawn simply because we see someone else do it. In fact, for many of us, yawning is a completely involuntary response.

This "social" yawning may have once been a protective device that is still functioning in many of the animal communities you'll see at the zoo. Ostriches, for example, will yawn as a group before settling down to sleep. The higher-ranking birds yawn first, a sign that the coast is clear and the lesser birds need not fret. A contagious yawn spreads through the herd, settling them down so they can sleep and be ready to feed together the next day. Without this behavioral sedative, the flock would be likely to scatter at the slightest noise, lose sleep, and not be able to maintain the security shield of group alertness the next day. After a full night's sleep, another bout of yawning wakes the troops and readies them for action. It's interesting to think that our own contagious yawning may have once promoted group togetherness around prehistoric campfires.

AVOIDING PREDATORS

Though zoo animals no longer have to worry about predators, their natural defenses are just under the surface and can easily be evoked by neighborhood crows or raccoons, hawks flying overhead, or even a visitor who suddenly pops into view. The common response to intruders is to freeze, flee, fight, or some combination of the three. Usually, animals that freeze have an uncanny ability to blend into their background. Their cryptic coloring fools the eye either by erasing or breaking up the outline of their body so predators, more attune to moving than still objects, may walk right by without noticing them. Once their cover is blown, however, even cryptic animals get the overwhelming urge to exit.

Fleeing has been perfected to a fine art, inspiring mythic levels of speed, endurance, and agility in prey species. Plains animals, such as antelopes, gazelles, and zebras, have also learned to measure their attackers' talents against their own. Knowing that lions, leopards, and cheetahs are capable of only short bursts of speed, the hoofed residents rarely panic at the sight of a cat as long as they have running room and a head start. The important thing is to keep an eye out so the predator doesn't "steal the bases" and get close enough for a deadly sprint. Against hunting dogs and wolves, however, prey animals know they can't depend on their endurance alone. Canines are not as fast as cats, but they can run for a long time, long enough to out-tire weak, old, or sick prey.

When straight-out running won't do the trick, some animals, such as rabbits, wind up zigzagging for their lives. They gain their advantage by being unpredictable, making trigger-fast turns, and outmaneuvering their less agile enemies. Grouse are the masters of another kind of escape behavior called dash-and-wait. The birds respond to danger by flushing dramatically, flying some distance, then dropping silently into the weeds to wait. They stay stock-still, hoping the predator strays off course while looking for them. If the predator picks up their scent again, they let it get close, then dash away at the very last moment. An embellishment to this technique is called flashing. Just before taking off, the grouse will flash a brilliant white underwing that is normally concealed. The predator keys in on this and is confused when the bird hides the patch of white and apparently disappears into thin air.

Other animals use the environment to protect them. If you've ever wondered why burrowing rodents stick so close to their tunnel entrances, consider the fact that their winged enemies "drop in" at speeds of up to 110 miles per hour. Without an underground bunker, prairie dogs, ground squirrels, and weasels would be easy pickings in the open fields where they live. Another way to escape is to head up above; birds have the friendly skies, and tree-dwelling species, such as squirrels and monkeys, high-tail it for the leafy branches. Finally, if you can't climb and you can't dig, it helps to be able to hold your breath underwater. Aquatic animals plop to safety, hoping their predators are not, like mink and water snakes, easily able to follow.

Sometimes the getaway strategies fail, and the predator comes within a ragged breath of its prey. Some animals still have one more diplomatic avenue available to them before they are forced to fight or surrender. If they can simply distract the predator, they may buy themselves some time. A sparrow in a domestic cat's jaws may suddenly go limp, for instance, prompting the cat to lay it down momentarily, long enough for it to flutter to life and fly away. Other prey species take the ruse even further, banking on the fact that if they look like they're dead, the predator will lose its appetite and its urge to kill, especially if it's an animal adapted to eating only fresh meat. Opossums curl up and go so rigid that predators can bite and maul them without eliciting any response. The Oscar for playing dead goes to the African wood snake, however, whose perfectly healthy eyes fill up with blood while a trickle of red drools from its limp mouth.

Other animals startle their attackers by flashing a frightening display of color, an unexpected move that makes the predator hesitate for a fateful second. The blue-tongued skink sticks out its vibrant tongue, and the frilled lizard pops open an umbrella of skin around its open jaws. Other animals flash spots of color on their bodies that look astonishingly like the eyes of a hawk or owl. It's easy to understand why tasty butterflies would sport these menacing eyespots, but why would cobras need to have them on the back of their raised hood? Cobras are abundantly equipped to win a fight,

STARTLE DISPLAYS *Owls can sleek their feathers to avoid being noticed, or fluff them out to intimidate enemies.*

DISTRACTION DISPLAY *Lizards known as skinks can drop their tail in an instant to throw a predator off track.*

but like other animals, they don't want to have to prove it. Spooking their enemy with owl-spots lets them save their venom for feeding time, when they need it to put struggling prey out of commission.

Animals that don't have venom to back them up will often act as if they do in order to fool predators. When cornered, a small cat (both wild and domestic) will spit and hiss, expose its "fangs," and switch its tail sinuously back and forth. When the snakelike sound is emanating from a dark cave or crevice, most predators will think twice before sticking their head in. A cut-throat finch in Africa does the same routine from its cavity nest, adding a snakelike body twist to its performance.

Another diplomatic way out is to offer the predator a part of yourself for consumption. It's best to offer something that you don't need for immediate survival and, better yet, something that can grow back. For many lizards, the peace offering of choice is the tail, sometimes brilliantly colored to provide a decoy target for the predator. Some lizards won't even wait for the predator to strike; by squeezing muscles at the tail's base, they can shed it themselves. By the time the duped predator subdues its wriggling, wormlike "prey," the real meal is long gone.

Snakes will also sacrifice their tail sometimes, even though they can't regrow it. To deflect the predator's attention from more vulnerable parts, they may turn the tail over, waving the red or pylon-orange underside like a target. The rubber boa has gone so far as to evolve a tail shaped exactly like its head. By the time a predator strikes down this weaving, taunting "head," the real brains are well on their way into the weeds.

When all else fails, some animals may try to fight back, all the while hoping their struggle persuades the predator that this morsel is not worth the trouble. The weapons brandished by cornered prey are impressive, ranging from the paralyzing saliva of shrews and the stinging spray of skunks to the sharp hoofs of giraffes. These showdowns provide some of the most dramatic moments in the natural theater. Though predator and prey clashes are a legitimate and necessary part of nature, many people would still rather not see them at the zoo. So until naturalistic exhibits become truly realistic, these scenes will be left to your imagination.

HOW ANIMALS INTERACT

Individual behaviors are fascinating and, for the most part, easy to see at the zoo. But it's social behaviors that really excite us, perhaps because we ourselves are social animals who have succeeded, in part, because we are able to communicate our thoughts, hopes, fears, and intentions. While we rely on the written and spoken word, language is only one of the ways we communicate. Gestures, facial expressions, and even a certain stance can reveal exactly how we feel and are many times more telling than mere words. "Body English" and our responses to it are genetically programmed in *Homo sapiens*. The laugh, smile, smirk, and look of surprise are performed and understood worldwide in cultures ranging from the aboriginal to the corporate.

Most animals you'll meet at the zoo practice some form of communication, which, though different from ours, is no less effective. Without communication, animals would not be able to form their elaborate societies, nor would we animal watchers have such fascinating social behavior to view. As you'll see in the following sections, animals have surprisingly similar ways of expressing themselves, regardless of whether they are in harems, hierarchies, or matriarchies. Learning the basic communication protocols will help you to "translate" almost any exchange you'll see at the zoo, be it friendly, feisty, sexual, or parental.

SOCIAL LIFESTYLES

To understand the behavior of zoo animals, it helps to know *what social lifestyle these animals practice in the wild*. In other words, do they normally live in herds, family groups, bachelor groups, harems, or off on their own? The interactions you will see in the zoo are drawn directly from these lifestyle scripts. For instance, you can count on pandas, which are essentially solitary in the wild, to all but ignore each other in their exhibit, unless it's breeding season. Monkeys, however, are constantly interacting—touching, hooting, playing, and grooming—proof that they have a more complex social culture. Each of these cultures has its own rules and reasons that ensure the survival of individuals and their genes. The cultures differ for birds, mammals, and reptiles and amphibians.

Reptiles and Amphibians

For reptiles and amphibians, the rule is to live a rather solitary life, interacting only to defend a territory, find a mate, or pass the winter in a huddle. Frogs sing, float, and fight in large communal groupings, but only when they're hoping to find a mate. In the winter, hundreds or even thousands of normally solitary snakes will mass together for warmth in underground pits called hibernacula. Once these events are over, amphibians and reptiles usually go their separate ways. For many, even parenting is a drop-your-eggs-and-run proposition, with little or no contact with the newly hatched young. There are notable exceptions, however, including the meticulous crocodile mothers described in this book.

Birds

Most species of birds are monogamous, which means one male and one female form a pair bond that may last anywhere from one breeding season to a lifetime. Without a mate, they would have a hard time feeding their young, sheltering them from weather and enemies, and teaching them the ropes before sending them off on their own. In some species, the parents get additional help from fledged juveniles that stick around to raise their younger siblings. These helpers presumably benefit by practicing their parenting skills and by helping their siblings (who share their genes) survive in the world.

Many birds, even if they are normally territorial, will let down their guard and group up for special purposes, such as migrating, roosting, or feeding in flocks. Some species, like flamingos, frigatebirds, and penguins, join flocks to breed, building their nests in noisy, crowded colonies or rookeries. Birds such as ostriches take the communal atmosphere one step further; up to 10 hens may lay their eggs in the same large nest.

Mammals

Mammals tend to belong to either (1) solitary societies, in which the most complex social unit is the female and her offspring, or (2) polygamous societies, in which one male monopolizes several females in a harem, tirelessly fighting off would-be suitors, or (3) permanent groups with even more complex structures. You'll most often find permanent groupings among the ranks of the marsupials, carnivores, hoofed animals, and primates. Many of these societies are organized matrilineally (as elephants are); mothers and offspring stay together, forming groups of aunts, sisters, mothers, daughters, and nieces. In other societies, such as gorilla troops and lion prides, adults of both sexes revolve around a dominant male leader. In wolf packs, a dominant male and female have ultimate control and are the only members of the pack that mate.

GROUP NESTING *Like many birds, magnificent frigatebirds nest in colonies where their eggs are less likely to be wiped out by predators.*

How did all these different strategies evolve, and how do they benefit their members? Part of the answer lies in the habitat: the type of food available, the terrain they must traverse, and the predators they must face. The most complex societies have developed among mammals that live in a relatively open habitat, especially one that contains large, dangerous predators. For these animals, it pays to have more than one set of eyes on the lookout and to be able to mount a group defense when predators attack. Part of the beauty of living in a group is simple statistics. If an animal is alone when a predator strikes, it will surely be eaten. If there's another animal nearby, it suddenly has a 50:50 chance of surviving. The more animals there are in a group, the more choices the predator has, and the better each animal's survival odds become. Birds flock together for similar protective reasons.

Members of a group also tend to eat better than loners do, especially if the food is scattered in patches over a large area. One bird on its own would have to search long and hard to scout out the best berry patch in a large ravine. Having a flock on the lookout cuts the search time way down; if one bird finds a rich store, the others quickly learn of it, and they cash in too. Some mammals also hunt cooperatively, allowing them to bring down prey much larger than themselves. With more meat on every carcass, lions, wolves, and hyenas can stock up so they don't need to hunt as often. Group membership has its advantages even when it comes to nonfood resources. A gang of elk that shows up at a salt lick, for instance, can easily drive out an individual animal such as a mule deer or a mountain goat.

Grouping up also means that males and females are at the same party when breeding time comes. This saves the animals the trouble of tracking down mates and helps them synchronize their mating cycles so all the young are born at once. A crowd of chicks or calves swamps the predator with choices, improving the survival odds for the vast majority of the group. Besides swamping predators, group breeding paves the way for communal day-care, as practiced by penguins, lions, giraffes, and bottlenose dolphins, just to name a few.

GROUP ALARM *When predators are afoot, pronghorns warn distant herd members by flaring the white hairs on their rump patch.*

When parents are free of the kids for a while, they can forage for better food, which in turn strengthens their young.

WAYS OF RELATING

Once you understand animals' social lifestyles, you're ready to move on to the next question: *What relationship do these particular individuals have to one another?* Are they relatives? Father and offspring? Mates? Friendly herd members? Or are they rivals? Without knowing what animals mean to one another, you have no context for their behavior, and your interpretations can miss their mark. For instance, mounting between two animals may not be sexual at all; between siblings, it may be a form of play. By the same token, what may look like fighting could actually be foreplay between lifelong mates. Your best bet is to ask the zookeeper to describe the family tree of the animals on exhibit. Once you know the relationships, you'll know what behaviors to be on the lookout for. The adjacent chart will get you started.

WAYS OF RELATING	
IF ANIMALS ARE . . .	LOOK FOR THESE BEHAVIORS . . .
Herd members	Friendly or aggressive (depending on rank)
Competitors (for mates, food, water, space, etc.)	Aggressive
Mates	Sexual
Parent-offspring	Giving or soliciting care
Siblings	Playing or fighting for rank

THE ART OF COMMUNICATION

When you watch two animals meet, the air seems charged with suspense. For each animal, there is a certain risk: Will this individual attack me, run from me, steal my food? Does it want to mate, groom, or play with me? Is it bringing food for my off-spring? Having no way to identify the animal or predict what it might do would be a recipe for disaster in the natural world.

Instead, animal greetings seem to have a deliberate formality to them, with both members exchanging information according to a set of agreed-upon rules. Their behaviors follow a predictable course: dogs smell each other, polar bears circle, zebras touch noses. These displays illustrate the art of communication, conceived and refined for eons by natural selection. Each animal shares information about who it is and what it is about to do next. The sender benefits by announcing its intentions, and the receiver benefits by being forewarned so it can decide how to respond.

The value of clear, unambiguous communication is obvious. Consider a human example: you are sailing a boat that has a 30-foot mast, and you must pass under a bridge that has only 20 feet of clearance. Obviously, it's important that you be able to signal clearly and understandably that you wish to sail through and you need the bridge-master to raise the bridge. Proper nautical protocol is to signal with three horn blasts as you approach a bridge. The signal is the three horn blasts; the message is "This sailor wants to sail under the bridge"; and the meaning is what the bridge-master understands, "This boat intends to sail under the bridge, so I'd better raise it." Communication occurs successfully when the sailor's action of blowing the horn three times causes the bridge-master to change his or her behavior and open the bridge. Without a signal that both parties understand, there would be chaos every time a boat wanted passage.

Of course, in nature, there are no official commissions to come up with these communication protocols. Instead, most signals are genetically programmed, in the same way as our involuntary smiles, frowns, and grimaces are. Natural selection rewards animals that can transmit unmistakable messages as well as those that can recognize a signal and apply the right meaning to it. These animals live longer because they know how to read a rival's mood, and they reproduce more successfully because they know how to neutralize a mate's aggression.

Natural selection also favors accessories that amplify a message. For instance, if light skin patches above the eyes accentuate facial expressions, then individuals born with eye markings will be the fittest, and this trait will spread through the population. Other traits such as crests, fancy plumage, and vocal control might also exaggerate or draw attention to the signal. The mere fact that animals are trying to draw attention to themselves tells you what a big deal communication is. Anything that prompts an animal to give up its usual quest to blend into the background must be worth the risk!

Besides calling attention to itself, the signal must be readily recognizable as an attempt to communicate. To help animals distinguish displays from ordinary movements such as feeding or preening, natural selection has exaggerated, abbreviated, or made the movements rigid through a process called ritualization. This ritualized quality also makes it easy for zoogoers to know when an animal is communicating.

A GUIDE TO COMMON SIGNALS
AND WHAT THEY MEAN

Signals or displays carry two basic kinds of information: nonbehavioral (who and where) and behavioral (what and how). Nonbehavioral messages identify the animal by species, gender, rank, and sometimes even individual identity. Examples of non-behavioral signals are distinctive striping patterns, hormone-laden odors, or territorial calls. Knowing the gender, status, and identity of the sender gives the receiving animal a *context* for future behavior. Since males behave differently than females, and dominants differently than subordinates, knowing gender and status will help the receiver know what to expect. A nonbehavioral message also tells the receiver *where* the sender is—in the next canyon, on top of the hill, or right over its shoulder. Behavioral messages, on the other hand, deal with action. They tell *what* an animal is likely to do next and *how* it is going to do it (weakly or vigorously, for example, and in what direction).

To put the nonbehavioral and behavioral messages in human terms, think of referees on a football field. Their uniform (nonbehavioral) identifies who they are, and blowing a whistle and throwing a flag (behavioral) signify that they have spotted an infraction of the rules and are about to call a penalty.

The Medium and the Message

It's easy for us to grasp that the referee is giving a signal because we naturally communicate in sounds and gestures. Animals have other ways to get their messages across, however, ways that may be too subtle for our senses to detect.

Communication signals have evolved to fit each animal's lifestyle and to carry effectively in its habitat. Animals with territories that take days or weeks to patrol, for instance, need to advertise with something long-lasting, something that will keep on broadcasting even when they are gone. Since odors have the ability to linger, animals often anoint their territories with scent marks that other animals can read as they pass by. Animals that operate in the dark depend on sounds as well as odors, whereas day-active animals are more likely to use visual means to communicate. You'll see the most dramatic visual signals exchanged between animals of the open plains, since, in their native habitat, they have the luxury of unobstructed views. Birds in dense forests usually can't see their receivers, however, so they have perfected a rich repertoire of songs instead. Low-frequency songs are preferred, since higher frequencies tend to hit and bounce back from objects such as tree trunks. Songsters usually perch in the tree canopy, where an "acoustic window" carries their messages most effectively. A similar sound channel is found in shallow seas, as well as in the surface layers of warm oceans. When whales broadcast on this channel, they can send their songs to other whales up to 2,000 miles away.

SCENT MARKING   *With surgical precision, a blackbuck inserts a twig into his facial gland, wetting it with a strong-smelling secretion that will mark his territory.*

For the sender, broadcasting a clear signal to the right audience is as critical to survival as webbed feet or powerful lungs. To complete the circuit, it is equally important for the receiver to be able to hear, see, smell, or feel the message coming through. However a species communicates, it is endowed with complementary adaptations that allow it to pick up the messages of its species. You can often see how animals at your zoo are tuned in to their particular communication channels by noticing their sensory organs. Visual animals have large, light-gathering eyes; tactile animals have sensitive whiskers; and vocal animals have large, twitching ears.

The Principle of Opposites

Here's a helpful concept to remember when you are trying to interpret signals: signals that carry opposite messages typically have opposite characteristics. Harsh, deep calls, for instance, indicate threat, while soft, high calls are used for appeasement or begging. In the same way, if a wolf raises its body, bristles its fur, thrusts its ears forward, and stares intently to show aggression, it does exactly the opposite to show that it is submissive. It lowers its body, droops its tail between its legs, averts its eyes, and flattens its ears back against its head. This principle of opposites is a universal rule that holds true for most species groups. Its main benefit is clarity; it dramatically differentiates the two behaviors so there is no mistaking them.

Facial Clues

While many social encounters occur over long distances (howl to howl or scent mark to scent mark), those that occur face-to-face are far more intense. For animals in the bird, reptile, or amphibian clans, a lack of facial mobility limits how expressive they can be. A gape may be all that these animals can muster. Since the gape exposes a mouthful of sharp teeth, however, it is often enough said.

Mammals, having the most facile faces in the animal world, are often able to move their lips, ears, eyes, and even noses with meaning. Many of these movements are characteristic not just of one species, but of an entire group of species. Primates, for instance, be they monkeys, orangutans, or humans, share a common repertoire of facial expressions. The human smile is a good example; both human and nonhuman primates bare their teeth in a grin when they are delighted or feeling frisky. The similarities in mammal expression don't end here, though. Carnivores, such as wolves and cats, and hoofed animals, such as antelopes and horses, use their eyes and lips and ears in many of the same ways that we primates do. In his excellent book *Animalwatching*, zoologist Desmond Morris explores the similarities in the facial expressions of primates, carnivores, and hoofed animals. Keep these principles of animal gesture in mind when you're watching not only the species in this book, but other mammals as well.

PRINCIPLE OF OPPOSITES *The dominant wolf's inflated posture contrasts sharply with the subordinate's quivering crouch. This clearcut difference helps wolves avoid misinterpretation.*

A QUICK GUIDE TO THE FACIAL EXPRESSIONS OF PRIMATES, CARNIVORES, AND HOOFED ANIMALS

SIGNAL	DESCRIPTION	MEANING	NOTES
EARS			
Relaxed ears	Slightly forward, with ear openings facing sideways	"No worries"	Neutral mode; ears can hear sounds over a wide range
Pricked ears	Stiffly erect, openings facing forward	"Periscope up"	Alert mode; ears and eyes focused on the source of sound
Twitched ears	Moving back and forth rapidly	"Stressed out"	Agitated mode; doesn't know whether to fight or escape
Rotated ears	Begins to flatten ears, openings facing backward	"Getting aggressive"	Halfway to a fight; look for conspicuous warning spots on backs of ears
Flattened ears	Flattened tight against head, openings facing down	"Attacking or being attacked"	Protective mode; keeps ears out of harm's way during fight
Airplane ears	Straight out on either side, openings facing down	"I mean you no harm"	Opposite of alert ears; sign of appeasement; also used in courtship

Pricked ears

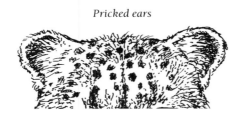

Rotated ears

Flattened ears

Airplane ears

EAR SIGNALS   *An alert cheetah pricks its ears, rotates them back when it anticipates trouble, flattens them for safety during a fight, or sticks them out to the sides to appease a partner.*

SIGNAL	DESCRIPTION	MEANING	NOTES
EYES			
Relaxed eyes	Open, but not completely	"No worries"	Neutral mode; no need for full-alert posture
Staring eyes	Wide open, focused forward	"Periscope up"	Alert mode; widening eyes expands range of vision
Blinking eyes	Rate of blinking increases	"Stressed out"	Agitated mode; blinks rapidly, perhaps to keep eye surfaces clean for action
Frowning eyes	Brow lowered and eyes half-closed	"In danger"	Protective mode; brow down to shield eyes, closed as much as possible without blocking vision
Glaring eyes	Brow lowered, but eyes kept as wide open as possible	"I'm angry"	A contradictory expression; wants to see all, but also wants to protect eyes
Closed eyes	Eyelids shut completely	"I give up"	In a struggle, a subordinate animal closes its eyes completely to switch off all stimuli

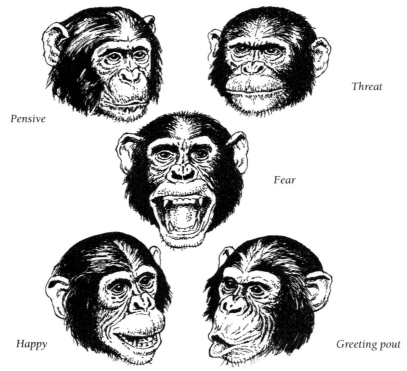

Pensive

Threat

Fear

Happy

Greeting pout

MOUTH SIGNALS *The chimpanzee's mouth is loosely set in repose, tightly tucked for threat, gaping when fearful, smiling when happy, and puckered during a greeting.*

SIGNAL	DESCRIPTION	MEANING	NOTES
	MOUTH		
Relaxed mouth	Lightly closed or slightly open; lips may cover teeth, but without tension	"No worries"	Neutral mode; baseline against which you can compare other expressions
Tight mouth	Closed tightly, lips pressed together in a short, hard line	"I'm concentrating"	Intense concentration; in encounters, a dominant, passively hostile expression
Gaping mouth	Wide open, lips tense, tongue withdrawn, interior of mouth showing; may be a tall snarl or long snarl (see Notes column)	"I will attack if pushed further"	A biting threat; two forms: (1) lips pulled back horizontally (long snarl) indicates fright; (2) lips pulled back vertically (tall snarl) indicates hostility
Play mouth	Open, but lips are covering upper teeth	"Just kidding"	A playful grin to show that mock attacks are not serious
Pouting mouth	Closed, puckered lips pressed forward as if offering a kiss	"Please be nice"	Used by primates in friendly greetings or for begging if the animal is in need of comfort
Lip-smacking mouth	Lips open and shut rapidly; tongue may be extended slightly	"Let's be friendly"	Used by primates in friendly, submissive greetings; derived from grooming actions
Teeth-chattering mouth	Mouth opens and shuts rapidly, but teeth chatter because corners of mouth are pulled back	"I'm afraid, but I still want to be near you"	Combination of lip-smacking and fear-gape; highly submissive signal, used by primates and young horses
Flehmen mouth or lip curl	Neck stretched forward, head tilted up, mouth open slightly, top lip curled up showing teeth and even gums	"What a fragrance!"	Inhaling deeply to "read" an odor, usually the urine of a female in heat

Defensive—long snarl

Aggressive—tall snarl

LONG SNARL AND TALL SNARL *When a wolf's aggression is diluted by fear, the long snarl (corners back) signals defensiveness. The tall snarl (corners forward), complete with open eyes and pricked-up ears, is a sign of pure aggression.*

HOW ANIMALS AVOID CONFLICT

1.	They keep a healthy distance away from other group members.
2.	They find a good excuse for being close, such as greeting, playing, or grooming. These behaviors benefit both partners and make close contact worthwhile.
3.	They avoid provoking extreme arousal or frustration.
4.	They use submissive displays.
5.	They behave predictably so as not to arouse fear or aggression.
6.	They divert their attack elsewhere. Many animals will basically "kick the dog," taking out their frustrations on an underling or an inanimate object rather than on an equal or dominant member of the group.

FRIENDLY BEHAVIOR

Animals that live in a group enjoy perks that they wouldn't have in a solo life: tighter territorial defense, early predator detection, and more efficient foraging. The cost, however, is the natural tension they must constantly overcome to live peaceably in close quarters. In the interest of harmony, animals have evolved ways to avoid conflict or tone down their neighbors' aggression.

Gestures of Goodwill

The most common friendly behaviors you'll see in zoos are greeting, submission, grooming, and playing. Animals use greeting ceremonies the way we use handshakes, to assure each other that neither intends to fight. The sniffing or touching that accompanies a greeting gives the animals a chance to collect clues about each other. One sniff, for instance, might tell them who their companion is, what sexual state they're in, and even what they've been eating lately.

Whereas greeting is a gesture between equals, submission is an attitude that inferior animals adopt to appease dominants and hopefully save their hide. As long as a subordinate stays submissive, the dominant can keep its cool, and harmony will prevail. A submissive act may be as subtle as moving out of the dominant's way or as blatant as groveling before a superior in a show of infantlike helplessness. Very often, a submissive individual will offer to groom a dominant one, a gesture that not only cleans the partner, but also seems to soothe it, dissolving its knee-jerk urge to attack. Grooming can also be a way to build trust between two parties, which is why petting our dogs and cats and brushing our horses work such wonders.

SOCIAL GROOMING   *Grooming can be a social lubricant as well as a cleaning exercise.*

Playing, though it is not always soothing, has its own set of rules that outlaw true aggression. Many animals invite a partner to play with a special invitation stance that says, "There's no telling what I'll do next, but I'm only playing." That's what your dog is saying when it lowers its forequarters on outstretched paws, bounces its hind end back and forth, and grins like a kid. Even fighting, when it's in the context of play, has special rules. Combatants pull their punches by biting without force or by swatting with their claws in. The wrestling matches of large animals like pandas seem to be in slow motion, with none of the intensity of a real duel. Nevertheless, these play matches help to establish a dominance order in certain species, because they give contestants a chance to measure one another's strength and stamina. Play also helps build alliances that may come in handy later in life.

For many animals, the urge to play wanes after adolescence, and by the time they reach adulthood, they're on to more solemn pursuits. Those that are intelligent and curious about the world around them, however, or those that have their hunting tactics down pat, usually have time

PLAY *As these young langurs play chase, they are perfecting the survival moves they will need later in life.*

to play even when they are adults. Otters are a perfect example. When we put playful animals like otters in zoos, we have a responsibility to indulge their exuberant, creative natures by giving them fellow playmates and a stimulating, ever-changing environment.

AGGRESSIVE BEHAVIOR

When people blame their violent natures on their animal ancestors, they're showing a basic misunderstanding of animal aggression. For all practical purposes, animals are diplomats, not warriors. They will go to great lengths to avoid a physical confrontation, and acts of extreme or fatal violence are rare.

It's not that animals don't have anything to fight about; in fact, their lives depend on tough defense of mates, shelters, hiding places, nests, food, and water—resources that are hotly coveted by those that don't have them. The real question is: How much fighting can the average animal afford? If a fight escalates to an exchange of blows, it can easily leave an animal wounded. Though the victor wins the scrimmage, it may lose the ultimate battle for survival if the wound prevents it from finding food or avoiding predators. Fighting, though it may be warranted, makes little biological sense: it's dangerous and time-consuming, and it wastes vast amounts of energy.

Property and Politics

This is where diplomacy comes in. Natural selection has favored systems that allow an animal to defend its rights without having to jeopardize its own life. In one system, animals divvy up the available space in a habitat into territories. Each territory holder advertises its claim with visual displays, scent marks, or sound, and patrols the borders to guard against trespassers. Neighboring claim holders may face-off at the "fence line" occasionally, but even then, they prefer to stay on their side. In areas where the habitat is in good shape, the system works beautifully; almost every animal has the privacy and space it needs to find food and shelter, court a mate, and most importantly, pass on its genes. The territory system is also flexible; some animals defend an area only when breeding or when they want to keep an abundant source of food all to themselves. The same animals can turn off their security alarms to join a herd at the water hole or when migrating in search of food.

A second way to avoid serious fighting, especially in species that travel in permanent groups, is to establish a dominance hierarchy. The best-known example is the linear pecking order of chickens, in which the flock members are arranged on the rungs of a social ladder, with each chicken superior to those below and subordinate to those above. The top animal in a hierarchy has priority access to the necessities of life: the best food, mates, nest sites, and so on. Subordinates get the leftovers and spend their time brownnosing the dominants—yielding to them, grooming them, and even helping to care for their offspring.

In many species, the organizational chart is more complicated than it is in chicken yards. Wolf packs, for instance, are led by two dominants and have three subclasses beneath them. Other systems feature a grand master at the top and a herd of equals beneath him or her. Within the ranks, animals often figure out who's who through threats and fights early in life or as soon as strange animals are thrown into an enclosure together. Individuals that have been beaten by a superior are usually reluctant to risk a rematch, so the ranking order generally maintains itself. There are so few overt signs of control, in fact, that you may not even recognize the dominance hierarchy until a squabble arises in the group. Then the order reveals itself, as one writer said, like an image on photographic paper dipped in developer fluid. Uprisings that lead to a shifting of power do occur, but they are the exception rather than the rule.

In more intelligent species such as apes, political factors beyond mere size and fighting ability begin to shape the hierarchy. Factors that affect dominance status include age, seniority, hormonal condition, personality, maternal lineage, and alliances with friends. But even here, clashes rarely erupt into serious battles, and disputes are quickly settled with harmless screaming and chasing. Given the laws of these systems, the aggressive behavior you can expect to see at the zoo will primarily be "safe aggression."

When Diplomacy Fails

The two basic motives for aggression are self-defense and property protection (property may include offspring as well as physical objects). Other aggressive encounters may involve mating rights and parent-offspring conflicts. No matter what the motive, however, most disputes follow a predictable course, designed to make aggression as survivable as it can be.

Step 1. Threat display: The first step is to issue a warning in the form of a threat display. Dominants warn subordinates, "I am your superior"; residents warn intruders, "This property is mine"; and prey animals warn predators, "I am willing to defend myself and my offspring." The threat displays usually show off the animal's weaponry, size, strength, and willingness to back up its claim (even though it usually doesn't have to). Most conflicts end here. A searing glance may be enough to make an animal think twice, turn tail, and run.

Step 2. Ritualized fight: If the threat doesn't register, the next step in the escalation is ritualized fighting. This is usually an abbreviated form of the real thing, with punches pulled and safeguards built in to avoid real injury. The ritualized fight is a taste of things to come, and it gives the opponents a chance to test each other's abilities. Once it becomes clear who is stronger or faster, the conflict usually fizzles, and the weaker animal slinks away.

Step 3. Actual fight: If either animal is in extreme pain or fear, the conflict may escalate to an all-out fight. This is especially true when one of the animals feels trapped, either physically or because it has a mate, offspring, or territory that it can't leave. Even in all-out warfare, physical adaptations and conventions make the fight as safe as possible. For instance, animals have reinforced skin or bones in the places where they are most likely to be bitten, rammed, or kicked. Animals also tend to fight defensively as well as offensively, instinctively protecting their tender spots.

THREAT   *A hippopotamus flashes its weapons in a threat gesture that looks like a yawn.*

While it's true that animals are rarely violent toward one another, all bets are off once they become overcrowded. The classic example of overcrowding is to place the nests of two territorial fish in the bottom of an aquarium. If you put the nests at opposite ends, the fish will meet at the center of the aquarium, perform their territorial threat displays, and peacefully swim back to their nests. But if you move the same nests toward the center, the fish no longer have a sphere of comfort and will fight constantly and unnaturally to their deaths. The lessons for humans are obvious, and, as we continue to grow at a rate of a quarter of a million people a day, ominous as well.

FIGHTING *In red kangaroo boxing matches, opponents lock forearms and shove each other backward, waiting for a chance to drive home a kick.*

SEXUAL BEHAVIOR

It was not too long ago that sex, the newfangled approach to multiplying organisms, burst onto the evolutionary scene. Before that, no partners were involved; organisms simply split in two and produced carbon copies of themselves. A colony of asexual offspring was the ultimate in harmony, since every individual had the same genetic makeup and, therefore, the same genetic interests at heart. Peace aside, cloning didn't win particularly high scores for innovation. Everything was fine as long as asexual organisms remained in a stable environment. If the environment suddenly changed, however, cloning didn't allow genetic mutations to happen often enough for the species to adapt.

It took sex, the combining of genes from two different individuals, to produce the novel combinations that allowed animals to change quickly. When sperm met egg, the random shuffling of genes was much more likely to produce a mutation—a new twist on an old theme. If this mutation proved beneficial to survival, it was *adaptive*. The lucky individual would survive and pass on the adaptive trait, and if it were truly a breakthrough, the trait would spread quickly through the whole population. In the same way, a bad mutation would be edited out of the population, since the tainted individuals would usually not survive to reproduce. This ability to adapt quickly allowed early organisms to emerge from the sea and colonize the land and the skies. Today there are millions of sexually reproducing species changing and evolving in response to their environments.

But sex, for all its advantages, also has some drawbacks. Primary among them is the fact that members of a sexual species can't produce offspring by themselves. There must be a union of egg and sperm, small cells that reside inside two unrelated individuals. To get the two together, an animal must find and attract a mate of the opposite sex, confirm that they are of the same species, reduce the mate's aggression, arouse its sexual interest, and then synchronize its partner's sexual cycle with its own. It's a lot to orchestrate, especially when you factor in the mates' conflicts of interest in both the genetic and the gender departments.

Strange Bedfellows

As we have seen, animals are more willing to cooperate with animals that share their genes. Since mates don't have any genes in common (a "rule" that prevents inbreeding), there's very little to temper their natural distrust for one another. As they get closer and closer to mating, the animals are held prisoner by three competing forces: sexual attraction, fear, and aggression. Sometimes the fear is on top, and the animal's retreating movements flash the signal "I am about to flee." Counteracting these are forward-approaching movements that reveal a strong sexual desire. As the animal approaches, it may suddenly express aggression, brought on by the mere proximity of a stranger. The three driving forces war with one another, like elastic bands pushing and pulling the animal in different directions. Though it may try to escape or launch an attack, its desire to stay pins it to the spot and holds its aggression in check. You can actually see the ambivalence as the animal bobs its head forward and back as if to say, "I'm staying, I'm leaving."

This visible ambivalence has been ritualized through eons of evolution until, eventually, the back-and-forth, flee-or-approach movements have become elaborate stereotyped displays. To the potential mate, these displays prove that the performer is in turmoil and that its motivation for staying is probably sexual rather than aggressive. Once fear and aggression are neutralized, the flames of arousal can win out in the war of emotions, allowing the pair to get close enough to copulate.

Opposing Game Plans

To further complicate sex relations, the male and female strategies for reproducing themselves are worlds apart. To understand the reason for this division, we have to travel deep inside the animals' bodies and take a look at the cells that do the work of reproduction—the female egg and the male sperm. Talk about differences! The human egg is 85,000 times larger than the sperm, and that outsizing is echoed in all species, from newts to blue whales. Naturally, it takes a lot more energy and resources to produce an egg than it does to produce a sperm. In the case of humans, a female will produce, at most, 400 eggs in her lifetime. The male, on the other hand, produces up to 300 million sperm with each ejaculation! Each sperm is so tiny, however, that the energy invested in it is minuscule.

Once the egg is fertilized, the female pours even more energy into the embryo, and, in many species, she continues to invest in the young even after birth. To protect her investment, it pays the female to find only the fittest father, perhaps one that will help defend or feed her offspring. For the male, whose initial investment is minimal, promiscuity pays. His strategy is "quantity not quality," and he seeks to impregnate as many females as he can, taking his chances that some of his sperm will find their mark and develop into adults that carry half his genes. The female, by contrast, puts her bets on only a few good mates.

These differences are most obvious when the female of a species is capable of raising the young successfully on her own. The male, assured that his genes are in good hands, dedicates his energies to mating with as many females as possible. This is especially true with mammals. Since the females have the exclusive patent on milk-producing breasts, many males need never look back on the raising of their offspring. These males belong to the 95% of all mammal species that are nonmonogamous.

In a scant 5% of the mammal species, however, it takes two individuals to raise the young successfully. In these species, males change their strategy and act more like females—staying close to home and lavishing attention on a few offspring. Male lions are needed, for instance, to protect the cubs from enemies, and male foxes are needed to bring the nursing vixen enough food. If these males were to leave the female, the chances of their offspring surviving would be negligible. Choosing quality over quantity, therefore, they opt to stay. Ninety percent of all birds are monogamous for the same reason; the female can't incubate eggs, brood the chicks, and catch food for the family all by herself.

In our Western culture, which values monogamy, people seem to approve of animals that form long-term pair bonds. It's important to remember, however, before we jump to biased conclusions, that these animals are not displaying "true love," but simply following the dictates of their genes. They are survival machines, and their mission is to multiply their own genes in the gene pool. If a male felt that his partner could raise young without him, he'd be off in a flash. But this would not be abandonment in our terms, and we need not feel sorry for the female. Both are pursuing their own game plan that will lead to the best positioning of their genes—a pursuit that is adaptive and, therefore, beautiful.

Courtship Displays

Now that you know how dicey the sex game really is, you can see why courtship displays are so important. Miraculously, these displays encourage the merger of sperm and egg, even between mates that have opposite agendas and a nagging reluctance to be together. As you watch these mating preludes at the zoo, try to see how well they perform their functions.

WHY ANIMALS COURT ONE ANOTHER	
FUNCTION OF THE COURTSHIP DISPLAY	HOW IT HELPS
Attracting	Brings individuals together, even if they are separated by long distances
Identifying	Ensures that only members of the same species mate, so they don't waste their energies on unions that are infertile
Arousing	Overcomes aggression and fear
Signaling quality	Helps female judge whether male is a good genetic specimen, or whether he will help raise the young
Synchronizing	Brings the male and female sexual cycles into sync
Signaling readiness	Lets individuals know that their partners are ready and willing to mate

FEMALE DISPLAYS

Signs of Estrus   Unlike humans, who breed throughout the year, many species have a select time during which they mate—usually when the female is in estrus, or in heat. In some species, estrus is a monthly event, while in others it occurs less frequently. The season of peak sexual readiness for pandas, for instance, may be only 2 or 3 days out of an entire year. Lest they miss this window of opportunity, natural selection has favored the development of unmistakable signals (unmistakable at least to other pandas) that tell a male when a female is in heat. Each species has its own special signals, which may include a flushed and swollen genital region, a beckoning invitation stance, or a chemical marker embedded in the female's urine. Some females mark their home range (or their enclosure) with these scents, leaving a fragrant calling card that indicates how close to estrus they are.

Playing Hard to Get   Even when a female is in estrus, she has good reason to take her time before mating. A female peacock, for instance, will let a male display his heart out while she casually feeds, giving him no more than a bored glance. With each refusal, she drives the male to improve his performance. His showmanship and stamina may tell her something about how healthy he is, and perhaps how fit his gene type is. In the meantime, the female may be buying herself time to become physiologically ready to copulate. By the time she acquiesces, the male's desire is likely to match hers, helping him overcome any fear he might feel.

COURTSHIP   *Arabian oryxes stiffen their forelegs to court their mates. Birds of paradise hang upside down and flaunt their beautiful plumage. Tortoise mates ram shells, trying to flip each other over.*

Surrender or Solicitation   A similar courting exchange occurs in many species; the male does his best to persuade, while the female evaluates his fitness as a sexual partner. Certain qualities of his antlers, mane, shoulders, or simply his superiority over other males tell her that he is a good genetic specimen. Signs such as nest building or courtship feeding may also tell a female something about the male's ability to provide for her offspring.

In many species, once the female has weighed the evidence, she makes her choice and sets her sights on becoming impregnated. She will first have to overcome her own fear, especially since her choice male may be considerably larger than she is (elephant seal males, for instance, are twice the size of females). When she has conquered these fears and is at last stimulated enough to mate, she indicates this to the male through a form of surrender or solicitation. Surrender may be as subtle as her stopping, ending what has been a steady chase-and-avoidance ritual. Or she may take a more active stance, soliciting the male by caressing him, backing up to him, presenting her rump, or crouching in place before him.

MALE DISPLAYS

Testing Female Readiness   In mammals, a female's urine often holds a bouquet of scents that tells the male what state she is in. Many hoofed species, including zebras, rhinos, and giraffes, practice a peculiar testing ceremony called *flehmen* or *lip curl*. After nudging the female's bladder, the male sniffs or drinks some of her urine. If the stream doesn't flow, he may simply fill his nostrils with the scented air around her genitals. To get the full effect, he leans his head back, points his nose skyward, and curls his lip up over his nostrils to trap the air so it can be analyzed by scent-sensitive cells. He wears a dazed, faraway expression as he samples, like that of a hungry person savoring the scent of cooking in the air. This analysis may tell him many things, such as the female's age, how far along in her sexual cycle she is, or if she is currently fertile. Knowing whether she will be receptive is important; a male that makes his move too soon may be greeted with sharp, snapping teeth or a well-placed kick.

Competing with Other Males   Because there are usually too few receptive females in a population to go around, sex becomes a competitive sport. Some males deal with the shortage by setting up a breeding territory and keeping all other males off-limits. Birds sing to announce their territory, frogs chorus, and many mammals mark their homelands with scent. These signals act not only to warn away competing males, but also to attract a female. Females that are lured to the area may mate with the resident and then take advantage of the food and cover that his territory affords.

A second type of territory, held by polygamous birds, is called a *lek*. Leks are small display stages of several square feet or yards, checkerboarded through a field like sites in a campground. Lekking birds, such as grouse, peacocks, and turkeys, fight to keep all other males off their small display stage. As the females wander by, the curtain goes up and each male performs, trying to outshine his opponent and win the female's eye. In these visual battles, the most impressive-looking males seem to have a certain edge over less impressive-looking rivals. The females pause to mate with the most sensational of the performers.

A third strategy in these sexual Olympics is to temporarily sequester a harem of females as your own personal mates and keep all other males away from them. Other males are of course anxious to cut in on the action, and so the harem master's job is strenuous indeed. When he is busy threatening or fighting with one male, another may sneak in for a furtive mating. If something happens to topple the harem master, such as disease or a thrashing by a rival, one of these males waiting in the wings will quickly take over.

Taking Chances Many of the male's displays seem to be a form of showing off: the prairie chicken's booming, the elk's antler-bashing in the bushes, or the common snipe's aerial daredevil stunts. But these displays are not about vanity. They act to dazzle the female, but also to say, in effect, "Look how fit I am," or, "Look how well I will protect our offspring from predators." Showing off his shiny plumage may be the peacock's way of boasting that he is free from parasites and disease. In the same way, when males build elaborate nests or bring females gifts of food, they are saying, "Look what a good provider I can be."

HAREM MASTER *A bull elk bugles during breeding season to advertise his territory and attract females.*

Besides promoting his strong points, the male helps allay the female's fears by signaling that he intends to make love, not war. Even though some of his courting moves may be borrowed from aggressive acts, he performs them with predictable and obvious restraint. This ritualized quality helps to calm the female and put her in the mood for mating. Besides changing her emotional state, the displays may actually change her physiology, bringing her into condition for copulating and perhaps inducing her to ovulate. Without this egg in the pipeline, mating might occur for naught.

All of this displaying carries a certain amount of risk. Birds with ornate, 5-foot-long tails and gaudy crests are as obvious to predators as they are to the females they are trying to attract. Sounds and smells may also give these showboats away. Besides being conspicuous, breeding ornamentation can make males less agile when it comes time to evade an enemy or hunt down prey. An elk's rack, for instance, adds as much as 22 pounds to his frame. A lion's bouffant mane, though it isn't heavy, can blow his cover in the grasses, making him look like a moving haystack.

Obviously, if animals are willing to take these risks, the payoff in terms of mating success must be worth it. It does seem that these handicaps help a female to evaluate a male's fitness. If a bird can evade enemies despite his long plumage, he must also have alert senses and quick responses. In the elk's case, only a strong specimen would be able to carry a horny candelabra through the woods and still find enough to eat.

But how far can evolution take this before the handicap becomes more of a detriment than a courting aid? Rather than walk this line, some animals have evolved a way to display their wares only during breeding season. Off-season, for instance, the brilliant plumage of some male birds fades. In an even more discreet design, animals can keep their sexual teasers concealed until the occasion warrants. The green iguana, for instance, is able to keep a low, green profile, expanding his colorful throat pouch for the female's eyes only.

The third way to manage risk is to not wear the bright colors of breeding at all, but rather to decorate a nest as an extension of yourself. The fascinating bowerbirds and weaverbirds take great pains to build a structure that will impress females and put them in the mood to mate. The satin bowerbird collects blue objects to decorate his palace, and even paints the walls with a stick dipped in homemade, berry-blue dye. The more garish the abode, the less important it is for the bird itself to be brightly colored. Indeed, the most elaborate builders in these groups have the drabbest (and most predator-proof) plumage.

To Fertilize an Egg

Copulation must be just as carefully orchestrated as courtship. This is the moment of ultimate fear/attraction, because the animals are actually touching, often in positions that leave them vulnerable to predators. The main goal of all this maneuvering is to transport a live sperm to a healthy egg. The safest deliveries occur in animals that have some way to plant the sperm inside the female's body, where it can't get lost, swept away by the current, or broiled by the sun.

One of the best designs occurs in mammals, where the males have penises and the females have vaginas, so the fragile sperm can be tucked close to the protected egg. Some male birds, such as ducks, geese, game birds, and ratites, also have a penis that resides in their cloaca, an opening which serves both excretory and reproductive systems. The penis emerges during copulation, and with it the male inserts the sperm safely inside the female's cloaca. Other species of birds and reptiles also fertilize internally through cloacas, but without the benefit of penises. When mounting from the rear, the male must twist his body under the female so the swollen lips of their cloacas can come into contact for the so-called cloacal kiss.

Salamanders (which are amphibians) have yet another method of sperm transfer. On the ground, the male deposits a sac of sperm perched on a gelatin stalk like a tiny golf ball on a tee. With his courtship conga dance, he leads the female over this stalk, so she can scoop up the sperm packet with her cloaca. Frogs have the least secure system of all. The female simply drops her eggs in the water, and the male follows behind, covering the eggs with a spray of sperm. The success of this method lies entirely in the timing of the dual release.

Once the deed is done, most animals quickly part. Fear and aggression reemerge to fill the void left by the waning sexual drive. Lionesses often get in one last snarl or swat, sending a message that the honeymoon, for the moment, is definitely over. (Part of the lioness's crabbiness comes from pain, no doubt; the male's penis has backward-pointing barbs that rake her vagina as he withdraws. This shock induces her to ovulate, as well as to protest.) The exception to this kiss-and-split strategy occurs in dogs, wolves, coyotes, and their relatives. The base of the male's penis swells into a bulb that locks into the female's vagina, causing the well-known copulation tie. As the male turns around to dismount, he stays connected and they wind up standing rump to rump for as long as half an hour. Though this looks awkward and uncomfortable for the couple, it is actually necessary for successful mating. The tie keeps the pair together long enough for the male to release the several ejaculations needed for fertilization to "take."

PARENTING BEHAVIOR

If we were to give out a Parent of the Year award to one animal in the zoo, we'd have a tough choice to make. The fact is, in the animal world, there's no such thing as bad parenting. When cowbirds dump their eggs in another bird's nest and elephant seals accidentally crush their newborns, they are being every bit as "good" as the elephant that feeds, protects, and tutors its young for years. Like other behaviors, each style of parenting evolved because it somehow helped the animal excel in its habitat. In the scramble called survival, the good parent award goes to every animal that manages to leave a successor.

The amount of care the offspring receive depends in part on how mature they are at birth or hatching time. Precocial young are precocious; they can run, feed, and even hide from predators at an early age. Altricial young, however, are born blind, naked, and helpless, needing their parents to shield them from the harsh outer world. It is for these altricial young that nests, burrows, and other shelters are most important.

Animal Nurseries

BUILT-IN SHELTERS

Kangaroos, opossums, and other marsupials give birth to extremely underdeveloped young. A newborn kangaroo joey, no larger than a kidney bean at birth, must climb from the vagina to the teats, which are hidden inside an expandable pouch. As the joey grows, it rides along in this pouch, eventually becoming large enough to peek out. Even after it leaves its mobile nest, it may clamor to get in again when it wants to nurse or be protected, no matter how large it has grown. The mother eventually has to forcibly restrict access to this comforting haven.

Other kinds of built-in papooses are not as well known. The Surinam toad of South America, for instance, carries its developing eggs in pits on its back, providing both moisture and protection from predators. Some seabirds, such as penguins, balance a single egg on top of their feet so it can't roll off the rocky cliffs of their rookery. For species that don't have these built-in nurseries, the next best thing is to create a shelter using natural materials.

BUILT-IN SHELTERS *The Surinam toad has an on-board hatchery where eggs are kept moist and safe from predators.*

BUILDING A NURSERY

Nest building is such an important chapter in some species' behavior that captivity will not squelch it. Zoo beavers, if they are provided with aspen branches, will work furiously to build a lodge, and foxes will dig a den in the dirt floor of their enclosure. Video cameras in these dens offer zoogoers glimpses of domestic life that are impossible to see in the wild. In the same way, a raised walkway may allow you to peer into the construction of a bird nest in the aviary. Nest building can completely consume the animals for a few days or weeks (usually in the spring), making this one of the most exciting times to visit your zoo.

Though we are not always aware of it, animals are raising their young all around us: on the ground, in the ground, in water, on branches, in tree trunks, and even on cliffs. Some occupy the abandoned nests of other animals or squeeze into natural tree cavities and rock crevices, while others build their homes from scratch. The building plans come in many shapes and sizes, from the hummingbird's thimble to the huge condominiums of the weaverbirds. Some dwellings, such as the underground burrows of prairie dogs, are real feats of architectural engineering. Others are no more than thrown-together piles of grasses, like those built by many species of waterfowl.

Nests, like paw prints, are so characteristic that you can often identify the species of nest builders even if they aren't in sight. Although the basics are the same within a species, individuals have some latitude in customizing their nests, and anything in the environment is fair game. An American robin on my college campus, for instance, had a penchant for the tear-off strips from the sides of computer paper. You may be able to see similar kinds of borrowed building materials in the nests of birds in your zoo aviary.

In addition to found items, animals may personally contribute their glandular secretions, feces, or saliva to help hold the nest together. (Gourmet food lovers may be surprised to learn that the main ingredient in the coveted bird's nest soup is the hardened spittle of the edible-nest swiftlet!) Gray tree frogs avoid the danger of ground predators by gathering in trees en masse and stirring the females' secretions into a foamy froth. Tadpoles hatch from the eggs laid in this froth and drop to the ground once they become frogs.

All nests—large, small, sloppy, or sturdy—are designed to protect the young until they are old enough to survive on their own. A nest can cryptically camouflage its tenants, place them out of reach of predators, and shield them from the elements. As you pe-

NEST BUILDING *Many aviaries provide visitors with a bird's-eye view of nature's master architects.*

ruse the nests at your zoo, consider what it is about the design that gives offspring a survival edge. By building a nest that works, the parents buy themselves a bit of genetic immortality—a fair trade, no matter how laborious construction may be.

Birth

Elephant and giraffe calves are in for a big surprise when they're born. Giraffes may drop more than 5 feet to the ground, a fall that gets them breathing, but doesn't seem to harm them. The young of aquatic mammals, such as whales and dolphins, have a much smoother exit; if they need help taking their first breath, their mother gently buoys them up to the surface. On land, mammal mothers usually lick the placental matter from their young, both to remove the odor of birth and to help dry the fur so that chilling won't set in. They may also lick the muzzle to encourage breathing or the anus to get the young to defecate. Where predation is a concern, the mother may eat the young's feces to keep telltale smells to a minimum.

Birds and most reptiles and amphibians develop outside the mother's body in protective eggs. Birds usually incubate these eggs with the heat of their own body, sometimes developing a brood patch—a thinly feathered patch on their chest or belly that gives the egg direct access to warm skin. Most baby birds are born with a sharp "egg tooth" on their beak that helps them cut their way out of the shell. Clearing this first hurdle in the outside world is exhausting work, and, depending on the species, the parents may or may not help.

Alligators bury their eggs in a mound of vegetation, which works like a compost heap to warm the eggs. The female checks the nest regularly, wetting, shading, or removing vegetation to adjust for rises and falls in temperature. This thermostat management turns out to be vital for balancing the sex ratio in the population. Eggs that are incubated below 86 degrees Fahrenheit will be all females, while those kept above 93 degrees Fahrenheit will be all males. Somehow, the mother keeps temperatures in the nest just right, giving birth to a balanced mix of genders. When it comes time for the young reptiles to hatch, they begin to cry out, prompting the mother to come running. She digs them out of their compost incubator, helps break them from their shell, and then carries them down to the water in her mouth, all without a scratch!

In contrast, most frogs, salamanders, and toads practice a lay-them-and-leave-them strategy. They deposit their eggs, coated with protective jellies or shells, in sheltered, moist areas. Since both eggs and hatchlings make great predator fare, they are usually produced in staggering numbers, in the hopes that at least a few will survive to adulthood. Snakes, depending on the species, either give birth to live young (tiny replicas of themselves) or else bury eggs to hatch on their own. Either way, snake parents are well on their way to other pursuits shortly after passing on their genes.

Getting to Know You

While some animals play the odds by producing vast numbers of offspring and leaving them to fend for themselves, others pour energy into raising just a few young. In birds, the relationship between mother and offspring often begins while the chicks are still in the egg. The young birds peep and the mother answers, as if urging them to come out. The two keep up this soft and incessant contact for days or even weeks.

Mammals are particularly interested in getting to know their newborns, and the mothers often use grooming as the vehicle. In cleaning and cuddling sessions, the female learns the sound of her young's voice, its smell, and perhaps even the taste of its skin. This initial getting-to-know-you period is essential, especially later in life when the mother and young must find each other in the confusion of a herd of animals that look very much the same.

In the first hours or days of life, the young of many species fixate on the first moving object they see, forever associating that figure with "mother." This process, called imprinting, ensures that the young animal will know exactly where to run when it's hungry or in trouble. Unfortunately, depending on what they see after birth, animals can easily imprint on a human, a dog, or even a truck or tricycle.

Imprinting comes into play again when adult animals are looking for a suitable mate. Studies have shown that swans seek out mates that look somewhat like their parents (this ensures a species match-up), but that also look different enough. Staying away from animals with the exact markings of their parents keeps them from inbreeding with close relatives. It makes you wonder about our human incest taboo; might our aversion to mating with relatives be biological as well as cultural?

A Parent's Job Is Never Done

KEEPING CLEAN

Mammals have neither Baby Wipes nor Q-tips, but nonetheless manage to keep their young clean, odorless, and free of parasites. Cats use their strong tongues to lick the fur clean, primates part the fur with agile fingers, while hoofed animals, such as zebras, nibble away impurities. For gregarious mammals, grooming is an important glue, cementing the mother-child bond as well as other social bonds later in life. Cleaning is especially important in the confines of a nest or burrow where diseases and parasites can multiply, and smells can tip off predators. Many mammals actually eat their young's feces, while birds carry the prepackaged sac of wastes away from the nest.

MOUTHS TO FEED

Mammals have a built-in food dispenser—the breast—that enables them to provide high-quality, fat-laced nourishment to their offspring without leaving the den. Black bears begin feeding their young even before they officially wake from hibernation. Having their food "on board" saves them a trip into the wintry outdoors.

Since birds don't have the luxury of breast-feeding, species with dependent chicks must frequently leave the nest in search of food (which is why two parents are needed in most bird families). During their rapid growth stage, chicks are essentially stomachs with gaping beaks attached. The beaks are commonly lined with bright colors or stripes that help guide the parent toward the all-important target. So powerful is the instinct to fill this gaping beak that a bird will work for hours on end to gather food. Many go out of their way to gather special baby food. Seed-eating birds, for instance, deviate from their normal diet and begin snapping up protein-rich insects. Raptors tear their live prey into extra small chick-size chunks. Seabirds make a blended fare by eating the fish themselves and then regurgitating it for the youngsters.

Eventually, both mammals and birds must wean their offspring away from breast-feeding and nest handouts. Wolves and foxes bridge the transition from milk to solid food by digesting food themselves and then regurgitating it at the den. Eventually, they will lead the young to carcasses and encourage them to participate in hunts. Birds wean their young by either ignoring the gaping youngsters, now rather large for the nest, or by threatening them, sometimes harshly, until they fly off.

This time of weaning accentuates the conflicts of interest set up by the "selfish genes" inside each individual. Parents that are starting their second brood can no longer afford to share resources. It's in the youngsters' best interest, however, to lobby as long as they can for handouts, even if they no longer need them. Eventually, the begging-and-refusal routine wears thin, and the parents force the young to earn their own living.

FIDDLING WITH THE THERMOSTAT

Many species of young are unable to keep themselves warm when they are born. Their feathers or fur may not be in place, and their internal furnace can't replace all the heat escaping from their skin surface, which at this age is quite large relative to their tiny body mass. At the other end of the spectrum, newborns are not yet adept at cooling off, and can easily overheat in direct sunlight.

In their "spare time," then, the busy parents must also create a comfortable nest climate. In cold weather, birds stick tight to the nest, using their own body heat like a top broiler to warm the nest below. They may also line the nest with downy feathers from their breast, helping to plug air leaks and seal in heat. In hot weather, watch for birds of prey, such as eagles, spreading their wings to shield the nest from sun.

HOLDING DOWN THE FORT

Even the mighty lion needs to watch out for predators when cubs are about. No matter how invulnerable the adult of the species may be, a cub, kit, or nestling is a perfect snack for some animal. Young are especially desirable because they can't run as fast, nor defend themselves as impressively as adults. All species experience some loss of young, held in check only by the evolution of a strong parental defense behavior. Pumped by hormones and instinct, even the most docile animal can be a formidable adversary when its offspring's life is at stake.

In birds, the defense campaign begins at the egg stage. In addition to sitting tight on the nest and warning away intruders with melodies, croaks, or cries, some birds perform a fascinating distraction display. Pretending to be injured, they make a target of themselves and divert the predator's attention away from the eggs or young. The predator, programmed to search for easy meals, eagerly begins to track the "injured" party. At the last minute, the trickster flies away, leaving the predator high, dry, and hungry, but still in the dark about the nest.

Once a predator discovers a nest, it may as well say *restaurant* in blinking neon lights. Predators will keep returning until they are able to slip by the parents and order everything on the menu. Once exposed, parents may either relocate their nests or abandon them completely. Keep this in mind when you clamor for a peek at bird eggs in your yard or woodland. Nest harassment, even with the best intentions, can put enormous stress on animal parents. Rebuilding and relocating drains precious energy that the birds need to gather food for their young.

Sometimes at the zoo, especially with endangered or very sensitive species, a mother-to-be will be taken out of the public eye and given a quiet, safe place to have her young. Round-the-clock watches are scheduled, and every effort is made to protect both mother and newborn. If all goes well, visitors may then have one of the most rewarding of all zoo experiences—the pleasure of seeing a healthy new member of a rapidly diminishing species.

OUTINGS WITH THE OFFSPRING

Sometimes, when parents are on the move, there's no time to wait for small legs to catch up. Traveling cats grab their kittens by the scruff of their neck, miraculously not harming them despite their scalpel-sharp teeth. The young instinctively fall limp in this neck lock. Voles also use their jaws, but they grab their infants by the belly rather than the neck. Bears have the most unusual mouth transport. They actually carry the cub's head in their mouth, a circus trick that seems to serve these gentle animals well.

In other species, the young do the work, saddling up and holding on to mom as she runs, swings through the branches, or paddles on a lake. Loons and swans ride on their mother's back, snuggled between her sheltering wings. Primate babies ride piggyback or cling to their mother's belly, occasionally bumping along the ground as she moves. Kangaroos and other marsupial young hitchhike in their mother's built-in pouch, even when they are ridiculously large. Ducklings have found that the best way of keeping in touch with mom is to line up behind her and follow her every move. Shrews also use a caravan approach, with each young shrew biting the tail of the sibling ahead. The single beluga whale calf "shadows" its mother, swimming in the slipstream right next to her body.

OUTINGS WITH THE OFFSPRING *Like first graders holding hands on a school trip, young shrews bite the tail ahead of them, forming a caravan behind their mother.*

HOME SCHOOLING

It's hard to say whether animals teach their young or whether they simply serve as examples. As the young watch their parents, they may learn which plants to eat, where to go for water, how to avoid danger, how to hunt, and even how to choose a place to sleep. In some species, however, the parents do seem more directly involved in teaching, as evidenced by the fact that they will discipline their young when they see them doing something dangerous. Parents also encourage their young to explore the world around them and to play with one another. This play, perhaps more than anything else, prepares a young animal for the complicated maneuvers of survival.

Communal Day-Care

As with many other human inventions, we come to find that nature invented group day-care long before we did. Flamingos, penguins, ostriches, giraffes, dolphins, crocodiles, and many other species leave their young in the care of other adults for a while. This gives parents the freedom to track down the most nutritious foods for their growing family. Just who are these surrogate parents that care for the young? The sitters may be parents taking random turns, or they may be nonbreeding individuals that are related to the parents.

Though it may look like altruism, the sitters are merely promoting their own genes tied up in the young nieces, nephews, or siblings that they are caring for. If their aim is to further their genes, you may ask, why not just have their own brood? In a stable, fully occupied habitat, there may not be enough nest sites or food available in a given year for new breeders to strike out on their own. Rather than be forced into a marginal nesting site, they might hold off for a year, learning tricks in the interim that will make them better parents.

The gesture of helping may also be good for an animal's "reputation." Species that live long lives and are intelligent may remember individuals that have helped them in the past and be more willing to come to their aid when times get tough. Human society is based on surprisingly similar principles. Even though we are no longer "primates in the mist," doing a good deed still makes sound genetic sense.

CAPTIVE BEHAVIOR

Although many of us equate the word "wild" with freedom, animals live in a different reality. Animals in the wild are not really free to go where they please. They are limited in their travels by competitors, predators, the amount of area they can defend, and by their own energy budgets. In fact, animals roam freely only if they absolutely have to; if they can satisfy their needs without going far, so much the better. Once they mark their territory as their own, they may be quite content to stay within these scented "bars."

In a zoo, quality of space (the plants, other animals, water sources, and other features that make up the exhibit) is usually more important than quantity of space. Unfortunately, however, there are still menageries, circuses, and so-called "zoos" that keep animals in cramped, lifeless exhibits that fall short in both quantity and quality categories. The animals in these places are the ones most likely to develop abnormal behaviors.

HOW ABNORMAL BEHAVIORS START

Many animals develop abnormal behaviors as (1) a way to deal with fearful stimuli that they can't escape, or (2) a way to liven up their existence. To understand these coping mechanisms, we first have to take into account how a particular animal makes its living in the wild.

An Inquisitive Nature

Animals that have a diet specialty, such as bamboo-chomping pandas and ant-eating anteaters, have something in common with premier predators like lions. They both have a well-defined niche and don't have to spend much time improvising when it comes to finding food. In between hunts, these specialists can lie back and take it easy. Since their nervous systems are designed for idling, they are quite content to laze away their days in the zoo.

Opportunistic species, however, make their living by their wits. They wander in search of food and mates, and, being naturally inquisitive, they leave no stone unturned. For them, every new discovery adds another feather to their survival cap. These neophilic (love of the new) animals include wolves, raccoons, coatis, martens, mongooses, monkeys, apes, and, of course, humans. Unlike the specialists, these animals are not always content to accept the easy life at the zoo. Their keyed-up nervous systems are constantly prompting them to patrol their territories and investigate their habitats.

Small, Sterile Enclosures

Patrolling in a stark cage leaves a lot to be desired. In the wild, every bend in the trail would be full of surprises, obstacles, and new objects that would guide the search and keep the animal on its toes. Cues are scarce in a sterile exhibit, however, and the animal soon reaches the end of its territory. It may move as if it intends to go farther, but to no avail. Eventually, the walking ritual becomes formalized and rigid in form. A bear may wear a rut in the exhibit floor, or stain the wall where it pushes off and turns around each time. It may walk in a figure eight or a circle, retracing its steps again and again. These behavioral stereotypes develop a rhythmic quality as the animal rounds off the jagged edges of its turns.

This stereotyping is similar to the development of display rituals in animals. Courting animals, as we saw earlier, are torn between flying away or staying. They make a move to flee, but their impulse is checked, until all that's left is an exaggerated bobbing of the head. When you see elephants weaving or bears swaying to and fro as if listening to music, you may be seeing the stylized versions of trying to leave when there is nowhere to go. Circus animals stuck in very small cages are especially susceptible to these stereotyped movements. When keepers put the animals in a larger exhibit or one with more interesting, naturalistic elements, these behaviors often disappear. Sadly, however, just as with many human neuroses, these scars of an unhappy past sometimes remain, no matter how plush the current accommodations.

Cabin Fever

When faced with boredom, some animals attempt to put complexity back into their lives. They may invent new motor patterns (like twisting for hours on a rope), or they may create stimuli to which they can react. The chimp that throws its own feces at the crowd is bound to create a ruckus to which it can then respond. A bear may beg for food not because it's hungry, but because it craves the interaction. A cat may throw a bit of meat into the air again and again just to put life back into it. A bored animal may likewise light a fire under its exhibit-mates by stalking, baiting, or sexually harassing them.

This craving for stimuli can cause animals to inadvertently harm one another. Parents may become too parental, overgrooming their offspring to the point of lacerating them. For that matter, bored animals may also overgroom themselves, biting or nibbling at an irritation until they wind up chewing off a bit of their own flesh. Birds that used to chew wood in the wild may chew at their feathers if they have nowhere else to direct their instincts. Hypersexuality can also develop, causing animals to rub their exhibit-mates the wrong way. Zoo personnel can often prevent these disorders by designing appropriate exhibits and then staying alert to the physical, social, and mental needs of their charges.

Bad Company

An inappropriate social setting may also spawn abnormal behaviors. Too much company can cause animals to become overly aggressive, especially during breeding season when competitive tides run high. If subordinate animals have nowhere to hide from the temper tantrums of dominant animals, they can die of stress and worry. Being totally alone, on the other hand, can depress gregarious animals such as primates and zebras. Zoo managers find that putting animals together in their natural groupings can alleviate problems such as overeating, undereating, loss of sex drive, or general apathy.

Upbringing may also affect an animal later in life. A young monkey raised without her mother's teaching, for instance, may not have a clue about how to raise her own young. She may neglect or even abuse her infant, foiling even the best-intentioned captive breeding schemes. Rescuing an animal and raising it by hand is not always the best thing humans can do; animals raised by humans may imprint on us, forever mistaking us for mother, rival, or even sexual mate.

WHAT YOU CAN DO

Good zoo professionals are well aware of the hazards of captivity-induced behaviors. If you see an abnormal behavior developing, ask your zookeeper what it means and what the zoo intends to do to correct it. Captivity need not be a prison term, and the more of us who push for stimulating, stress-free exhibits, the better off zoo animals will be.

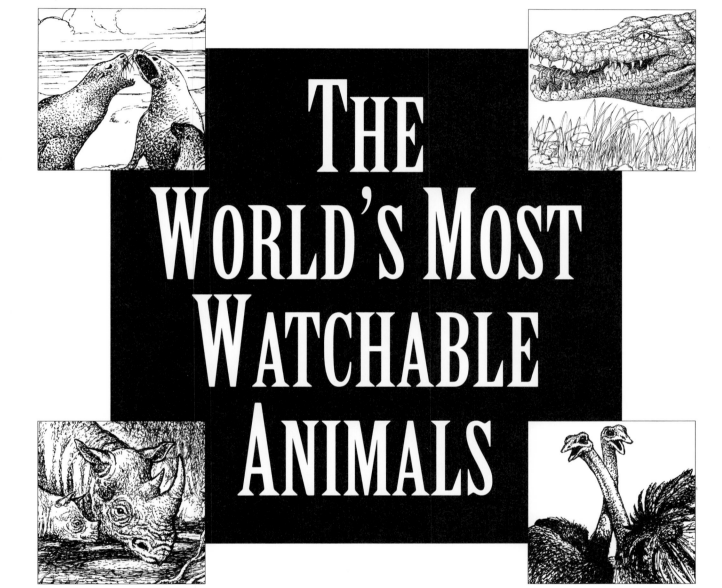

THE WORLD'S MOST WATCHABLE ANIMALS

African Jungles, Plains, and Waterways

CARING FOR THE YOUNG   *Gorilla mothers spend hours examining their infants.*

GORILLA

Watching gorillas in a naturalistic outdoor exhibit is myth-shattering. Despite what Hollywood would have us believe, gorillas are rather shy, ponderous, reclusive creatures, not at all interested in snatching humans, climbing towers, or swatting planes from the sky. To find out how miscast they have been, set your alarm some morning and get to the zoo as soon as it opens.

In the warmth of the morning sun, you'll find the gorillas in slow-motion, patiently combing the grasses for scattered food and murmuring belches of contentment to one another. Here is a family that still eats together! Even the much maligned male gorilla is relaxed, knowing that his family is safe and within sight. Despite his imposing size, the youngest gorillas clamor to be near him, and he endures their clinging and hair tugging with gentle good humor. (In the wild, this tolerance boosts survival by placing the most vulnerable group members within the male's sphere of protection.) Later on, as the midday sun shines hotter, the giant beasts will quiet one by one, stretching, yawning, arranging a comfortable daybed of straw, and easing into a light sleep.

Don't you wonder how anyone could have portrayed this peaceable animal as "nature's most savage beast" for so many years? The answer, of course, was in the profits. Circuses that housed a "dangerous killer" drew record crowds, and safari hunters who killed a "bloodthirsty beast" seemed that much braver to their friends back home. As a result, fears and myths about gorillas became deeply embedded in our culture. Not too long ago, a survey taken among British schoolchildren showed that gorillas ranked right up there with snakes and rats as the kids' most hated animals.

Today, however, thanks to some insightful research and the education efforts of zoos, public attitudes toward gorillas are changing. The pictures of Dian Fossey (researcher and author of *Gorillas in the Mist*) surrounded and accepted by a dozen wild mountain gorillas have helped amend some of the ferocity myths. In the same way, the controversial

VITAL STATS

ORDER: Primates

FAMILY: Pongidae

SCIENTIFIC NAME: *Gorilla gorilla*

HABITAT: Lowland or mountainous tropical rain forests

SIZE: Height, 4.6–6.2 ft   Chest circumference, 4-5.7 ft

WEIGHT: Male, 298–606 lb Female, 154–308 lb Heaviest in zoo, 772 lb

MAXIMUM AGE: 50 years

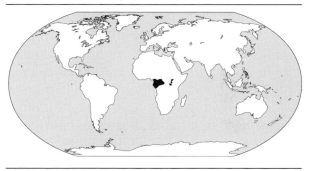

work of Dr. Francine Paterson with a young female named Koko has forced us to re-think our opinions about these so-called brooding beasts. Dr. Paterson taught Koko a form of the American Sign Language used by the deaf. By her seventh birthday, Koko had used 645 different signs in dialogue with various researchers (and occasionally when "talking" to herself or other gorillas). According to Dr. Paterson, Koko not only described current events and objects, but was also able to communicate about past events—something that only humans were thought to be able to do.

Though the results of her studies are still being debated, they suggest that the gorilla staring at us through the greenery might have more on its mind than we imagined. Gorillas that have taken the Stanford-Benet IQ test (in sign language) have scored 85 to 95—nearly identical to the scores of an average human child. This is particularly remarkable when you consider the heavy human bias built into these tests. When gorillas were asked, for instance, "Where would you go in the rain?" *house* didn't even occur to them. Though it may have lost them a point, *tree* was a much more logical and gorilla-correct response.

The public debate about gorilla intelligence is a welcome one. When scientists began to investigate whale intelligence, it did wonders for the movement to stop whale hunting. People tend to be much more empathetic if they think an animal knows what is happening to it. Perhaps the gorilla's gaze will elicit similar emotion, hopefully in time to save these beautiful, yet vulnerable, creatures.

Basic Gorilla Behaviors to Watch For

LOCOMOTION

Gorillas are shaped much like we are, except their arms are proportionally longer and their upper bodies—chest, shoulders, neck, and head—are much larger and heavier than ours. Gorillas walk on all fours, using their long arms as front legs. Rather than rest on their palms, they curl their fingers under to **knucklewalk**. Using a diagonal stride, they can walk, **trot**, or even **canter** to move quickly. At the zoo, you can often see younger gorillas **climbing** trees; as they get older and heavier, however, gorillas are less inclined to hoist themselves aloft.

FEEDING

The various subspecies of gorillas (western lowland, eastern lowland, and mountain) haunt many elevations, from moist valleys to thin-air altitudes where only stunted bonzai-like trees grow. They frequent lowland rain forests, mountain rain forests, or bamboo forests, preferring the young, open sections growing back after fire or windstorms. The abundant sun in these openings nurtures succulent leaves, stems, shoots, and fruits, all within easy reach of a sitting gorilla. To garner sufficient nutrients, these 440-pound vegetarians spend more than half their day **foraging**, eating up to 70 pounds a day. Thankfully, they are surrounded by food, and need not waste too much energy traveling. They typically cover less than half a mile a day, munching as they meander or sitting in the middle of a verdant patch and pulling in as much as their large arms will allow. The lush greenery contains so much moisture in each bite that they rarely need open water for drinking.

Gorillas don't disguise their love of eating; listening to their loud smacking and contented chewing noises is likely to make you hungry for a snack as well. Selective shopping is the gorillas' secret to remaining in balance with their habitat. Instead of stripping an area bare, they eat only certain parts of certain plant species and then move on, allowing the browsed jungle to rejuvenate from their pruning.

As you watch the gorillas at your zoo manipulate their food, pay close attention to their facile, humanlike hands. Since their thumbs are opposable like ours, they can grasp everything from bamboo stalks to the hands of other gorillas. Gorillas' thumbs are also more in line with their fingers than those of many other primates, another development that shows how closely related they are to us.

SOLITARY PLAY

Like humans, gorillas (particularly the young ones) find endless ways to occupy themselves. They escape into their own world, **spinning** on ropes or vines in the exhibit, turning **somersaults**, **climbing**, **jumping**, **batting vegetation**, **running** with arms akimbo, **sliding** down hills on their stomach, and **examining** their own body parts in detail. If they have water in the exhibit, you'll see them **wading** and slapping the surface to splash one another. Novel objects in the exhibit encourage play among the gorillas, and, as with human kids, the simplest toys—sticks, branches, bamboo stalks, banana leaves, ropes, and burlap bags—are often the best. By watching a gorilla's play habits, you can get a good feel for how content it is. If there is something in its environment that stresses or worries it, a gorilla will be less likely to play.

LISTENING TO GORILLAS	
(Sounds are in order of most to least frequently used.)	
GORILLA SOUNDS	WHAT THEY MEAN
Belch	Contentment when feeding
Chuckle	Social play
Pig grunt	Mild aggression; when group is moving
Hoot bark	Initiating group movement
Hiccup bark	Very mild alarm or curiosity
Cries	Infant in distress or separated
Question bark	Very mild alarm or curiosity
Scream	Aggressive encounters in group; female, when copulating
Pant series	Mild threat; female
Copulatory panting	Copulating; male
Growl	Mild threat
Wraagh	Sudden alarm; unexpected contact, loud noise
Roar	Strong aggression; male

SLEEPING

Don't be disappointed if the gorillas at your zoo are sleeping when you arrive. Even in the wild, gorilla families take lots of naps between feeding bouts, not because they are bored, but because they are full and content. Good zoos provide live vegetation or other material that gorillas can use to **build sleeping nests**.

Day nests are usually a simple affair—bits of straw arranged into a makeshift mattress. About an hour before dusk, however, gorillas build a more complicated structure that provides a dry, springy buffer from the moist ground. They reach out in a circle around themselves and bend down the stalks of large-leafed plants and grasses, taking care to avoid anything with thorns. By intertwining the branches in the center, they form a bowl-shaped bed as large as 5 feet in diameter. They may also climb into bushes or low trees to sleep, bending the branches together to form a loft bed.

SLEEPING NEST *Gorillas bend branches toward themselves to create a springy sleeping nest.*

HOW GORILLAS INTERACT

Gorillas live in small groups headed by a dominant male called a silverback (referring to the silvery "saddle" of hair that males acquire between the ages of 11 and 13). The group typically includes a few blackbacks (younger males), as well as several females and their offspring. Depending on subspecies, the number of individuals in a group ranges from 2 to 35, with 6 to 16 being the typical size.

For the most part, group living is peaceful and uneventful. The reigning silverback need not compete with the younger males of his group for the privilege of mating with the adult females. As leader, he decides which way to travel, when the group will rest, and where it will sleep each evening. As they steamroll through the jungle in single file, the silverback leads, followed by the blackback males and then the females and their young.

When the blackbacks become silverbacks, they may leave the group and travel alone, or they may remain with their family of origin, helping the dominant male fight off intruders. These groups with more than one silverback present a more formidable challenge to intruders, and, since the younger silverbacks tend to take the front line, the dominant silverback can hang back and avoid some of the danger. In return, the second-fiddle silverbacks occasionally get to mate with the younger females, or if they bide their time, they may get a chance to take over when the leader begins to lose his power.

Female offspring rarely stay with their natal groups, however. As soon as they reach puberty, they leave to join lone males or established groups. Each adult female that enters a group brings new genes, ensuring that the genetic mix stays fresh. Since the females in a group have come from other groups and are not related, their connections with one another are rather weak. Consequently, the real glue in gorilla society is the bond that each female has with the dominant male.

It pays for a female gorilla to attach herself to a male for several reasons. First, he can use his gargantuan size and strength to defend her against predators, keep competitors out of her feeding area, and break up fights within the group. Second, he can protect their offspring from invading males that may want to take control of the group. These takeover artists sometimes eat the offspring of the former silverback so that the females will be free to become pregnant again, this time with the new male's young. Third, having a live-in bodyguard and baby-sitter allows the females to wander off and forage for the best foods.

FRIENDLY BEHAVIOR

Coordinating Group Movements

When feeding, the gorillas naturally space themselves out a little, but keep in vocal contact through contented **belches** and grunts. When on the move, it's the silverback's job to keep the group together. He uses **pig grunts** to keep them assembled

FAMILY GROUP   *Gorilla families feed, rest, groom, and play together, all under the protective eye of the silverback male.*

and issues the hoot bark (sounds like a dog bark, except the *woof-woof* is replaced by a *who-who*) to speed them along.

Social Grooming

Grooming gorillas use their fingers, lips, and eyes to systematically examine and clean the skin, hair, or nails of their grooming partner. They work hardest on the upper part of the body: the shoulders, head, face, neck, and arms. Although social grooming is not as prevalent in gorillas as it is in some primate species, it's worth watching for. Seeing who grooms whom can tell you a lot about the structure of gorilla society.

Adult females, for instance, rarely groom the silverback, perhaps because they are already in his good graces and don't need to improve their bond with him. Instead, they spend most of their time grooming their infants, both for utilitarian purposes and to strengthen the mother–offspring bond. Unlike adult females, young animals don't have an automatic "in" with the silverback and must work to build a relationship with him. By grooming the silverback, they are investing in their own future; they not only soothe his immediate aggressiveness, but also bond with an ally they might need in the future.

Social Play

Young gorillas spend a fair share of their day playing, which is, by definition, doing vigorous things that seem to have no definite purpose (yet another human parallel). Besides maintaining social bonds, play serves to entertain these intelligent animals, just as it does us.

A gorilla often sports a huge, open smile (no teeth showing) called a **play face** when trying to entice a companion into a bout of **mock biting**, **chasing**, **lunging**, **tackling**, or **wrestling**. The play partners usually sit face to face or walk upright toward each other with arms waving slowly and alternately. When they meet, they may grapple in a slow-motion wrestling match that one researcher compared to the ritualized behavior of Oriental sumo wrestlers. Their favorite spot for a play bite is right where the neck meets the shoulder. After a gorilla scores a bite, you may hear something familiar: **chuckling**.

CONFLICT BEHAVIOR

Threat Gestures

The elaborate threat gestures of gorillas are like Geiger counters: they warn receivers that the mood of the performer is approaching a danger point. An annoyed dominant, for instance, need only threaten once to send group members scattering in submission. Threat displays directed

SOCIAL PLAY *This "conga line" is typical of the games young gorillas create to amuse themselves.*

at intruders can be even more dramatic, and if the response is anything less than retreat, the threat may escalate into an actual fight.

One of the most common signs of displeasure is the **tight lip expression**, surprisingly similar to the one we humans use when we're steamed. The gorilla tilts its head downward, tucks its lips in a tight line, and looks out under its furrowed brows with a hard, steely stare. **Growling** may accompany this show of dominance. Another way to express anger is with a **series of pants**—rapid bursts of air that warn other group members that the gorilla is teed off. As it gets more annoyed, the gorilla may **jerk** its head toward the opponent with a loud snap of its jaws. Another form of threat is the **strutting walk**. The gorilla shuffles in front of its opponent with a stiff-legged, exaggerated gait. It holds its upper body erect, keeps its arms bowed out at the elbows, and tilts its head to one side, averting its eyes and glancing up only occasionally. As the aggressive mood develops, the gorilla may push forward the corners of its mouth and retract its lips to display its toothy arsenal in the aggressive **bared-teeth threat**.

The **chest beating** for which gorillas are known occurs in many different contexts, including play, sexual excitement, frustration, and alarm at the sight of intruders (including zoogoers). It is often part of a longer display (see the three charges described below), but it can also occur all by itself. The loud *pock-pock* sound that you hear occurs because the gorilla cups its hands as it beats, trapping air against the naked skin of its chest. A dilated sac in the larynx acts as a resonator. Females may thump their thighs instead of their chest.

When used in a threat situation, chest beating is a brandishing of the fist, so to speak; it allows the gorilla to advertise his or her strength and size while avoiding an actual attack. Within groups, the dominant male may use chest beating to rally the troops to move on, to get the female's attention, or to stop squabbles that arise among members. The display sometimes spreads contagiously throughout the group, as if members are communicating their positions to one another. In a wild population, chest beating may help to space groups apart so they don't compete for food sources.

Threat Charges

If gestures alone don't do the trick, the gorilla may resort to one of three threat charges: the diagonal bluff charge, the rush charge, or the slam charge. The gorilla begins all three from a standing position or from a stiff, elbows-out, four-legged stance. Listen for a soft hooting, which gets faster and less distinct toward the peak of the dis-

CHEST BEATING *A gorilla cups its hand against its chest to make a loud* pock-pock *sound when alarmed, excited, frustrated, or angry.*

play. During the display, the gorilla may kick out one leg, roar, throw a handful of vegetation or rocks into the air, or beat its chest. In the **diagonal charge**, the gorilla runs sideways on two or four legs *past* the adversary for up to 66 feet. In the **rush charge**, the gorilla rushes *directly toward* the adversary, stopping just short of a collision. In the **slam charge**, the most intense of the displays, the gorilla *makes contact*, shouldering the adversary or slamming an exhibit wall as it charges by.

Fighting

Typically, the only contact occurring during threat displays is the shouldering or smacking during the charge sequence. All-out fights between gorillas are extremely rare, usually occurring only between resident males and their male challengers. You know a fight is in the making when both parties are displaying their teeth, screaming, and roaring in fury.

Submission and Fear

STRUTTING WALK   *To remind other gorillas of his dominance, the silverback stands and strides in a stiff, exaggerated gait.*

Gorillas that want to keep out of this kind of trouble have a variety of humble-pie displays. To show submission to a dominant, a gorilla will **cower** on the ground with its abdomen protected and only its broad back exposed. Sometimes it may use its arms to cover its head. When a gorilla wants to interact, but is uncertain, it folds its lips in against its teeth and averts its gaze. This **lip tucking** is an expression similar to human lip biting. Another telltale mark of nervousness is repeated **yawning**.

Gorillas use several sounds to express alarm or outright terror. The **question bark** (sounds like "who are you?" because the middle note is higher) and the aptly named **hiccup** are signs of mild alarm or curiosity, often heard in response to a sharp sound in the distance. A really fearful troop of gorillas will fall absolutely **silent**, as if paralyzed with listening. When faced with extreme danger, a gorilla may also open its mouth wide, throw back its head, and **scream**. Another form of alarm is not as shrill as a scream or deep as a roar, but sounds like its name—**wraagh**. Listen for this abrupt outburst at the zoo when a sudden noise surprises the gorillas.

SEXUAL BEHAVIOR

Another myth about gorillas is their insatiable sexual appetite, often the subject of grisly tales about women being dragged away from jungle camps by lustful beasts. In reality, male gorillas have rather lackluster libidos, probably because they have no need to woo or fight for females that are already securely bonded with them. In fact, females often wind up playing the role of initiator, soliciting the male when they are receptive. Once females are pregnant, they become even more sexually active than when they are in estrus, but not with the silverback. Gentle, leisurely homosexual pairings (among females) may occur, along with masturbation and dalliances with the younger males in the group. The silverback tolerates these extracurricular activities as long as the offspring of the pregnant females are sure to be his.

Courtship

The female gorilla is in heat for 3 days every month. Shortly before estrus, her odors change, and the male spends more time **sniffing** her armpits and genitals. She also becomes visibly attentive to him, **presenting her rump** or simply **staring** fixedly at him with stiffened limbs. She may lie on her belly or back in front of him, making suggestive **pelvic movements** and extending a hand in invitation. Throughout these solicitations, she presses her lips together and draws in the corners to create a bulge. The male also makes this tense face during courtship, nervously glancing at the female with quick quarter-turns of his head. When he is trying to get her attention, he may also adopt a **strutting** posture or perform a **chest-beating** display.

A rare form of courtship that you may see is **back riding**. The female rides the male like a horse, and the male plays the part by bucking up and down. The pair's excitement becomes obvious, and copulation soon follows.

PRESENTATION   *The friskier female gorilla blatantly solicits the male's sexual attentions.*

Copulation

In addition to the more traditional mounting posture, the pair can also copulate face-to-face. Scientists say that the gorilla's ability to vary its sexual regime is related to its relatively large brain size. While copulating, both partners display the **copulation face**: with eyes squeezed shut, they press their lips together and draw in the corners to form a bulge. In the final phase of copulation, these expressions grow more intense, and the male gets a dazed look on his face. **Copulation sounds** are also characteristic. The male utters staccato panting sounds, the female may scream, and both partners grunt and growl a bit. These sights and sounds often attract young males that try to touch the pair's genitals, sniffing their own fingers afterward. These curious onlookers are presumably learning sexual techniques that they will use later in life.

HUMANS AND GORILLAS

There are three subspecies of gorillas, each found in a different part of Africa: mountain gorillas, eastern lowland gorillas, and western lowland gorillas. Mountain gorillas are by far the most endangered, and at last count, only 400 or so were surviving in the wild. Eastern and western lowland gorillas are faring slightly better, but their fate is shaky as well. Although zoos and museums have stopped taking gorillas from the wild, new threats loom just as large. Like so many other species, gorillas are competing with humans for land and losing. The more acres of forest that we bulldoze for agriculture, lumber, and living space, the less wild country the gorillas have.

Even in the few sanctuaries that remain, gorilla families are not safe. Poachers go to great lengths to hunt down the remote, shy gorillas, despite the fact that gorilla poaching is a crime. No matter how strict the laws, the killings will continue as long as people are willing to pay large sums of money for gorilla skulls, hands, and feet (used as good-luck charms), or for gorilla infants to be raised as pets. In the process of getting one gorilla, poachers invariably kill the entire group that rushes to defend its comrade or young.

Tragically, gorillas sometimes lose their lives in snares set for other animals. They also die at the hands of farmers who are worried about losing their crops. Although gorillas have been known to pilfer crops, it doesn't happen often, and it seems a small crime compared to the amount of gorilla habitat we humans have destroyed. Arguments like these lose their power in the face of hunger, however. Whether the local people are defending their crops or simply feeding themselves on gorilla meat, the results are the same: entire families of the slow-breeding gorillas are lost.

In recent years, research on mountain gorillas has made people in this country more aware of the plight of our beleaguered relatives. We can only hope that the opportunity to see these primates in near-natural settings in zoos will arouse even more public sentiment. If this tidal wave of concern for gorillas is to be effective, however, it can't stop at the zoo; it must be felt across the ocean, in the form of better research, better protection against poachers, and more local education. Only then will the bloodshed pause long enough for the gorilla population to stabilize and, maybe, in protected habitats, to even stage a comeback.

PARENTING BEHAVIOR

Gorillas give birth only once every 3 to 5 years, devoting the intervening years to child rearing. Infancy and childhood are when gorillas absorb much of the culture and experience of their group, and proper socialization at this time translates into better success in adult life.

Birth

You can tell when a female at your zoo is pregnant by looking for a swelling of the nipples and abdomen that starts showing about 3 months before birth. (This latter clue is tricky since even nonpregnant females tend to have a large stomach.) The female also starts sobering noticeably, playing with other gorillas less often, and becoming, in the words of one zookeeper, temperamentally flat. As the birth approaches, she grows irritable and nervous, occasionally touching her vulva and licking her fingers as if gauging how long it will be until labor.

Caring for Young

Immediately after birth, the mother begins a **grooming** ritual that will become commonplace. She licks, nibbles, picks, scratches, and rubs her newborn, turning it in a series of seemingly uncomfortable positions to reach all parts of its body. The young may protest with wriggles, kicks, and cries that sound eerily human. Infants also cry when left behind, caught up in a tree, or otherwise troubled or uncomfortable. Like human infants, their cries are likely to build into shrieking tantrums that stop the moment they get their way.

When the female walks off, you'll see her infant instinctively **clinging** to the fur of her underside. As it gets older, you'll see it **riding** atop her back. The lactating female goes through 2 to 4 years of celibacy while caring for her newborn. As soon as she sends the young out to fend for itself, her menstrual periods return, and she is once again ready to prod the male into action.

GORILLA BEHAVIORS TO LOOK FOR AT THE ZOO OR IN THE WILD

BASIC BEHAVIORS

LOCOMOTION
- ☐ knucklewalking
- ☐ trotting
- ☐ cantering
- ☐ climbing

FEEDING
- ☐ foraging

SOLITARY PLAY
- ☐ spinning

- ☐ somersaulting
- ☐ climbing
- ☐ jumping
- ☐ batting vegetation
- ☐ running

- ☐ sliding
- ☐ examining body parts
- ☐ water play

SLEEPING
- ☐ building nests

SOCIAL BEHAVIORS

FRIENDLY BEHAVIOR
Coordinating Group Movements
- ☐ belching
- ☐ pig grunting
- ☐ hoot barking
Social Grooming
Social Play
- ☐ play face
- ☐ mock biting
- ☐ chasing
- ☐ lunging

- ☐ tackling
- ☐ wrestling
- ☐ chuckling

CONFLICT BEHAVIOR
Threat Gestures
- ☐ tight lip
- ☐ growling
- ☐ pant series
- ☐ head jerk
- ☐ chest beating
- ☐ strutting walk
- ☐ bared-teeth threat

Threat Charges
- ☐ diagonal charge
- ☐ rush charge
- ☐ slam charge
Fighting
Submission and Fear
- ☐ cowering
- ☐ lip tucking
- ☐ yawning
- ☐ question bark
- ☐ hiccup bark

- ☐ silence
- ☐ screaming
- ☐ wraagh

SEXUAL BEHAVIOR
Courtship
- ☐ sniffing
- ☐ rump presenting
- ☐ staring fixedly
- ☐ pelvic movements
- ☐ strutting
- ☐ chest beating
- ☐ back riding

Copulation
- ☐ copulation face
- ☐ copulation sounds

PARENTING BEHAVIOR
Birth
Caring for the Young
- ☐ grooming
- ☐ crying
- ☐ clinging
- ☐ riding

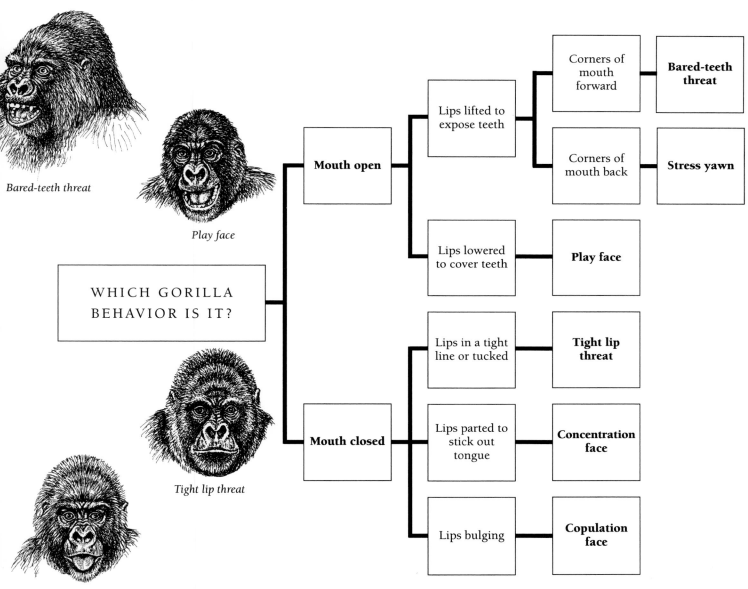

Bared-teeth threat

Play face

WHICH GORILLA BEHAVIOR IS IT?

Tight lip threat

Concentration face

Mouth open

Lips lifted to expose teeth

Corners of mouth forward → **Bared-teeth threat**

Corners of mouth back → **Stress yawn**

Lips lowered to cover teeth → **Play face**

Mouth closed

Lips in a tight line or tucked → **Tight lip threat**

Lips parted to stick out tongue → **Concentration face**

Lips bulging → **Copulation face**

NURSING AND PLAY   *A cub is a hungry bundle of energy. Nursing nourishes its body, while play prepares its muscles for the sprinting ambushes it will need later in life.*

LION

If your delicate Siamese cat seems content to get along without you, can you imagine how independent and aloof a 550-pound lion must be? Lions must not need a soul to help them survive, right? In reality, lions are a particularly social species, dependent on one another to catch prey and to pass on healthy offspring. To understand how this system of cooperation evolved, it's helpful to look at the forces that shaped the lion.

Lions evolved in a land of opportunity. Hoofed prey streamed across the African savanna by the millions, giving rise to a variety of predators, each with a slightly different hunting strategy or niche. Most carnivores concentrated on the old, sick, or very young members of prey populations, zeroing in on species that weighed 220 pounds or less. Lions took a grand detour in feline evolution, however, and began to hunt together in small groups. Through cooperation and stealth, they were able to exploit perfectly healthy prey over 550 pounds, giving them access to kongoni, wildebeest, zebra, warthog, topi, impala, and even the massive giraffe! Cornering this market made them the moguls of their habitat.

While it's possible for a single lion to bring down one of these animals on its own, it's not exactly a sure bet. In one study, only 17% of all solo hunts yielded dinner. When two or three lions joined forces on a cooperative hunt, however, their success rate doubled. This advantage, more than any other, led to the evolution of a social lifestyle in Africa's savanna-living lions.

Lions live in groups called prides, composed of 3 to 12 females of breeding age, 2 to 4 (usually 2) breeding-age males, and cubs of various ages. The size of the pride varies with locale, but can range from 5 to 37. Although pride members share a common territory and are friendly when they meet, you'll rarely find them all together in one place. They spend most of their time in hunting groups of twos and threes. When these subgroups happen to meet in the field, it's easy to tell which lions belong to the same pride and which do not. Outsiders are not tolerated in the territory; both male and female lions act quickly to chase away intruders.

VITAL STATS

ORDER: Carnivora

FAMILY: Felidae

SCIENTIFIC NAME: *Panthera leo*

HABITAT: Steppe, bush, or savanna

SIZE: Male length, 5.5–8 ft, shoulder height, 4 ft
Female length, 4.5–5.5 ft, shoulder height, 3.5 ft

WEIGHT: Male, 330–550 lb   Female, 265–395 lb

MAXIMUM AGE: 30 years in captivity; average is 15

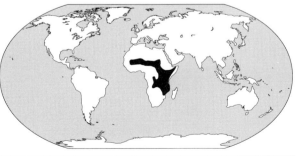

Group living, as it turns out, is well worth defending. Besides providing strength in numbers when it comes to hunting and defense, the close family ties also make cub-raising a whole lot easier. Females in a pride come into estrus around the same time, and, down the road, births are synchronized as well. Nursing, rearing, and protecting cubs are communal endeavors, with all females caring for all cubs, including their own. This day-care allows females to leave periodically for hunting without worrying about their young being harassed by predators or intruding male lions.

Intruding males are part of the nomad class of lion society—animals that live on the outskirts of prides, waiting for their chance to be part of a group. All young males and some surplus females become nomads when they are forced to leave their family pride to find territories of their own. This self-regulating mechanism keeps prides at a reasonable size so their appetites won't outstrip the hospitality of their habitat.

On the Serengeti Plain, nomads form 20% of the lion population. They hold no territory and must travel long distances to find food, working that much harder than pride members because they are hunting alone. Pregnant female nomads (usually mated by a male nomad) also have a harder time raising their cubs to maturity because of the lack of communal day-care and protection by males. The only way for nomads to improve their lot is to create their own pride or to join an already established group. Where pride rosters are filled to capacity, understudies must wait for a pride member to die before they can fill in the gap. If a natural gap doesn't become available, male nomads often attempt to force their way into a pride. Despite the danger, a coup d'etat is the only guaranteed way for nomad males to pass on their genes.

Ousting the dominants in an established pride is not an easy task, especially for one lion. Related males that leave a family pride at the same time often form gangs in order to improve their chances of cutting in on resident lions. If a gang does manage to invade, its members can look forward to only a 2- to 3-year tenure in their new pride. After that, they are likely to be driven out by yet another marauding gang or a solo male that is younger or in better shape than they are. After being driven out, the defeated males again become part of the nomad population. This revolving door of personnel in and out of lion society is one of the more interesting aspects of their social behavior. Even a lone lion has a life to live, however, and their individual behaviors are worth studying as well.

BASIC LION BEHAVIORS TO WATCH FOR

LOCOMOTION

Lions can **sprint** at speeds up to 36 miles per hour, a rate that would make human runners look as if they were standing still. If you could slow down a sprinting lion, you would see it plant its forepaws, then surge forward, placing its hind paws far ahead and outside of its forepaws. Repeated over and over, the motion is fluid and mesmerizing. Even when **walking**, the lion is a pleasure to watch. Its silent, floating gait is driven by a supple musculature that gives the lion extraordinary predatory power.

FEEDING

In a long chase, however, a prey animal blasting along at 50 miles per hour would easily leave a lion in the Serengeti dust. Undaunted, lions have evolved a different style of approach: the agonizingly slow and deliberate stalk.

A stalking lion looks a lot like your neighborhood tabby sneaking up on a sparrow. The three basic stalking moves are easy to pick out: the stalking walk, crouching walk, and crouch. The **stalking walk** is a quick walk or a trot with the legs slightly bent, head lowered, and ears cocked forward. Using the land-

THE STALKING WALK *With head lowered and body sleeked, the lioness glides incognito below the grass tips.*

scape to hide itself, the stalker slinks along termite mounds, road cuts, slopes in the terrain, or bushes. As it gets closer and is in more danger of being seen by its prey, it drops its abdomen, holds its head and neck almost to the ground, spreads its ears out to the sides to lower its silhouette, and literally disappears below the tips of the waving grasses. In this **crouching walk** the lion moves forward only when the prey is inattentive and stops dead at the first hint of suspicion. By freezing and advancing in this way, the lion is able to creep to within 30 to 60 feet of its quarry. In the final stage, the **crouch**, it waits with muscles tensed and quivering, calculating the perfect time to pounce. A human watching this crouch tingles with suspense, fear, and an inner empathy for both predator and prey. Perhaps we remember, in some instinctive way, the days when we hunted on prehistoric savannas, creeping ever closer to prey while predators, undoubtedly, crept toward us.

Lions are probably a hundred times better at this than we humans ever were. Before the wild-eyed zebra or warthog has a chance to react, the lion explodes from its hiding place, pulls the animal down with a swipe of its paw, and bites deep into the prey's neck or muzzle. Holding on like a bronco rider, the lion stays with the frenetic animal until it suffocates and ceases to struggle. The lion's head is perfectly adapted for this type of killing: the skull is large, the muzzle is short, and the powerful jaws are studded with massive teeth. Even the tongue gets in the act. The backward-curving horny bumps on its surface give the lion good traction on slippery meat.

Most zoos still hesitate to feed live animals to their carnivores, at least when visitors are around. Until that happens, you may have to depend on nature films or safaris to watch lions hunting. As you tune in, keep in mind that there are as many as eight hunting styles, but the most common are stalking and communal hunting (see Cooperative Hunting, p. 91).

ELIMINATION

Unlike your fastidious house cat, lions scatter their feces anywhere on the landscape and don't bother to bury them. Both males and females urinate from a crouched position unless they are trying to broadcast their scent as a way of marking territory.

When urine is being used as graffiti, lions will stand and eject a spray of urine back-wards towards a tree trunk, an exhibit wall, or even a row of zoo visitors (see Scent Marking, p. 92).

SELF-GROOMING

Males do quite a bit of self-grooming and are meticulous about keeping their front paws clean, as well as the front parts of their mane and chest. When it's time to wash their face, however, they enlist the help of a grooming partner. Lions will also rub against an object or roll from side to side to scratch an itch or rid themselves of in-sects. Zoo researchers suspect that lions groom themselves more often when they are under stress, which can occur, for instance, when an individual is not well integrated into the group. Watch to see if the lions at your zoo are comfortable by noticing how often they groom themselves.

COOLING DOWN

To beat the tropical heat, a lion will often rest on its back, with its limbs akimbo, let-ting the air circulate and cool its sparsely furred underside. Cubs are not as tolerant of the sun as adults are, so they usually head for shade cast by a tree or an obliging adult. The whole family will hit the shade after large meals, when, with very full stomachs, they lie panting and drooling to cool themselves down.

SLEEPING

When you see a lion sleeping in your zoo exhibit, remember your indolent house-hold cat. Have you ever noticed how many hours it spends sleeping? For starters, it

HUMANS AND LIONS

Powerful, invincible predators like lions are in many ways the most vulnerable of all species. They are the animals of choice for human hunters seek-ing to prove how powerful they are. The machismo attraction of lion killing was once so strong that white hunters were willing to pay exorbitant fees to "bag" a lion, and hunting parties would kill as many as a dozen in a day. Today, though protective legislation aims to control the hunting of lions, this horrible test of manhood continues illegally. Lions that live near human settlements are also at risk, since most farmers shoot to kill when a lion comes anywhere near their livestock.

Even more threatening to lions is the slow, in-exorable loss of habitat. Lions depend on acreages that are vast enough to encompass the feeding ranges of their prey. Whenever a loss of land causes prey populations to dwindle, lion popula-tions plummet too.

There is hope for lion protection in national parks, however, thanks to the lion's star status. In the late seventies, a biologist in Kenya's Amboseli Park estimated that the male lion earns $1 million in his lifetime because lions are the single biggest draw to wildlife watchers. Unfortunately, these dollar figures are often the only way to convince people of the lion's worth. Hopefully, zoo visitors who happen to lock eyes with this magnificent an-imal will understand the lion's inherent value in a much more personal and powerful way.

sleeps when you sleep, but it also sleeps when you're at work. In fact, it's only really active a few times during the day (like in the morning when you're trying to sleep in). Lions are just as fond of naps. In fact, these animals of the hot plains are inactive for up to 21 hours a day! In zoos, they are likely to sleep for 10 to 15 hours a day.

Though it looks like sloth, it's shrewdness on the lion's part. After all, each hunt brings in enormous servings, so they don't need to chase prey very often. In between hunts, why waste energy in idle motion? Wasting energy would force lions to hunt more often, a trade-off that this species is obviously not willing to make.

YAWNING

Lions yawn after waking from a rest period, or just before feeding time, as a way of getting their blood well oxygenated for the activity to follow. Yawning seems to be a synchronizer as well, since one lion yawning will set off a whole chorus of yawns (sound familiar?). Perhaps this ensures that no member of the pride will be unduly sluggish if the need to flee, hunt, or simply greet the keeper should arise. In short, yawning helps put all the pride members in the same physical mood.

STRETCHING AND SCRATCHING

Lions stretch with their forepaws straight out and their back arched in, or with all their paws together and their back bowed out. They stretch after resting or before an activity period. Also watch for lions placing their forepaws on a tree and rhythmi-

Yawning

Grooming

Scratching

Roaring

CONTAGIOUS BEHAVIORS   *Once one lion starts, the rest of the pride can't seem to resist.*

cally extending and retracting their claws (like your favorite cat does on your good couch); this seems to work the muscles of their back, forelegs, and paws.

How Lions Interact

FRIENDLY BEHAVIOR

Greeting

When you're a member of a pride, you rely on other members to help you rear cubs, hunt, and defend your territory. To keep your relationships in good working order, you have to defuse tensions and constantly reinforce social bonds. One of the ways lions do this is by giving each other a proper greeting when they wake up, when reuniting, or before embarking on a communal activity such as a hunt. As you watch the greeting ceremony, think how similar it is to your own cat's hello.

Greeting lions **rub their heads** and foreheads together, sometimes quite vigorously, humming and moaning as they do. In more elaborate greetings, they **circle** their rubbing partner, **rubbing against its whole body**, **flicking their tail** up and toward their partner's back and **swiveling their rump** slightly toward its head. A small lion may rub its head or body under a large lion's chin and then lean earnestly into its chest as if to get as much contact as possible. An abbreviated greeting may be as subtle as a slight bending of the head toward the other without touching.

These greeting ceremonies express peaceful intentions and may even spread a common odor through the pride. You'll notice that females and cubs rub one another, and they rub males quite often, but the males rarely reciprocate. In their superior position, it seems that males don't have to worry about building bonds or smoothing any ruffled fur.

Another kind of greeting to be on the lookout for is **anal sniffing**. One lion sniffs the anal area of the other, or both take a sniff at the same time. Lions that are unsure of each other's identity are most likely to do this as a way of "checking ID." A male may also sniff a female, especially if she greets him with rubbing, circling, and tail raising. Her gestures are reminiscent of sexual moves, and he might be sniffing to see if there is more to her attentions than mere friendliness.

Social Grooming

In social grooming, lions **lick** each other's face, neck, and shoulders—areas they can't reach for themselves. The abrasive bumps on their tongue dig deep in the fur to remove skin irritants. Interestingly enough, it seems that social grooming dislodges more parasites than self-grooming because social partners lick against

GREETING *Like domestic cats, lions greet each other by rubbing their heads and bodies together.*

the grain of the fur, while solo groomers lick along the lay of their own fur.

Social grooming is more than just hygienic, however; it also acts as a social lubricant, reinforcing mutual dependence and friendliness. The dawn and dusk grooming and head-rubbing sessions are especially social in nature. Notice who grooms whom: in the wild, females groom males, cubs, and other females, but males don't bother to groom anyone. The only time they come close is during courtship, when they may give the female some token grooming attention. As with rubbing, dominant males may not feel the need to appease anyone.

PLAY INVITATION *Cats will be kittens; to say, "Let's play," the lion lowers its forequarters in a friendly bow.*

Social Play

Though cubs are famous for mischief, play is not the sole domain of the young. Lions maintain their playful natures throughout life, and it's not unusual to see two females chasing, biting, and jumping on each other in a buoyant ruckus. Motherhood can also bring out the kitten in the mature cat. Watch how good-natured a lioness can be, even when cubs are batting, nibbling, and crawling on her. If the action gets slow, she may twitch her tail up and down, as if inviting the cubs to catch the end if they can. This game, besides being irresistible, may serve to teach the cubs prey-catching behavior.

Cooperative Hunting

One of the most common and by far the most successful forms of hunting is the communal hunt, an exciting practice that you can see in nature films, if not at the zoo. The lion that first senses the prey cocks its ears and opens its eyes wide, staring fixedly in the prey's direction. The other lions notice, and they too redirect their gaze. Once synchronized, one of the lions will set off in a straight path, while its partners set off at an angle, fanning out to encircle the prey. When the prey animal spots one of the lions, it runs the opposite way, often into the waiting jaws of the others.

CONFLICT BEHAVIOR

Roaring

The lion's best-known trademark is its full-throated, thundering roar, a long-distance telegram that tells other lions in all directions, "I am here, this is my territory!" In the zoo, you'll often hear a lion roar in response to a loud sound, triggering a series of contagious roars among its exhibit-mates.

The carrying power of the roar is evolution's answer to the vastness of the lion's wild habitat. Even over a distance of 5 miles, neighboring lions may hear and answer the call. This communiqué helps lions in two ways: by warning intruding males that they are trespassing and by reuniting scattered pride members. Roars can also be effective at close range, especially when an aggressive lion wants to make itself look and sound larger than life.

Scent Marking

Scent marking is an attempt at long-distance communication of a more lasting nature. Even after the lion leaves, its scent will continue to tell others what, when, and perhaps even who left the mark. Lions, like dogs, tend to mark anytime they come across a previously used scent post. They may also scent mark when approaching one another and during aggressive encounters.

On the veldt or at the zoo, lions mark by backing up to a tree or bush, raising their tail, and emitting a backward **spray of urine**. By wetting a tree trunk or a bush, lions can post their message on a large evaporating surface and at a convenient nose height for other lions that pass by. Males spray most often, but females also get in the act when they are in heat.

Certain lions in zoos have found that one way to activate crowds is to back up to them and unexpectedly shoot a jet stream of urine 10 to 13 feet, far enough to cover the front row. Now that you know about the lion's tendency to mark, however, you can watch for the raised tail and give them plenty of firing room. Also keep your eye on the other lions in the exhibit; both urination and defecation appear to be contagious behaviors.

Lions can also mark when urinating from a crouched position. They wet their hind legs and then **scrape** them backward to leave a visual addendum to their chemical message. Lions will also mark by rubbing their head and mouth into the branches of a bush or tree, leaving saliva and skin gland secretions.

If there are trees in your exhibit, look for signs of **claw raking**—scratches or grooves running vertically down the trunks. Lions rake by gripping the trunk with their extended front legs and then drawing the claws down. The claws snag in the bark, helping to remove the loose claw sheaths and also leaving a visual and possibly olfactory message having social significance to other cats.

Low-Intensity Threat

Roaring and scent marking are both mechanisms that keep lions well-spaced so they avoid confrontation with one another. Whenever lions do cross over boundaries and are suddenly face-to-face, they switch to a visual communi-

URINE SPRAYING   *Resident lions spray their personal scent at nose height so lions that pass by will not miss it.*

cation mode to express threat. Lions use these same threat gestures to settle everyday disputes within prides, so there's a good chance you'll see them at the zoo.

The most subtle threat gesture is the **direct stare**. Within eyeballing range, this silent form of dominance gets the point across without wasting precious energy. Lesser lions, if they know what is good for them, will simply move aside and give the staring lion the right-of-way. A male lion will also assert his rank by **strutting**—raising himself up to his full stature on rigid legs, holding his neck bolt upright, and turning sideways so the opponent or the female he is trying to impress will see his inflated profile.

High-Intensity Threat

If a lion needs to raise the ante on its staring or strutting threats, it may move to the high-intensity **aggressive threat**. Facial expression is all important here: the aggressive lion sinks its head between its shoulders, stares directly at its opponent, and opens its mouth slightly. It pulls the corners of its mouth so far forward that the lips form almost a straight line, nearly hiding the teeth from view. It rotates its erect ears so the openings face backward and the dramatic black spots on the backs of the ears face the opponent. To add a menacing touch, the lion may **growl**, **cough**, or **lash its tail** up and down. You can often see these aggressive threats at feeding time when hungry neighbors get too close.

When fear is mixed into the equation, a lion will issue a defensive rather than an aggressive threat. **Defensive threats** differ from aggressive ones in that the corners of the mouth are pulled back, the teeth are visible in a "long snarl," the ears are laid back flat against the head, and the eyes are nearly shut. This is a cat that expects to be scratched and doesn't want its ears or eyes hurt.

When fear begins to take the upper hand, the defensive lion turns away from its opponent, using the **head-twist posture** to avoid a direct stare. At its highest intensity, the fearful yet aggressive lion opens its mouth to expose all its teeth, then backs up its threat with lunges, bites, and slaps. This display is performed by females before copulation, by males during copulation, and by both sexes during tense encounters, especially over food.

If a fearful lion wants to avoid risky contact altogether, it will use the stalking walk, the crouching walk, and the crouch to sneak out of sight of rivals, predators, humans, or unexpected guests.

SEXUAL BEHAVIOR

When a female comes into heat (every 3 to 4 weeks on the average), the first male that finds her becomes her temporary consort. The other males in the pride usually respect his right and wait their

FIGHTING *Two males struggle for control of a pride.*

turn without challenging him. Besides draining their precious energy, a fight at this point would be bad politics. Males need one another to fight off marauding gangs and are therefore hesitant to tarnish their esprit de corps or disable a team member. Besides, the male that is waiting his turn is usually related to the consort, so even though he's not the one mating, some of his genes are still being passed on.

The best reason for waiting without a fuss, however, is that opportunity eventually knocks for all the males. It would be impossible for a single male to keep up the sexual pace set by the estrous female. Once she comes into estrus and for the next several days, she will mate every 15 minutes on the average! At the same time, other females in the pride are coming into estrus, making even more matings possible. This phase of the cycle is phenomenally active. In one study, a male mated with two females, copulating a total of 157 times in 55 hours. He mated every 21 minutes on the average, with a breathing space ranging from 60 seconds to as long as 110 minutes, leaving little time for normal pursuits, including eating.

Courtship

Courtship usually takes place in the open, sometimes away from the rest of the pride. The male and female circle each other restlessly, performing a typical sequence of displays. The male signals his intention with a **mating grimace**, a sneezelike snarl in which he wrinkles his nose, pulls back his lips to expose his teeth, opens his mouth slowly, and rolls his head from side to side without a sound. It is similar to the defen-

Mating grimace

Soliciting

Neck bite

Post-copulatory threat

SEXUAL BEHAVIOR *An aroused male signals his intent by grimacing and rolling his head from side to side without a sound. Females do their own soliciting, circling the male suggestively. While mating, the male anchors the female with a soft neck bite. Afterward, the female snarls and snaps as the male withdraws his barbed penis.*

sive snarl except he makes no sound and opens his mouth quite slowly. If the female doesn't immediately crouch in the mating posture, the male may move to her and begin to lick her fur, as if to groom her. If she still isn't ready, she may get up and move away a few steps, tail looped high, with the male following her in a tight **mating chase**. Eventually, the female turns her rump toward the male, lowers her forequarters, and elevates her hindquarters in a crouched **presenting posture**.

Females are by no means passive in the courtship sequence. In one study, they initiated courtship in 57% of the cases by circling the male, rubbing against him, curling themselves around him, or crouching in front of him. Females tend to **solicit** males in captivity even more frequently than they do in the wild. You'll see them presenting repeatedly, often treading with their forepaws as they do. Zookeepers have also noted that estrous females often roll on their back, grabbing their own hind paws and biting them. If you see a female doing this, you may want to stick around to see what happens next.

Copulation

During copulation, the male grabs the female's neck and shakes it, but with no force behind his **neck bite**. The female **growls** throughout mating, with her face tensed in an **aggressive threat expression**, except that her ear openings are facing forward instead of back. The male **miaows** during mating and wears the **defensive threat expression**. He lets out a drawn-out **yowl** as he ejaculates and then quickly dismounts.

Post-Copulation

Before he has a chance to get away, the female will turn, snarl, and even snap at him with an **aggressive threat**. You'd snarl too if you were a female lion. Males' penises are covered with backward-pointing barbs that strongly stimulate the female as they withdraw, a shock that induces her to ovulate. Once away from the male, she takes a few steps and then **rolls** sensuously on her back, resting up for the next go-round.

PARENTING BEHAVIOR

Caring for Young

In the wild, a cub's chances of surviving to age 2 are only one in five. The two or three littermates weigh only 3 pounds each and are blind and helpless at birth. In the early weeks before the mother takes them to live within the security of the pride, she is their sole caretaker. When she leaves to hunt or to socialize, she hides the cubs in the bushes as well as she can, but they are still vulnerable to hyenas and leopards. When she returns to the area, she **grunts** or **roars** softly as if to say, "Come see me." She may **move** them once or twice to different hiding spots, picking them up by the scruff of their neck and carrying them the way your house cat carries its kittens. For the first 2 months, she **feeds** them a diet of milk supplemented with meat.

At first glance, some of the lioness's behaviors seem negligent, at least from a human perspective. If, for instance, there is only one cub, she will invariably abandon it to die, a move which, though it seems heartless, makes good ecological sense. By abandoning her single cub, she will be able to mate again and increase her chances of having a larger litter. Mothers can also seem rather selfish at the dinner table. A hungry mother will always satisfy her own hunger before she brings her cubs to the kill, and depending on the size of the prey, there may not be enough left to keep all the cubs alive. This is especially true during the dry season when most prey species migrate north, and Serengeti lions must make do with the small Thompson's gazelle. Once again, it's a matter of investing in her own life, which is a surer bet than that of her cubs. Cubs are notoriously vulnerable to mishaps in their first few months, and the lioness needs to keep up her own strength if she is to make up for these losses and maximize her reproductive potential.

Communal Care

If the cubs do manage to survive their first 6 to 8 weeks, the mother introduces them to the pride. Here they are communally raised, along with the litters of other females born at the same time. Because they can suckle from any of the females, they are likely to live even if their mother perishes. They start following mom to the kill as soon as they are big enough, and by 11 months, they are actually hunting with other lions. During these outings, they enjoy the protection of the males in the group, who are always on the lookout for predators or hostile takeovers by other males.

Despite this communal care, cubs retain a special relationship with their own mother, a tie that lasts for 2 years until she has her next litter. She does her best to keep them from dangerous situations (like getting on a male's nerves) and brings

A MANE IS A MIXED BLESSING

Human rulers have long used a lavish mantle around their neck and shoulders to signify power and majesty. The lion's mane has that same look of royalty, and for the male lion, it's a necessary symbol of dominance. At communal feasts, for instance, the lion with the largest mane (and the largest body to go with it) has first dibs and can clear a group of lesser individuals just by showing up. Manes can also save lions the trouble of fighting. A lion that spots a large mane on the horizon wisely keeps his distance, while at closer range, lions size each other up mane to mane before coming to blows. If they decide they are closely matched enough to fight, it's their manes that buffer them from most slashes and bites. Females

seem to appreciate a great mane as well, and males are not at all shy about parading their locks back and forth so females can gauge their suitability as partners.

The best thing about a mane is that it's mostly fluff, not half as heavy as it looks. Weight may not be crucial to males in a pride (the sleek females do their hunting for them), but it is to nomads that have to chase down their own prey. As light as it is, the mane still cuts down on the lion's attempts at subterfuge, however. Male lions on the hunt look like "large haystacks moving through the grass," said one researcher. Even so, the protection and status enhancements they get from their manes must make them worth the trouble.

only her own offspring to a kill site. You'll have no trouble picking out mother-cub pairs at your zoo. The mothers are the ones being tugged and hugged and pounced on by hyperactive bundles of fur doing their best to find a free nipple and shatter the limits of her patience.

WHAT DO MALES DO, ANYWAY?

On the face of things, the division of labor in a lion pride looks grossly lopsided. Even though the males are dominant, females are in charge when it comes to moving the pride, deciding where to sleep, finding water sources, and so on. A lioness always makes the first move, and the males, usually roused from slumber, take up the rear, trotting along after the cubs. Lionesses also do most of the hunting, which makes sense since their bodies are lighter and less conspicuous than the males'. After hours of stalking and taking down a big kill, however, the females stand back while the males saunter over and help themselves.

Despite appearances, the male lion *is* doing his part, and without him the pride would not be as fruitful in the offspring department. For one thing, the pride's well-being depends on its access to abundant prey. Since a section of land can support only so many lions, it's important to defend the hunting territory against competitors. It's the male's job to remain alert to the surroundings, to scent mark, roar, and let other lions know that he is on guard. If lions do trespass, the male and female will both chase them, but it's up to the male to take on other males, which the female, weightwise, would not be able to do. Second, the male lion, playing "sweep" anytime the pride moves, protects the cubs from predators.

Third, and even more important, he protects the pride from strange males that may try to usurp his role. A new male coming into the group would want to pass on his genes as quickly as possible. One sure way is to kill the offspring of the previous male so the females can stop nursing, cycle into estrus, and become pregnant with his cubs. Knowing this, the resident male works feverishly to protect his genetic investment, a service that the rest of the pride benefits from as well.

LION BEHAVIORS TO LOOK FOR AT THE ZOO OR IN THE WILD

BASIC BEHAVIORS

LOCOMOTION
- [ ] sprinting
- [ ] walking

FEEDING
- [ ] stalking walk
- [ ] crouching walk

- [ ] crouch
- [ ] ELIMINATION
- [ ] SELF-GROOMING

- [ ] COOLING DOWN
- [ ] SLEEPING
- [ ] YAWNING

- [ ] STRETCHING AND SCRATCHING

SOCIAL BEHAVIORS

FRIENDLY BEHAVIOR
Greeting
- [ ] head rubbing
- [ ] circling
- [ ] body rubbing
- [ ] tail flicking
- [ ] rump swiveling
- [ ] anal sniffing
Social Grooming
- [ ] licking
- [ ] *Social Play*

- [ ] *Cooperative Hunting*
CONFLICT BEHAVIOR
- [ ] *Roaring*
Scent Marking
- [ ] urine spraying
- [ ] scraping
- [ ] claw raking
Low-Intensity Threat
- [ ] direct stare

- [ ] strutting
High-Intensity Threat
- [ ] aggressive threat
- [ ] growling
- [ ] coughing
- [ ] tail lashing
- [ ] defensive threat
- [ ] head-twist posture
SEXUAL BEHAVIOR
Courtship

- [ ] mating grimace
- [ ] mating chase
- [ ] presenting posture
- [ ] soliciting
Copulation
- [ ] neck biting
- [ ] growling
- [ ] aggressive threat
- [ ] miaowing
- [ ] defensive threat
- [ ] yowling

Post-Copulation
- [ ] aggressive threat
- [ ] rolling on back
PARENTING BEHAVIOR
Caring for Young
- [ ] grunting
- [ ] roaring
- [ ] moving young
- [ ] feeding
- [ ] *Communal Care*

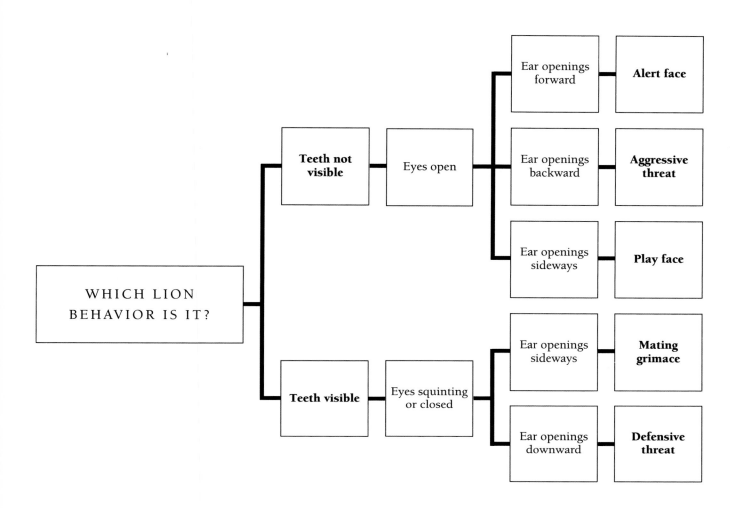

GREETING   *Elephants "shake hands" by putting their trunk tips in each other's mouths, perhaps picking up telltale tastes and smells this way.*

AFRICAN ELEPHANT

You would think an animal as colossal as an elephant would be something of a brute. Certainly, if we humans tipped the scales at 12 to 14 thousand pounds apiece, we'd find a way to throw our weight around. But not elephants; despite their awesome payload, the world's largest land mammals are downright gentle. Their weight is magically distributed on four squishy footpads that leave relatively dainty tracks. Their javelin-length tusks, sharp enough to puncture the sheet metal of a Land Rover, are more often used for peaceful pursuits such as root digging, peeling the edible bark of trees, drilling for water, and suggestively nudging the female before mating. Even their skin, at $1\frac{1}{2}$ inches thick, is really very fragile—so much so that zookeepers must scrub the elephants daily to keep their skin clean and well-conditioned.

Elephants lead gentle lives as well, characterized by co-operation rather than conflict. In their 50 to 70 years of life, elephants come to know one another intimately; every birth is a cause for celebration, and when a fellow member dies, all the elephants in the group struggle to lift the fallen one with their trunks. Their bonds are so strong, in fact, that when hunters shoot at family groups, separated members will run *toward* their imperiled family, not away.

The basic unit of this close-knit society is the cow-calf group, made up of a matriarch or leader (usually the oldest cow), her adult female daughters, and their immature offspring of both sexes. The matriarch's sisters, cousins, and

VITAL STATS

ORDER: Proboscidae

FAMILY: Elephantidae

SCIENTIFIC NAME: *Loxodonta africana*

HABITAT: Prefers the edge between grassland and forest, especially near rivers; also found in deep forests, open savannas, wet marshes, thornbush, and semi-desert scrub

SIZE: Length, 20–24 ft   Shoulder height, 10–13 ft

WEIGHT: 4,850–16,534 lb   Heaviest ever recorded, 22,050 lb

MAXIMUM AGE: 70 years in captivity

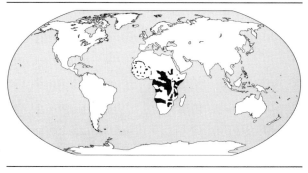

TEA-SPOUT POSITION *Curious or concerned, elephants "periscope up" to sniff out danger.*

their offspring sometimes join in as well, bringing the total to as high as 24 members, though the average is fewer than 10. The social hierarchy in cow-calf groups is based on size and age, with the largest and oldest females at the top and the smallest and youngest coming in last. Adolescent males determine their own ranking order through head-butting contests, where strength and temperament are as important as size and age.

Most females will stay with their original or natal group for a lifetime, except for the occasional few that develop a following and leave to become their own leaders. Males, on the other hand, are destined to leave their natal group as soon as they mature (at 10 to 20 years). Some don't even make it that long; females that grow tired of the males' rowdy behavior may oust them at any time. So begins the wandering life of the bull elephant. It's a solitary life for the most part, spent crisscrossing the range in search of female mates. The only time you'll see a bull with a family group is when he finds a cow in heat, or if he happens to be near the group when they are feeding or migrating. The rest of the time, lone bulls pal around only with other bulls, forming loose bachelor herds whose members change from day to day.

Besides the cow-calf groups and bull groups, there are composite aggregations called kinship groups that may contain as many as 70 animals. These kinship groups or clans are made up of two to five splinter groups, subsets of the family that have broken off from the original cow-calf group but remain in the neighborhood, intermingling from time to time. After the rainy season, clans from miles around may come together in large herds of several hundred to as many as 2,000 individuals to feast in areas of new growth.

Whether it's the five animals at your zoo or a herd of 2,000 on the veldt, the same rules of congeniality and loyalty apply. You'll notice that as family members feed, bathe, or move from place to place, they act as one, remaining within 50 yards of the matriarch. Even when bushes separate them, they keep in contact with special calls and probably scent cues. When danger threatens the elephant group, the members bunch together to present a united front, with the matriarch front-and-center and the calves tucked behind the adults. All the members, large and small, hold their trunks in a tea-spout position, sniffing the air for danger. If the situation warrants it, the lead cow may charge, or the whole group may stampede at once, trumpeting and shaking their great gray heads.

This furious mob can also show unusual tenderness. When a calf falters and falls, for instance, the whole herd pauses until the mother and various helpers have raised the calf and steadied it. This communal defense and care of the young, sick, and injured means that elephants have more in common with primates (and therefore with us) than they do with species in their own order.

LOCOMOTION

Walking leisurely, an elephant can easily cover 2 to 4 miles in an hour. When **running** in a dead heat, however, the same elephant could beat even the fastest human sprinter, reaching a blistering 24 miles per hour.

FEEDING

Life is one long meal for elephants in the wild. They are on the move for 16 hours a day, **collecting plant food** from all levels—root level, ground level, shrub level, high up in trees, and anywhere else their agile trunk can reach. If they can't reach the crown or get at the roots, they merely **overturn trees** or pluck them right out of the ground. Their diet is at least 50% grasses, supplemented with forbs, leaves, twigs, bark, roots, and small amounts of fruits, seeds, and flowers. Some of the more exotic items on their menu are desert dates, black plums, raspberries, ginger, wild celery, olives, figs, and coffee berries.

On a good day, an adult elephant can eat 300 to 600 pounds of food, making up in volume for what they lack in efficiency. Unlike cattle, which rechew their food at least four times, elephants use only 40% of what they eat, and the rest leaves their body undigested. Luckily, they process vast quantities of food very quickly with the help of enzymes produced by bacteria in their stomach. The results are what you see being swept up at the zoo: an impressive amount of manure deposited 10 to 30 times in 24 hours.

DRINKING

All life on the African savanna rides an annual roller coaster of wet and dry, feast and famine. During the dry season, the elephant is essentially tethered to water holes, and may have to travel several miles from hole to hole, drinking as much as 24 gallons a day. As you'll see at the zoo, they suck up 2 to 3 gallons at a time with their trunk and then spray it into their mouth. When elephants pull up to a water hole in the wild, most other species bow out of their way. A subtle and lordly wave of their trunk is all it takes to make giraffes, buffaloes, rhinos, and even lions shrink back, awaiting their turn.

For some animals, there would be no turn at all if not for the elephants. When rivers dry up on the surface during severe droughts, elephants can reach down to the sunken water level with their tusks. They **dig** a hole several yards deep and then wait for water to seep slowly into the sand. No matter how long they've traveled or how thirsty their journey, the elephants wait patiently for the sand to clear before

DIGGING   *Tusks can be used to mine the earth for salt or drill for water during dry times.*

taking a good long drink. Other animals flock to these holes once the elephants have left. This well-digging service is essential during the downside of the savanna cycle and would be sorely missed if elephants were to fade to extinction.

BATHING

One of the most joyful sights at the zoo is a poolful of elephants boisterously dunking themselves in **water** and shooting great fountains of spray with their trunks. If the elephants at your zoo are lucky enough to have a **mud wallow**, you may see the whole family luxuriating in the ooze, rolling side to side and even spraying it on themselves. Rather than just beading up and sliding off, water and mud are trapped deep in special fissures in the elephant's skin. As this moisture slowly evaporates, it releases energy in the form of heat, leaving the bathers cool and comfortable.

GROOMING

Besides lowering the thermostat, mud and water baths keep the elephant's delicate skin from drying and cracking. Ticks and other annoying insects don't appreciate the treatment either. After wallowing, the elephants often find a convenient tree, rock, or termite mound to **rub** against, dislodging these hitchhikers from their favorite hiding places. You may also see elephants blow **dust** on themselves as an added abrasive. The dust and mud that remain after a rubbing session paint them the color of their habitat (or exhibit) and may help protect their skin from too much sun.

COOLING DOWN

One disadvantage to being so big is not having enough skin to release all the heat that the elephant's portly body generates. Think, by way of comparison, of how much skin your neighborhood squirrel has in proportion to its body volume. When it heats up, there's plenty of surface area through which the heat can escape. Elephants have compensated by evolving several square feet of extra surface area in their enormous ears. These ears are heavily veined, especially on their rear sides, and act as natural radiators for excess heat. When elephants **spread their ears and flap them** slowly, air flows over these veins, cooling the blood before it courses through the rest of the body. When temperatures really soar, an overheated elephant can always go for a dip.

BATHING   *Elephants bathe to clean and cool themselves. Their trunk holds up to 3 gallons of water and makes a fine shower spout.*

SLEEPING

Elephants sleep deepest when most of us do—in the early hours of the morning just after midnight. They may also take a siesta during the hottest part of the day, for a daily total of 4 hours' sleep. When a group decides to sleep, they all go down, flopping one after another as if on cue. If one elephant shifts its weight in sleep, the rest follow suit, sending a gigantic wave of tossing and turning down the line until all is peaceful again. People who have seen a sleeping herd have been struck by the same

WHAT ELEPHANTS ARE KNOWN FOR

A **nose that swings**: The organ that the elephant uses like an arm is actually a modified nose and upper lip, complete with nostrils, antennalike hairs, and 40,000 muscles. Although this trunk is powerful enough to lift a tree out roots and all, it is also a delicate instrument loaded with touch and smell sensors. The two fingerlike projections at its tip are so facile that an elephant can pick up a single blade of grass. Trunks also come in handy for drinking, greeting, caressing, "snorkeling" across rivers, squirting water, blowing dust, making threats, and placing elephant-to-elephant calls.

Crown jewels: The beautiful, curved tusks are overgrown teeth made of simple dentine just like our own. They grow throughout a lifetime, so that by the age of 60, a bull's tusks may average 132 pounds each (a cow elephant's tusks are only 20 pounds each). Record tusks on very old individuals have measured 7.7 feet long and weighed about 287 pounds. You may notice that the tusks on the elephant at your zoo are not always the same length. This is because elephants favor one tusk over the other in the same way we favor our right or left hand. The favored tusk tends to wear down faster. Also, living ivory is a flexible material, and though it bends under pressure, it may break off if stressed too far.

Unfortunately for elephants, tusks have a unique, diamond-shaped core that makes them more lustrous than ordinary teeth. When a ravenous demand for this "white gold" drove prices up in the 1970s, poachers armed with automatic ri-

fles went on a slaughtering spree, killing nearly 100,000 elephants each year. Today, despite recent trade bans and consumer boycotts, elephants are still losing their lives. If the elephant at your zoo has long, beautiful tusks, be sure to admire them closely. It's a sight you're not likely to see in the wild anymore, since older "tuskers" are the primary targets of this kind of greed.

Tie an elephant around your finger: Besides being valued for their tusks, elephants are known for their legendary powers of memory. Although we have no way of knowing whether an elephant's memory is superior to that of any other species, we can point to a physical trait that may give them a leg up on learning. When an elephant is born, its brain is only 35% developed, leaving ample room to grow. The human brain has an even greater growing capacity, starting at 27% development. In contrast, most other animals come fully equipped with a brain that is 90% finished, leaving only 10% to hold new or learned information.

Elephants probably learn through a combination of trial and error, observation and insight. By watching older group members, younger members reap a lifetime of learning experiences. They in turn pass this information on to the next generation, perpetually replenishing the population's information pool. A critical link in this process is a mature leader, usually the matriarch, who has no doubt lived through severe drought and will be able to lead the group to dependable water holes when the next cycle of drought occurs.

thought: Only an animal with a secure, queenly position in the animal empire could sleep so peacefully through an African night.

How African Elephants Interact

When you see elephants interacting, pay special attention to each animal's head, ears, and trunk—the most visually expressive organs. The rules are simple: (1) the higher the head, the more excited the elephant, (2) the wider the ears are spread, the more aggressive the elephant, (3) trunk out in front signals confidence, while (4) a trunk curled toward the body shows hesitancy or fear, unless the elephant is simply tucking it away during a serious charge. (See Which African Elephant Behavior Is It? at the end of this chapter.) As you watch, don't forget that elephants are also communicating by means of scent, sound, and touch.

FRIENDLY BEHAVIOR

Keeping in Contact

You'd think it would be hard to lose sight of anything as big as an elephant. But in the wide open spaces of Africa, a restless male can quickly put distance between himself and other elephants. Even at close range, thick vegetation may screen one browsing group from another. Out-of-sight elephants are rarely out of touch, however, thanks to their low-frequency calls and system of scent marking. The most common call you may feel throbbing in the air at the zoo is a duet—a **contact call** and a **contact answer**. Another way to keep the group together is the **let's go rumble**, a kind of reveille issued by the leader when it's time to go somewhere new.

Greeting Ceremony

The friendliest thing you can do if you are an elephant is to walk toward another elephant with your ears high and folded and touch trunk tips. You might even put your trunk in the other elephant's mouth. This greeting ceremony gives family and kin group members a chance to reaffirm social bonds, especially after being separated. The longer the separation, the more intense the greeting. An excited reunion among long-lost family or kin groups may include urinating, defecating, trumpeting, screaming, loud rumbling for several minutes, and lots of physical contact. Besides building goodwill, the rambunctious participants may also be swapping information that they need to survive.

Mating Pandemonium

The same sort of hoopla often occurs when two elephants are copulating. The family group rushes up to the mating couple, vocalizing, urinating, defecating, and secreting from their temporal glands (see Pre-Courtship, p. 111) in what seems to be a contagious show of emotion.

Group Defense

When threatened, a family unit bunches together and faces the enemy en masse. They shake and toss their heads, trumpet, and launch short charges. During all this commotion, they are constantly reassuring one another with caresses and even placing trunk tips in each other's mouths. At the lower levels, young calves find their own asylum behind the stout legs of the elders.

Social Play

Well cared for and well defended, young elephants are free to spend their calfhoods playing. Watch for them **sparring** head to head, **trunk wrestling**, **chasing**, and **rolling** on one another. Elephants may also perform play **mountings** and other sexual gestures that prepare them for their lives ahead. As the males get older, their sparring will take on more significance as they start to vie for rank order.

CONFLICT BEHAVIOR

For the most part, females are model members of a cooperative society, and you'll see very few instances of conflict within their ranks. Males, on the other hand, can be antagonists, especially when they are young and their dominance ranking is still undecided.

Once they know their place, however, all-out fights between adult bulls are rare. Instead, males use threat displays to prove their might, especially when testing a strange male's rank, keeping other bulls away from their female consort, or simply clearing a water hole that is too crowded for their liking. Females will also perform these threat displays and charges, but they direct them only at intruders that threaten the safety of their family.

Threat

There is nothing quite as intimidating as an angry elephant. When 6 tons of pachyderm towers above you, ears spread wide, forefeet scraping the ground, tail twitching, body weaving from side to side, and head raised so high that it is looking down over its sharp tusks at you, every instinct tattooed on your genes tells you to turn tail and run.

Submission

It seems that many elephants feel the same way. A submissive or vanquished elephant will demonstrate its fear in gestures that are exactly opposite to those of the aggressor. Instead of standing tall with its ears spread wide, it will lower its head, hold its ears back, and curl its trunk inward. It may even run away or simply present its rump to show submission. In the all-male groups, the winner may then mount the loser, especially where there is no room for the loser to escape.

If an elephant is equally matched to the pow-

THREAT   *As a warning to would-be rivals, the elephant stands tall, spreads its impressive ears, and pointedly aims its tusks.*

erhouse, however, it may not back down. Instead, its show of confidence may trigger the next stage of aggression—the charge.

Charging

When a threat display fails, an elephant may rush toward the intruder with ears flapping and trunk raised to sniff the air. Though the shrieking and trumpeting sound fierce, this **mock charge** is primarily a warning, and the elephant will stop well short of its target. To top off the display, it vigorously **shakes its head**, causing the ears to crack like bullwhips as they flap against the body.

The next stage of irritation is the **serious charge**. The elephant brings in its trunk, aims its forehead like a fist and its tusks like lances, and rushes toward the enemy, trumpeting, growling, and screaming. Researchers have documented quieter and more direct versions of the serious charge, however; one watched an irate bull walk up to a parked van and unceremoniously drive a tusk through the door.

THE ELEPHANT CHANNEL

Anecdotes about elephants abound with examples of their loyalty and group cohesion. Maintaining this kind of togetherness calls for a good system of communication. We are only now beginning to appreciate how complex and far-reaching this system is. Researcher Katharine Payne first started to delve into elephant communication after a visit to Portland's Washington Park Zoo. Standing in the elephant house, she began to feel throbbing vibrations in the air, and after a while realized that they were coming from the elephants. What Katharine felt, and later went on to study, is a low-frequency form of sound called infrasound.

Infrasound is the same type of sound that earthquakes, wind, thunder, volcanoes, and ocean storms make. Our world is full of it, and though we humans can't hear much of it, other animals may depend on it daily. Whales, for instance, may listen for the infrasound of breaking waves to avoid stranding on shallow beaches.

Depending on how close you're standing to an elephant, how loudly it's calling, and how low the frequency, you may or may not hear the rumbling or feel the throbbing. Nonetheless, you can often tell when an elephant is communicating by watching its body language. An elephant receiving a message has an attendant posture—ears out, trunk and head raised, body frozen—as if listening. Calling elephants open their mouth and flap their ears, then pause stock-still as if listening for a response.

Researchers in Africa noticed that whole herds of animals would sometimes freeze like this and then suddenly head off in a new direction, as if being summoned. Perhaps infrasonic calls could explain why various herds coming from different directions often arrive at water holes within minutes of one another, as if they had planned the rendezvous in advance. Researchers have recently begun to record and interpret this "elephant talk" using high-tech microphones and their own powers of observation. They have distinguished many kinds of rumbles, some operating at close range, others reaching at least 2.5 miles. These calls may help aggressive musth elephants avoid one another, help mates find one another, and keep groups together in their wide-ranging savanna habitat.

After a threat or a charge, it takes a while for the aggressor's tension to dissipate. Still keyed up, the elephant might redirect its energy, much as we do when we pound our fist on a table instead of punching an opponent. It may rip up vegetation with its trunk or tusk, throw dirt or grasses onto its back, peer over its shoulder and jerk its head, or simply weave from side to side until the tension drains away.

Fighting

Although tusks are occasionally used to drive home a point, serious fights between bulls are rare. Instead, when two males are on a collision course, both heading for the same water hole, for instance, their intentions are clearly signaled through subtle movements. The dominant elephant need only spread its ears or barely raise its head, and the submissive elephant usually gets the idea and stands aside. This exchange of signals is of course adaptive, saving both elephants energy and avoiding the risk of a fatal wounding.

The largest and oldest male elephant is normally dominant over all other males in the area (except during the time of musth, as you will see). This dominance hierarchy is usually worked out in the bulls' adolescence, when struggles are common. These fights are essentially **head butting** and **trunk wrestling** matches in which the two young bulls stand head to head, with trunks entwined, alternately butting, shaking, and nodding heads to throw each other off balance. Once the loser surrenders, the bulls are likely to remember each other's positions for a lifetime.

Every now and then, however, strange bulls that don't know each other (such as new exhibit-mates) may go head to head to determine status. Since the strategy in an elephant fight is to get above and push down on your opponent, each bull tries to move upslope or climb a mound. They use their tusks (which grow to 7 feet) to keep their head from slipping sideways and to lever their opponent's head up or down. Both parties parry as best they can, taking pains to avoid the business end of each other's tusks. A tusk will occasionally pierce through hide, and if the momentum of a fast charge is behind it, the **stabbing** can be fatal. The fact that these deaths are rare, however, shows that fights are more ritualized than bloodthirsty.

One of the safety valves in a fight is called **tusking.** This is when one of the opponents takes a break from the fight and directs his tusks—and his frustrations—elsewhere. He may even get down on his knees to tusk up clods of dirt, mud, grass, or logs, and then toss them at any convenient target.

SEXUAL BEHAVIOR

Female elephants may come into estrus at any time of year, making breeding a possibility year-round in the zoo. In the wild, however, most breeding activity occurs during and following a rainy season, which places the birth—22 months later—at a time when trees are putting on fresh green growth. Because males and females normally travel in separate social and spatial spheres, getting together to breed requires special orchestration. As you will see, a fascinating array of signals and sounds has evolved to tackle this long-distance dating dilemma.

For thousands of years, primitive humans and elephants struggled to survive side by side in Africa. Through the years, elephants adjusted to and occasionally foiled humans' attempts to hunt them and then passed these adaptations on to their kin. Unfortunately, the contest has become terribly unbalanced, and elephants as a species cannot evolve fast enough to escape our modern assaults.

The first clue that something is amiss in the world of elephants is the lack of older group members. In a species whose wild longevity is easily 50 or 60 years, very few elephants in Africa reach the age of 35 before being slaughtered for their tusks. The animals that manage to survive poaching are likely to be orphans, belonging to shattered family units that no longer have a matriarchal leader. These leaderless groups tend to aggregate, perhaps harking back to their instinct of ganging up to defend themselves against primitive humans. In ages past, these aggregations were temporary, and elephants dispersed when the danger was gone. In modern-day Africa, however, where elephants are under continual stress, the groups no longer break up.

Besides the battering they take from hunters, farmers, and ivory poachers, elephants are also experiencing the poverty of closed-in places. The parks and reserves that house most of the remaining elephants are mere remnants of natural Africa, surrounded by farms and other tamed landscapes. As it turns out, elephants marooned here are wreaking their own brand of havoc on this limited haven.

It's completely natural for elephants to uproot, knock over, and trample bushes and whole trees if it helps them to get to the foliage that they love to eat. After these natural bulldozers move through, the area becomes ripe for grass fires, which burn so hot that they kill any woody seedlings trying to come back in. The ultimate result is a landscape changing from woodland to grassland. This affects the food supply and shelter of all woodland species, including the elephants themselves.

In the days before political boundaries and fences, elephants could bulldoze a small area and then move on to allow the habitat to recover.

Today, national parks are surrounded by development, and the hemmed-in elephants are having a massive effect on the parks' vegetation. Some people advocate killing "excess" individuals, while others believe that the elephant population will eventually come to equilibrium with its habitat. At what cost this will be to other species, no one can say.

Even when left to its own devices, the natural checks-and-balances system is no longer free of extraneous human-caused effects. In the seventies, when ivory (used for jewelry, signature seals, billiard balls, and carved ornaments) experienced a 10- to 15-fold increase in price, the rate of illegal hunting skyrocketed to satisfy the demand for tusks. To a herder or subsistence farmer, an elephant was a walking fortune, its tusks worth more than a dozen years of honest toil. Biologists estimate that Kenya lost more than half of its 120,000 elephants between 1970 and 1977. (Today there are only 16,000 in that country.) As the big tuskers started to disappear, poachers had to kill more elephants to get the same amount of ivory. In all of Africa, twice as many elephants were killed in 1988 as in 1979, and in the 9 years in between, the total population dropped from 1.3 million to about 735,000.

By 1990, 105 of the 110 nations party to the Convention on International Trade in Endangered Species (CITES) had agreed to ban the raw ivory trade. Nevertheless, poaching continues, and experts believe the well-armed poachers may be stockpiling tusks until they can establish new black-market trading networks.

Even if elephants survive poaching, they will still be crowded and threatened by agricultural expansion on their remaining ranges. Today it's not a question of *whether* the elephants are in danger, but simply *which* danger is greatest: overcrowding, woodland conversion, or illegal poaching? The most immediate effect we see is the unraveling of the elephant's well-woven social networks. Already, most of the 625,000 elephants that remain are being raised without the wisdom of their elders. No matter how well protected, this generation and the generations to come are already impoverished in ways that cannot be mended.

Pre-Courtship

BULL BEHAVIOR

One of the main reasons that zoos hesitate to keep male elephants is a phenomenon called musth. Musth, which comes from an Urdu word meaning intoxicated, is a period of heightened sexual and aggressive energy that mature males (over 30) experience once a year. As their testosterone level multiplies by fiftyfold, they become touchy and unpredictable, and this state may last anywhere from 1 to 127 days. During this time, they are temporarily catapulted to the top of the dominance ladder. Other male elephants, even those that are larger and normally dominant, make it a point to keep out of the musth male's way. Two males in musth can be a deadly combination when an estrous female is present, but luckily for them, run-ins rarely occur.

The male in musth has a mission in mind: find an estrous female, guard her from the amorous attentions of lower-ranking males, and mate with her himself. He is usually successful, both because of his elevated dominance status and because females prefer musth males over all other sexual partners. In fact, estrous females normally run from most males, but stay to solicit the attentions of those in musth.

How do females know that a male is in musth? There are plenty of signals, and though your senses may not be as keen as a cow elephant's, you can spot a musth male by looking for the following signs:

- **Sticky Streak on Side of Face**: A sweet-smelling secretion flows from openings called the temporal glands, located on either side of the head between the eye and the ear. Somewhat confusing is the fact that females and nonmusth males also secrete a liquid from their temporal glands, especially when alarmed or excited by social interactions. There is a difference, however: the nonmusth secretion is watery and tends to evaporate quickly, while the musth secretion is sticky and long-lasting.
- **Urine Dribbling**: Musth males continuously dribble a trail of strong-smelling urine from a sheathed penis (nonmusth males unsheathe their penis when they urinate). This sends a visual as well as a scent message and is a telltale sign of musth.
- **Musth Rumbling**: Here's one of the low-frequency rumbles that you can hear. Listen for a pulsing or throbbing sound, and at the same time, watch for an ear fold. Musth males rumble several times an hour, ending each call with a cracking ear flap. Researchers suspect that rumbling elephants may be communicating over long distances, because even when they are alone, they show listening behavior before and after rumbling, as if engaged in a dialogue that we can't hear. Perhaps they are tuning in to the challenges of other musth males or to the solicitations of estrous females miles away.
- **Trunk to Head**: In this behavior, the male stretches his mouth open, lifts his head high, and reaches his trunk up to touch his temporal gland. Using his trunk like a paintbrush, he can spread the scent on any vegetation he touches.
- **Marking**: A musth male can also mark vegetation in his area by rubbing his temporal gland on it, smearing it with mouth mucus, or visually scarring it with his tusks.

- **Musth Walk**: Even across a large, grassy exhibit, you can tell when a male elephant is in musth by watching him walk. First notice how a nonmusth male walks: his head, held at shoulder level or below, is thrust slightly forward; his ears are laid back; and his gait is relaxed. By contrast, a musth male will raise his head high above his shoulders and tuck his chin in. His ears are rigidly spread high and wide, and he walks with a controlled swing of his head and tusks.
- **Head Swinging**: In the wild, researchers have noticed standing musth males swinging their heads in a powerful figure-eight motion, sometimes lifting a leg to keep balanced. This normal response is similar to the abnormal, stereotypic head-swaying behavior that some captive elephants exhibit. To verify that what you're seeing is musth, look for the other indicators.
- **Tusking**: Frustrated musth elephants use their tusks to take out their aggression on inanimate objects, especially during a fight.

COW BEHAVIOR

The female plays her part by advertising her sexual state so that musth males can find her. Here are some infrasonic (very low frequency) calls that she may use:

Testing

Chasing and butting

Caressing

COURTSHIP   *To find out whether a female is ready to mate, the male touches and sniffs her genitals with the tip of his trunk. If she is in heat, he pursues her and nudges her rump with his tusks. Just before copulating, the male lays his head along the female's back, grasping and caressing her with his trunk.*

- **Female Chorus**: The females in a family group give this short-range call in response to a musth male coming into their area, rumbling nearby, or sniffing them. They may also chorus after smelling a male's urine on the ground.
- **Estrous or Post-Copulatory Call**: Females use this low-frequency call when they are in estrus, most commonly just after mating or occasionally while mating.

Courtship

When a male finds a female, he **tests** her by putting his trunk tip to her vulva, presumably tasting and smelling the bouquet of hormones that will tell him about her sexual state. If she is ready to mate, he may begin courtship by briefly **chasing** her or **butting** her rump with his tusks. Eventually she holds still while he **lays his head**, tusks, and trunk along her back, grasping and **caressing** her. As she turns to face him, they may entwine their trunks together for a while. Eventually, after a little more nudging, the female **presents** her hindquarters to the male, and they mate. A sexual encounter may also begin in the female's court when she backs up to a musth male and rubs her hindquarters against him to **solicit** attention.

Copulation

It's amazing (and a little worrisome) to see a 12,000-pound elephant rear up and lean his front legs on a 6,100-pound female. The family seems to love it, however, and they get in the act with the mating pandemonium described under Friendly Behavior. Mercifully, copulation lasts a short 2 minutes. Afterward, the pair separates to browse casually, sometimes mating again after 10 or 20 minutes.

PARENTING BEHAVIOR

Birth

Before giving birth, the mother-to-be and another female usually move away from the group and prepare a soft birth spot by loosening the soil with their forefeet. The calf drops head first into this soft spot, falling from the birth canal of its squatting mother. As it drops, the umbilical cord snaps, it takes a startled first breath, and the mother whirls around to begin caring for her infant.

Caring for the Calf

Even at a whopping 260 pounds, a new elephant calf looks tiny and fragile beneath the grand bulk of its mother. The mother usually helps the doll-like form to its feet by gently putting her foot under it and then lifting, steadying it with her trunk. It takes only a few seconds to see how devoted the new mother is. She instinctively hems her calf in toward her, **tucking** it under her chin and **fondling** it with her trunk.

For the first 6 months, the mother follows her calf everywhere, never letting it out of her sight. After that, the calf follows the mother, and the two stay in contact with a continual volley of **contact calls** and, probably, scent cues. The calf will have its mother all to itself for at least 4 years before the female gives birth again and begins to wean the older one.

NURSING AND TRUNK WRESTLING *A young calf uses its mouth, not its trunk, to noisily nurse. In the background, two young bulls vie for their place in the dominance hierarchy.*

Teaching

Even after weaning, however, the calf stays on and spends the next 8 to 10 years learning and developing. It watches its mother carefully, even placing its trunk in her mouth, perhaps to see what kinds of plants she eats. There is much to learn about surviving in their range: the location of good water sources, mud wallows, shade trees, and migration routes, as well as how to harvest the tastiest food.

Communal Care

Calves in an elephant family are also fortunate to have helpers—immature females that help the mother look after, teach, and protect the calf from predators and physical hazards such as heat and cold. This **allomothering** is a favor that is returned again and again, thus building long-term relationships and enhancing the stability of the group. By the time an allomother has her own young, any female calf that she took care of will be 5 to 10 years old—old enough to return the favor.

It's not just allomothers that are solicitous of calves, however. If a calf gives the slightest indication of distress, all members of the group gather around to investigate. This **communal defense** is one of the distinct advantages of a social lifestyle. An entire group is much more of a deterrent to lions, hyenas, and other predators than one mother elephant on her own would be. This system probably evolved not only to outwit these predators, but also to stave off the attacks of early humans.

AFRICAN ELEPHANT BEHAVIORS TO LOOK FOR AT THE ZOO OR IN THE WILD

BASIC BEHAVIORS

LOCOMOTION
- ☐ walking
- ☐ running

FEEDING
- ☐ collecting plant food
- ☐ overturning trees
- ☐ **DRINKING**

BATHING
- ☐ water bathing
- ☐ mud wallowing

GROOMING
- ☐ rubbing
- ☐ dusting

COOLING DOWN
- ☐ spreading ears
- ☐ flapping ears
- ☐ **SLEEPING**

SOCIAL BEHAVIORS

FRIENDLY BEHAVIOR
   Keeping in Contact
- ☐ contact call
- ☐ contact answer
- ☐ let's go rumble
- ☐ *Greeting Ceremony*
- ☐ *Mating Pandemonium*
- ☐ *Group Defense Social Play*
- ☐ sparring
- ☐ trunk wrestling
- ☐ chasing
- ☐ rolling
- ☐ mounting

CONFLICT BEHAVIOR
- ☐ *Threat*
- ☐ *Submission Charging*
- ☐ mock charge
- ☐ serious charge
   Fighting
- ☐ head butting
- ☐ trunk wrestling
- ☐ stabbing

- ☐ tusking

SEXUAL BEHAVIOR
   Pre-Courtship Bull Behavior
- ☐ secreting
- ☐ urine dribbling
- ☐ musth rumbling
- ☐ trunk to head
- ☐ marking
- ☐ musth walk
- ☐ head swinging
- ☐ tusking

   Pre-Courtship Cow Behavior
- ☐ female chorus
- ☐ estrus call
   Courtship
- ☐ testing
- ☐ chasing
- ☐ butting
- ☐ head laying
- ☐ caressing
- ☐ presenting
- ☐ soliciting
- ☐ *Copulation*

PARENTING BEHAVIOR
- ☐ *Birth*
   Caring for the Calf
- ☐ tucking
- ☐ fondling
- ☐ contact calls
- ☐ *Teaching Communal Care*
- ☐ allomothering
- ☐ communal defense

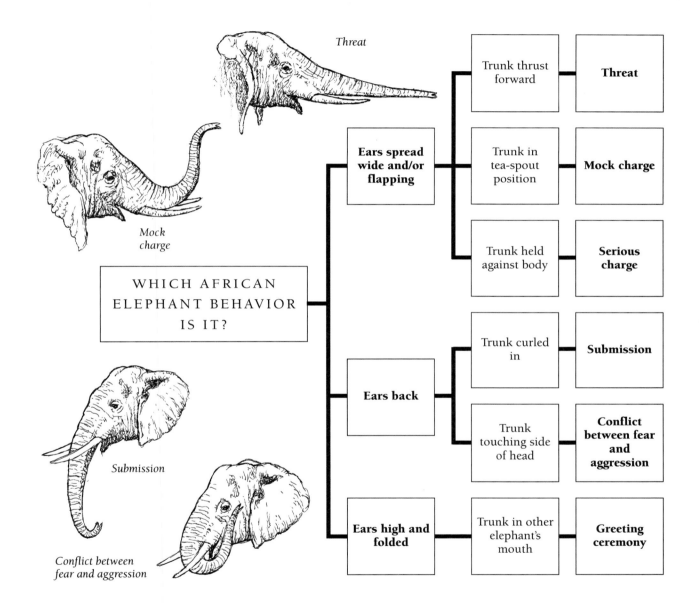

Threat

Mock
charge

Submission

Conflict between
fear and aggression

WHICH AFRICAN
ELEPHANT BEHAVIOR
IS IT?

Ears spread
wide and/or
flapping

Trunk thrust
forward — **Threat**

Trunk in
tea-spout
position — **Mock charge**

Trunk held
against body — **Serious
charge**

Ears back

Trunk curled
in — **Submission**

Trunk
touching side
of head — **Conflict
between fear
and
aggression**

Ears high and
folded

Trunk in other
elephant's
mouth — **Greeting
ceremony**

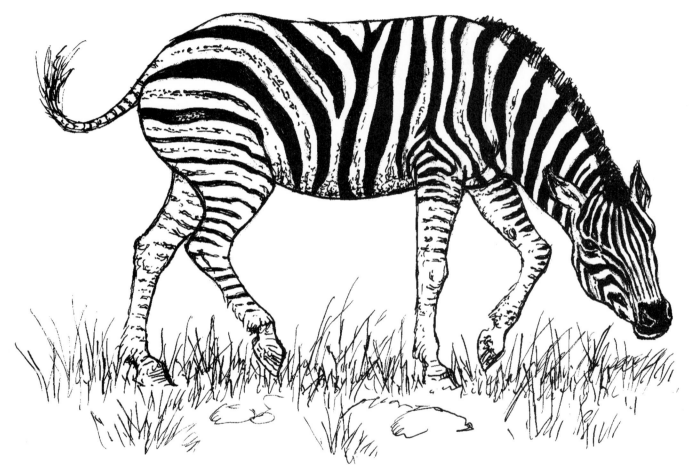

HERDING THREAT   *The dominant stallion flattens his ears and lowers his head to herd the group in a new direction.*

PLAINS ZEBRA

Living on the open plains is like being a delectable fish in a glass bowl constantly watched by a hungry house cat. In this case, the cat is the African lion, and the brilliantly striped plains zebra is its main temptation. The fact that zebras are flourishing rather than being decimated by these stealthy predators is a tribute to their well-adapted bodies and behaviors.

Sly Stripes At first sight, you would think the black and white zebras were colored all wrong to live and hide against a backdrop of golden grasses. Actually, you're probably just standing too close. To get the full effect of their camouflage, try watching the zebras at your zoo from hundreds of yards away, preferably on a hot, hazy day that is reminiscent of African summers. Notice how the stripes seem to break up the outline of individual animals in the wavy air. World War II ships were painted with this same striped pattern to help them "disappear" on the horizon.

Now imagine that you're a lion that has finally spotted the herd. You've got your eyes on the slowest, weakest, most vulnerable member, and you're inching in for the kill. As you get closer, the zebras begin to circle together until, in the psychedelic swim of stripes, you lose the individual you're supposed to be looking for. Rather than risk an energy-draining attack on the wrong zebra, you decide to retreat. In fact, the stripes can be confusing no matter who you are. Even tsetse flies, the blood-sucking parasites that plague most plains animals, can't see the zebra for the

VITAL STATS

ORDER: Perissodactyla

FAMILY: Equidae

SCIENTIFIC NAME: *Equus burchelli*

HABITAT: Grasslands

SIZE: Length, 7.5 ft   Shoulder height, 3.6–4.7 ft

WEIGHT: 385–848 lb

MAXIMUM AGE: 25 years in captivity

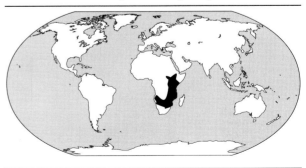

stripes. Instead of homing in on a large white expanse of belly, they see a broken pattern and don't recognize it as a single animal.

Though it may seem ingenious, camouflage may not be the only, or even the primary, function of the zebra's stripes. The real secret to stripes may be the way they captivate the eye and attract zebras to one another. Besides being compelling, the striped pattern is also unique to the individual, allowing zebras to identify one another in the midst of a dizzying crowd. Both seductive and familiar, the bold pattern is part of the glue that holds zebra society together, allowing it to prosper in the shadow of the crouching lion.

Hiding in the Crowd   The lesson of living well in the fishbowl of the savanna is simply this: you can't have too many friends. By keeping more than one set of eyes on the lookout, the zebra herd has a greater chance of spotting a predator before it pounces. A herd also looks more intimidating to the potential marauder, and, if tested, a herd can mount a stronger defense than a lone zebra could. Even when a predator does manage to make a kill, the law of averages still favors individuals in a group; with more prey available, each individual is less likely to be the one eaten.

A willingness to live in groups is also good for the zebra's stomach. Since they are grazers (and not very efficient ones at that) they need large quantities of grass to get enough nutrition. To take advantage of grasses when they are at the peak of their nutritional content, zebras "follow the rains" to the latest flush of growth. Naturally, thousands of other nomadic zebras key in on these feasts too. The fact that they are willing to endure lots of company lets them all partake in prime grazing.

But zebras do more than simply endure one another. They show tremendous loyalty and true affection for fellow members of their family group. Within huge wild herds, or even within the smaller herds at your local zoo, it's easy to pick out these closely knit family groups. They're the ones cleaning one another's hides, nuzzling and playing, feeding, and napping together.

Surprisingly, though, family groups are not automatically aggressive toward other family groups in the herd. In fact, relations among lead stallions are downright amicable, which is somewhat of a phenomenon among hoofed animals. The males of most other species are worn ragged trying to fight off the affronts of rival males that want to take over their territory or harem. In plains zebra culture, other males rarely contest the family stallion's right to mate with his mature mares, so there's no need for mistrust (except when it comes to young mares, as you'll see below). Even the eventual transfer of dominance, when a stallion gets old or sick, usually occurs without a fight.

This harmony both within and between groups enables plains zebras to peacefully share the wealth at prime feeding and watering spots. In a region where these places are few and far between, social bonding, rather than territorialism, makes great sense.

BASIC PLAINS ZEBRA BEHAVIORS TO WATCH FOR

LOCOMOTION

When a predator gets too close for comfort, the zebra's saving grace is its well-muscled legs. In brief, dead-heat getaways, zebras can accelerate to 50 miles per hour. You may see these amazing bursts of speed in the zoo when young zebras are racing or playing games of tag. As they run, look past the stripes and you'll see the strong family resemblance between zebra foals and their relatives—our thoroughbred race horses. Like all horses, zebras **walk** with a diagonal stride, **trot** to gain speed, then **gallop** to get away. And despite their weight and size, zebras are not stopped by water; if it's too deep to wade, they are surprisingly strong **swimmers**.

FEEDING

If you see no other behavior at the zoo, you're sure to see feeding. Unlike ruminants, which have four stomachs to break down cellulose in grasses, zebras have only one simple stomach. As a result, they must spend 50% of their time **grazing** and eat twice as much as other grazers to glean the same amount of nutrition. Luckily, they aren't picky eaters and will go after dry, coarse grass as well as more succulent fare. Zebras are also content to eat all parts of the grass, from the tip of the blade to the soil line. To find the acres of pasture land they need, zebras have evolved nomadic habits, moving ahead of the herds of other grazers on the plains to avoid competing with them.

When the first green flush of growth arrives at the start of the wet season, zebras congregate in vast herds of up to tens of thousands to partake of the plenty. Because they are the only plains animals that have both upper and lower incisors, they can bite grasses in two rather than just pluck them. By snipping the taller and coarser grasses, zebras uncover the smaller grasses for the succession of animals that follow. "Preparing" the range in this way is just one of the many indispensable and irreplaceable roles that zebras play on the African veldt.

DRINKING

Zebras also congregate near water sources when the dry season takes its toll. If drought sends rivers underground, zebras may dig into the sand with their front hoofs to unearth some liquid refreshment. As with all plains animals, zebras are especially wary when drinking, since hungry predators make a habit of shopping at water holes.

SELF-GROOMING

If you've ever tried to scratch an itch between your shoulder blades, you'll understand why zebras are so crazy about rubbing posts. When workers at a zoo in Switzerland tried to install an artificial termite mound in the African plains exhibit, they were hounded unmercifully by appreciative zebras. As soon as the workers stepped away, the entire herd descended on the mound, **rubbing** blissfully as if making up for lost time. Zebras also rub against trees to remove annoying insects, loose hair, or dandruff. For full-body

RUBBING   *Termite mounds provide a convenient post for rubbing impurities from the skin.*

rubs, they **wallow** in sand and dust or even go belly-up in the grass for between-the-shoulders relief. To reach their head and neck, they may **scratch** with a hind leg like a huge dog. They'll also twitch, shudder, and shake to rid themselves of insects and dust or to dry off after a cloudburst. And, finally, if a grooming partner is not available to do the honors, they'll **nibble** at disturbances anywhere their teeth can reach.

SLEEPING

When zebras sleep, one member of the herd usually stays awake and alert to do their watching for them. For deep sleep, they lie down with their legs tucked beneath them so they can jump up quickly should the sentinel sound the alarm. You can tell the younger animals in the herd because they lie on their side with their legs stretched out to one side. When dozing rather than sleeping, half-awake zebras stand with their head drooping, their eyes half closed, their ears pointing out to the sides, and their tail chasing flies back and forth.

SLEEPING *One zebra acts as sentinel, keeping watch while the others sleep.*

HOW PLAINS ZEBRAS INTERACT

The two basic units in zebra society are family groups and bachelor groups, both of which you can see if your zoo herd is large enough. Family groups consist of a stallion and as many as 6 mares and their foals, for a total of 2 to 16 animals (the wild average is 7). The stallion is the dominant figure in the group and the only male that mates with the females. Young males leave their family group when they are between the ages of 1 and 3 to join bachelor groups. A bachelor group (stallions only) may have as many as 10 animals, although the average is 3.

Zebras remain in family groups not because the dominant male stallion is forcing them together, but because group members have phenomenally tight bonds with one another. Friendly behaviors such as grooming are social lubricants of sorts, keeping the connections in good working order. Ironically, even conflict behaviors help ease

LISTENING TO PLAINS ZEBRAS		
WHAT YOU'LL HEAR	WHEN	WHAT IT DOES
2-syllable *i-ha* or loud snort	■ when danger approaches	■ warns
loose-lipped blowing	■ when content	■ signals well-being
contact call—*ha,haha, hahaha*	■ when member is lost ■ when crowded with other family groups	■ maintains or reestablishes contact
high-pitched squeal	■ when in pain or distress	■ solicits help
long squealing	■ when foals are in pain or distress	■ solicits help

tensions by establishing a hierarchy that all members seem to know and respect.

To see the hierarchy in action, watch the zebras at your zoo when they travel. They march in single file with the dominant or alpha mare in the lead and the other mares lined up according to their social standing. Foals follow behind their mom, with the youngest sibling coming first. As they mature, these foals will assume their mother's social rank. The stallion walks behind or off to the side of the troops and is usually content to let the mare decide where they will go to graze or drink. Every so often, the stallion might show his preference for a direction by displaying a herding threat.

GROOMING PATTERNS IN FAMILY GROUPS	
Most Frequently	Mares and youngest foals
↓	Mares and the next oldest foals
↓	Stallions and mares
↓	Stallions and foals
Least Frequently	Mares and mares

In some groups, researchers have noticed that the ranking among mares shifts every few months. You might be able to observe changes in the group at your zoo or animal park over a series of visits. Try singling out the alpha mare, and then watch for her every time you come. If it's a bachelor group you're watching, try picking out the adults; they're the ones pulling rank on the adolescent stallions in the crowd.

FRIENDLY BEHAVIOR

If you watch zebras for any length of time, you are bound to see plenty of friendly gestures. These behaviors bolster trust among members of a group and reaffirm the relationships that keep them together.

Social Grooming

Grooming can occur at any time of day, and a single session may last up to half an hour. Standing face-to-face, grooming partners begin by **nibbling** each other's neck and back. They work their way down toward the tail on one side and then walk around to the other side and start again. As they nibble, their upper incisors scratch against the lay of the coat, helping to extract loose hair and clean the skin.

Every zebra in the group engages in mutual grooming, even foals that are only a few days old. Watching who grooms whom can help you discern relationships between individuals. The animals with the strongest bonds groom each other most frequently. See the table above for grooming relationships.

You'll notice, after watching a herd for a long time, that the stallion plays favorites when it comes to grooming the mares. He grooms certain mares more often, and if there is a young mare in heat, he will pay most attention to her. In bachelor groups, all the males groom one another, but without noticeable preferences.

SOCIAL GROOMING   *Zebras run their teeth against the lay of the hair to nibble irritants from each other's skin. Grooming also builds social bonds; individuals with the strongest bonds groom one another most frequently.*

It's interesting to note that humans have become social "cleansing buddies" for domestic horses. Even a shy animal will eventually learn to trust the handler who consistently brushes and scratches it. Zookeepers report that this sort of grooming can earn a zebra's trust as well, which can be important when it comes time for immunizations and other kinds of veterinary care. A zebra that is trusting enough to be handled won't have to undergo the trauma of being tranquilized.

Social Play

Young zebras from a few weeks to several years old bond with one another in much the same way that human children do—through play. They spend their endless font of energy on **racing games**, **mock fights**, and other imitations of adult behavior. These bouts of neck wrestling, biting, rearing up on their hind legs, and chasing only look serious. Even the most vigorous row usually ends with the exhausted partners resting their heads on each other's backs in a friendly show of bonding. If they can't find cohorts their own age, young zebras will play chasing games with a willing adult or even with other animals such as gazelles, mongooses, and birds.

Young stallions in bachelor groups are especially frisky and fun to watch. In a moment of excitement, they may suddenly break into a full-speed gallop, racing neck and neck toward some imaginary finish line. As if to celebrate the victory, they often end with a **greeting** ritual, enacted with the comical looseness of play. These games sharpen the colts' physical and social reflexes, preparing them for the days when survival will depend on fast getaways and loyal relationships.

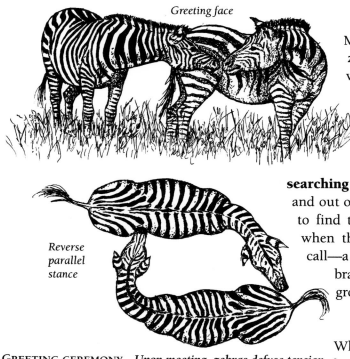

Greeting face

Reverse parallel stance

GREETING CEREMONY   *Upon meeting, zebras defuse tension by sniffing each other and wearing an appeasing expression. Standing head to tail, they rub each other's flanks and sniff each other's genitals.*

Keeping in Contact

Maybe our pundits should go to the zoo and study the zebra's version of a "kinder, gentler society." They would notice, for instance, that traveling zebras don't leave young, old, or lame animals behind. Instead, the pace of the group is determined by the slowest rather than the fastest member. Furthermore, if a clan member happens to get separated, the stallion goes to great lengths to find it, even if it means **searching** through a herd of 10,000 other animals. Weaving in and out of the milling crowd, he'll use his eyes, ears, and nose to find the lost member, periodically giving a **contact call** when the trail grows faint. Stallions also use this contact call—a combination of barking and inhale-exhale donkey braying—to get in touch with stallions from other groups.

Greeting Ceremony

When stallions from different groups meet, they enact a beautiful ritual that has both sexual and aggressive elements. To begin, both zebras stretch their heads forward and **sniff** each other's noses. Next, either facing each other or standing side by side, they stretch their faces into the **greeting face**. With ears bent forward, lips pulled back, and mouths slightly open, they make subtle chewing motions. This greeting face is very similar to the estrous face described in the Copulation section on page 129. Some researchers believe both expressions are forms of submission that say, in effect, "I do not intend to fight."

If the two stallions are not of equal rank, only the lesser one will display the greeting face and the ceremony will end there. If the stallions *are* of equal rank they will continue the ceremony by standing head to tail in a **reverse parallel stance**. Together they push their heads against each other's flanks and vigorously **rub** up and down, stopping every now and then to sniff each other's genitals. Finally, they smell noses and perform an abbreviated version of the **farewell jump**. The full jump involves standing up on the hind legs and springing apart, but more often, you'll see a kind of shorthand—a leg thrown out or a head thrown back—which represents the jump.

You may also see the greeting ceremony conducted within a group of male bachelor zebras at your zoo, but it will not be particularly formal or ceremonial. At the end of the greeting, one partner may put his head on the other's back, in the same way a dominant male may end an aggressive encounter with a submissive male. All of these gestures reassure the two zebras that though they have noticed each other, neither intends to fight.

Communal Defense

The zebra's social network works particularly well when predators, such as lions, hyenas, wild dogs, leopards, and cheetahs, are lurking in the grasses nearby. When a sentinel zebra sees a predator, or something suspicious at the zoo, it gives an **alarm call** to warn the rest of the group. It may be either a loud snort and a stamp of the foreleg, or a raspy braying (*i-ha, i-ha*) produced by breathing in and out with the

HUMANS AND ZEBRAS

Flying above or driving through a herd of boldly graphic zebras can be dizzying. They seem so numerous; you can't believe that they could ever dwindle the way many of the savanna species have. In reality, however, two subspecies of zebras have already been exploited to extinction. Burchelli's zebras (*Equus burchelli burchelli*) were wantonly destroyed by colonists who used their skins to make connecting bands for machinery. In the same way, colonist farmers and trekkers slaughtered quaggas (*Equus quagga*) relentlessly, eating their meat and using their skins as carrying pouches and their hocks as shoe soles. In less than 40 years, the once numerous quagga was completely exterminated.

Today, only 500,000 plains zebras remain, and these are still threatened by our hunger for land and places to raise our livestock. One subspecies of plains zebra is classified as vulnerable, two are endangered, and the rest are considered to be threatened. So, though the herds may look inexhaustible from the air or in nature films, we have to keep reminding ourselves that, just like the optical illusion caused by their stripes, appearances can be deceiving.

mouth open. In unison, the zebras stop what they're doing and **stare** at the source of danger. If the predator continues to approach, the group forms a **semicircle** and faces the predator with heads raised and ears cocked forward.

Another form of defense is to **walk closer** to the predator and continue staring. Zebras have a 100-foot flight distance; if the predator comes within this invisible sphere, the zebras will turn tail and run. As they flee, the family stallion acts as a rear guard, flailing backward with his sharp, powerful kicks and turning occasionally to bite. One group's reaction to a predator serves as an early warning for all the other groups in the herd.

When zebras that are separated from the group get in trouble, the group may attempt a daring **rescue** mission. One biologist tells the story of a group of 10 zebras that came galloping back in full force to rescue two foals and a mare being harassed by hyenas. The zebras surrounded the distressed trio and galloped off with them, kicking backward to discourage any followers. Although zoos don't let hyenas or lions into their zebra exhibits, you may see these protective moves in nature films, at open-air wild animal parks, or on safari.

CONFLICT BEHAVIOR

Plains zebras are not territorial, so you won't see them defending a particular area. They will, however, exert their dominance when it comes to being first in line for food, water, or a rubbing post. If the zebras at your zoo eat and drink from communal troughs, for instance, you're likely to see even more squabbling than you would at a wild water hole, where there's more room to spread out. Rather than endanger themselves with kicking and biting fights, however, the squabblers usually start and end with threat gestures.

Threat

Typically, a threat gesture is an abbreviated version of an actual fighting move—a taste of things to come. Symbolically, the zebra is shaking its fist and saying, "Don't go any further with this. I am willing to fight with you if need be."

At its most subtle, a threat gesture may be as simple as lowering the head and laying back the ears. A stallion that wants a lineup of zebras to change course, for instance, will nudge them from the sidelines with this **herding threat**. Females may use the herding threat to intimidate lesser mares that try to cut ahead in line.

For a more intense threat, the zebra may stretch out its neck, point its head at its opponent, and lay back its ears. To show that it is willing to bite if need be, the zebra opens its mouth and even bares its teeth in a highly agitated **bite threat**. Also symbolic of things to come, the zebra may **kick**, **rear up** briefly, or **scrape** the ground with its foreleg. Mares typically make their point using a hind-leg kick.

Threat displays and an occasional **chase**

KICK THREAT   *The female typically threatens with a hind-leg kick.*

are usually enough to decide an altercation. If a true fight erupts, chances are it will be between males, and the prize will be the right to mate with a young mare. If your zoo keeps stallions together in one exhibit, be on the lookout for these sexual battles.

Fighting

A zebra fight looks almost like a martial arts performance. It begins with **circling**, in which the two animals, standing head to tail, try to bite each other on the back and legs while trying to avoid being bitten. They go around and around, crouching lower to protect their hind legs until they are almost pivoting on their haunches. After a bout of circling, the zebras usually start to **neck wrestle**, jockeying for position the way humans do in a game of thumb wrestling. One zebra places his neck across the top of the other's and pushes down while the zebra beneath pushes up with all his might. Suddenly the one on the bottom may drop down, pull his head out, and rush to cross over the neck of his opponent.

When fights escalate to serious conflicts, you may see the animals **rearing up** on their hind legs and **kicking** each other with their hoofs. Hoofs are the zebra's only real weapons, and when wielded with enough force, they can be sharp enough to leave a nasty gash. As the zebras strike downward, they also **bite** at their opponent's neck, ears, and mane, sometimes clamping down for a good hold. Luckily, their teeth, most of them worn smooth by grinding grass, rarely penetrate the tough skin. The real object in all this drama seems to be knocking the opponent off balance; once he stumbles, the steadier zebra has won the upper hand.

Circling

Neck wrestling

Kicking and biting

FIGHTING   *Opponents snap at each other's back legs while trying to avoid being bitten. Going neck to neck, they pour on the pressure, each trying to place its neck on top. Their kick, powered by thoroughbred muscles, is far worse than their bite.*

Zebra fights, like most conflicts in the animal world, are rarely fatal. Although kicks and striking blows may leave some wounds, they rarely debilitate the contestants. Before they come anywhere close to that intensity, one opponent usually waves the surrender flag and runs away.

Fighting for mates serves an ecological purpose beyond the winning of a particular female. Because a male must win a fight before he can mate, only the strongest and healthiest specimens in the population wind up passing on their genes. This selection process helps to improve the entire species.

SEXUAL BEHAVIOR

There is no one best time of the year to see sexual behavior in zebras; foals can be born throughout the year. The only requirement is that mares be in estrus—a monthly event lasting only about a week.

Courtship

The details of a zebra's sex life depend entirely on who they are and how old they are. For adult mares that have bred before and are comfortably ensconced in a family clan, estrus comes quietly. The stallion, who has undisputed sexual privilege in the group, learns of her condition through hormones in her urine and dung. After **sniffing** her rear and inducing her to urinate, he takes a good strong sniff and raises his head in a **lip curl** gesture—nose thrown to the sky and lips curled back. This gesture seals the nostrils and helps the odor travel more efficiently to scent receptors, called Jacobson's organs. He then defecates on her dung and sprays urine on the spot where she has urinated. Researchers suggest this may be a way to conceal her condition from neighboring males, and thereby keep the family group together.

For young mares coming into estrus for the first time, courtship has a more dramatic flavor. The dominant male, usually her father, pursues her in earnest, lavishing her with attention (**grooming**) for long periods of time during the week she is in estrus. With these first-time mares, the dominant male does not enjoy his usual undisputed claim. The young mare comes into estrus between the ages of 13 to 18 months old and begins to advertise her sexual state months before she is ready to mate. In the unmistakable **invitation stance**, she stands with her hind legs apart and tail raised at an angle of about 45 degrees. Her mouth is usually open in a gesture that looks like a threat.

Males from miles around notice her display and begin to gather in hopes of mating with her. Though she repels all mating attempts for weeks, the bevy of suitors waits anxiously at the fringes. In what must be a stressful time, the family stallion fights off the suitors one by one, while still trying to win her for himself. The hopeful males watch closely, waiting for their chance to **abduct** the mare, usually while the stallion is tied up in battle. The filly's first abductor is not necessarily her final mate, however. She may move on several more times before she becomes a permanent member of a new group when she is about 2½ years old.

What appears to be a loss for the family stallion can be a gain for the species as a whole. Consider how tight the bonds between family members are. If young mares didn't attract suitors to steal them away, they would probably never elect to leave, and incest and inbreeding would eventually weaken the population. By joining new groups, females spread their genes around, and the entire genetic pool is richer.

Copulation

It's only the youngest mares that use the invitation stance to beckon to strange males. When an older mare assumes the invitation stance, the family stallion **mounts** her immediately, leaving little opportunity for other males to see and be attracted to her. This inconspicuous intimacy helps maintain the stability that is so important in zebra culture.

Mares display a typical expression called the **estrous face** before or during copulation. They usually hold their ears back and downward, draw back the corners of their lips, and chew with slightly exposed teeth and open mouth. They may even drool. (Submissive males also wear this expression when confronted by a dominant male; see Greeting Ceremony.) Occasionally, the mare turns her head back toward the stallion, throwing him a few symbolic **snapping** movements. Some researchers feel that the estrous face may have been derived from defensive biting.

PARENTING BEHAVIOR

Birth

As the time of birth approaches (after a gestation period of 1 year), males become ever more protective of the female. In fact, if your zoo has a pregnant mare, watch the stallion closely; his excited whinnying is a telltale sign that the birth is about to begin. The female also shows signs: a distended belly, swollen vulva, raised tail, and a difficulty moving her hind legs.

The female lies on her side during foaling, expelling her newborn hoofs and head first, embryonic sac and all. Although the mother doesn't assist, the foal shakes itself free of its covering, and after only 20 minutes, it stands up, breaks the umbilical cord, and begins taking stock of its world.

Caring for the Foal

For several days after the birth, the female keeps all the members of the group, including the father, away from the newborn foal. **Isolating** the foal in this way prevents it from accepting another animal as its mother, which can happen during the critical imprinting period when foals follow the first object they see. The female needs to bond with the newborn and encourage it to **suckle**; otherwise, the foal would try to get nourishment from anything larger than itself, including humans, gnus, and even Land Rovers. Once the mother and her offspring can identify each other by scent, voice, and stripe pattern, they can rejoin the group without fear of losing touch.

Communal Care

A zebra group is a safe, nurturing place for a foal to grow up. If it gets into any danger, the foal need only give a drawn-out **squeal** to trigger a group **rescue**. Considering the many predators that enjoy zebra meat, surprisingly few foals are lost, thanks to this unique group care.

PLAINS ZEBRA BEHAVIORS TO LOOK FOR AT THE ZOO OR IN THE WILD

BASIC BEHAVIORS

LOCOMOTION
- ☐ walking
- ☐ trotting
- ☐ galloping
- ☐ swimming

FEEDING
- ☐ grazing
- ☐ DRINKING

SELF-GROOMING
- ☐ rubbing
- ☐ dust wallowing
- ☐ scratching
- ☐ nibbling
- ☐ SLEEPING

SOCIAL BEHAVIORS

FRIENDLY BEHAVIOR
Social Grooming
- ☐ nibbling
Social Play
- ☐ racing games
- ☐ mock fights
- ☐ greeting ceremony
Keeping in Contact
- ☐ searching for lost members
- ☐ contact call
Greeting Ceremony
- ☐ sniffing
- ☐ greeting face
- ☐ reverse parallel stance
- ☐ rubbing
- ☐ farewell jump
Communal Defense
- ☐ alarm call
- ☐ staring
- ☐ grouping in a semicircle
- ☐ approaching predator
- ☐ rescuing

CONFLICT BEHAVIOR
Threat
- ☐ herding threat
- ☐ bite threat
- ☐ foreleg scrape
- ☐ symbolic kick
- ☐ rearing up
- ☐ chasing
Fighting
- ☐ circling
- ☐ neck wrestling
- ☐ rearing up
- ☐ kicking
- ☐ biting

SEXUAL BEHAVIOR
Courtship
- ☐ sniffing
- ☐ lip curl
- ☐ grooming
- ☐ invitation stance
- ☐ abducting the female
Copulation
- ☐ mounting
- ☐ estrous face
- ☐ snapping

PARENTING BEHAVIOR
Birth
Caring for the Foal
- ☐ isolating the foal
- ☐ suckling
Communal Care
- ☐ squealing
- ☐ group rescue

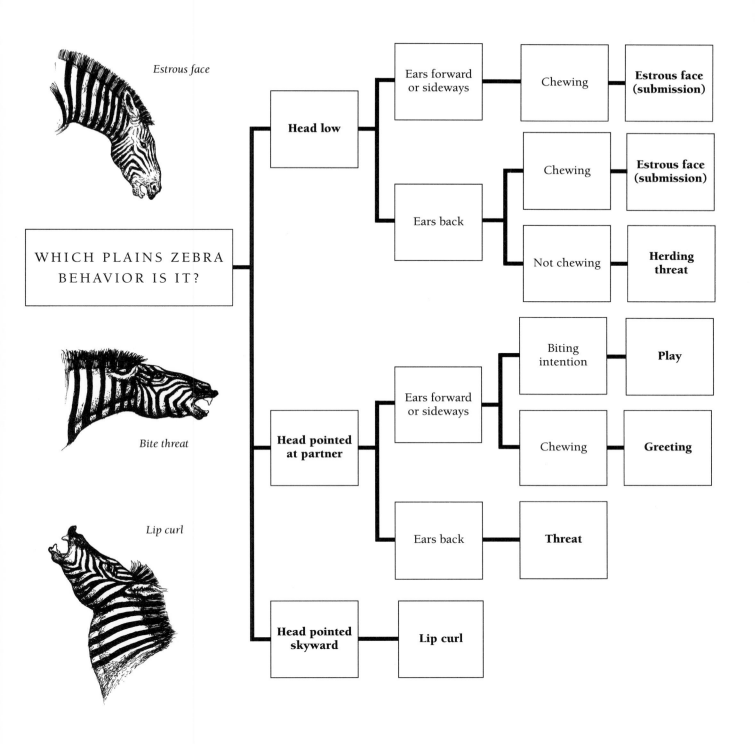

Estrous face

Bite threat

Lip curl

WHICH PLAINS ZEBRA BEHAVIOR IS IT?

Head low

Ears forward or sideways → Chewing → **Estrous face (submission)**

Ears back → Chewing → **Estrous face (submission)**

Ears back → Not chewing → **Herding threat**

Head pointed at partner

Ears forward or sideways → Biting intention → **Play**

Ears forward or sideways → Chewing → **Greeting**

Ears back → **Threat**

Head pointed skyward → **Lip curl**

MUD WALLOWING *A mother and calf dip themselves in mud to cool off and to rid their skin of parasites.*

BLACK RHINOCEROS

Though not exactly glamorous, rhinoceroses do have a certain mystique, and it's made them a favorite with the folks on Madison Avenue. The latest ads featuring rhinos use the endangered animals to sell vinyl conditioners, alarm systems, a yuppie saloon, and even beers "brewed the old-fashioned way." With their solid stature and their prehistoric good looks, rhinos are the Brink's armored cars of the animal world, epitomizing the durable and indestructible.

When you think about the rhino's life off the set, however, their maximum security gear seems a bit overdone. After all, rhinos are the second largest land mammals next to elephants, the adults have no natural enemies, and most animals (besides human hunters) give them a wide berth. Then why all the combat insurance? To answer that question, we have to admit that in one way the advertisers are correct: the rhino design has lasted a long time. Many evolutionary moons ago, rhinos were probably hounded by some pretty formidable predators. Though these giant carnivores no longer exist, the rhino's behaviors and physique are reminders of the days when they needed all the help they could get.

Long ago, rhinos learned that the best combat plan of all is to stay home from the war. Rather than take on the predators of those days, rhinos developed a keen sensitivity that allowed them to hear and smell danger before it got too close. Today, their cupped ears still swivel in all directions to catch the slightest sound, and their nose can read

VITAL STATS

ORDER: Perissodactyla

FAMILY: Rhinocerotidae

SCIENTIFIC NAME: *Diceros bicornis*

HABITAT: Transitional zone between forest and grassland; in thick thornbush, acacia shrubs, or more open country up to 9,000 ft altitude

SIZE: Body length, 9–12 ft Shoulder height, 4.5–5 ft

WEIGHT: 2,204–3,968 lb

MAXIMUM AGE: 45 years in captivity

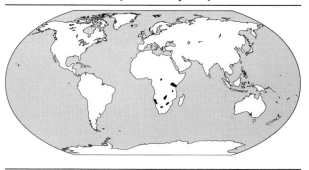

the air like a book. In fact, the nasal passages in a rhino's head take up more space than its entire brain. When sufficiently alarmed, they can pivot on a dime and gallop into the brush at 35 miles per hour. Their bulldozer of a body can either mow down vegetation or negotiate a mean slalom, thanks to a muscle structure that twists and turns with precision.

With all this talk of retreating, whatever happened to the snorting, paw-scraping, charging straight-at you rhino we've all seen in the movies?

The two-horned black rhino spends most of its life alone.

Rhinos do get this worked up, but only on rare occasions, like when they are caught unawares or when a male decides to push his luck against a rival. It's times like these that their horns and tough hide come back into fashion.

Loners by Design   In between times, rhinos live rather solitary lives, ranging over an area that is large enough to provide them with all the woody plant food they need. Although they're not strictly territorial, they do discourage intruders by blanketing the area with "no trespassing" scent marks. This insistence on elbow room, besides keeping them out of one another's hair and away from one another's horns, is a blessing for the habitat as well. If rhinos didn't space themselves out for feeding, they would quickly deplete and overbrowse their home range.

There are a few exceptions to the "alone and apart" rule. The rhino's longest-running social relationship—between the mother and calf—lasts up to 5 years before the mother gives birth to a new calf. For a while, she may tolerate the presence of both an older calf and the newest arrival, leading to sightings of groups of three rhinos. Sexual trysts are another reason for togetherness, but these are relatively short-lived. The only thing that brings solitary rhinos together on a regular basis is the need to drink and wallow each day. Animals that use the same water hole get to know one another, but are quick to chase or threaten any new rhino that intrudes on their party.

Even though these water-hole clans (or the groups at your zoo) tolerate one another, you won't see them going out of their way to be friendly the way zebras and giraffes do. That's because rhinos simply don't need one another the way some hoofed animals do. Unlike plains animals that are always in the predator's eye, black rhinos live in a forest full of camouflage and don't need to shield themselves with allies. Keep this in mind while watching the individuals at your zoo. Besides the occasional conflict over spacing or perhaps mating overtures, the bulk of the rhino show is a solo performance.

LOCOMOTION

Despite their short, squat legs and boxy bodies, rhinos are surprisingly nimble. Their normal gait is a **fast walk**, but, when frightened, they may **trot** with head raised and tail up over their back or straight out behind them. They can execute hairpin turns at a thundering **gallop**, but they can also step lightly, in a delicate dance step, when courting females. In the dirt of their exhibit, look for the ace-of-clubs track left by their three-toed feet. You'll no doubt find the tracks in well-worn paths, since rhinos, both in the zoo and in the wild, are notorious creatures of habit. Like humans, they cling to their routines and prefer to take familiar paths.

FEEDING

At the tip of the rhino's solid, prehistoric snout, you'll find loose, labile lips. They use these lips almost like fingers to grab and **browse** twigs and leaves from a wide variety of shrubs in the acacia woodland community. They may also lend their horns to the task, yanking branches down to lip level. One of the more unusual food items that rhinos indulge in is the feces of other animals such as the wildebeest. **Dung feeding** often occurs when browse is in short supply and when grass, a supplementary food, has been either grazed down to the nub or scorched by fire. With no fresh grass to graze, rhinos eat the partially digested grasses from other animals as a way to recoup needed minerals.

DRINKING

In dry seasons, rhinos are virtually tethered to their water holes. Though they wander, they stay in a circular area where no point is more than 3 miles away from liquid refreshment. When surface water is not available, rhinos may dig their own wells. Their forelegs send the dirt flying beneath them until they hit a source of underground water, where they can finally drink their fill.

MUD WALLOWING

In the sweltering African heat, water becomes as important for dunking as it is for drinking. The rhino's voluminous body generates a lot of surplus heat, but the surface area of its skin isn't large enough to radiate and release it all. Instead, rhinos must rely on the power of evaporation to pull heat from their body. Since they have no sweat glands to provide moisture, they coat themselves with moist mud which cools them as it dries. The mud mask may also give the rhinos some relief from biting flies, and when it cracks and falls off, ticks and other parasites may fall off too. Even zoo rhinos seem to appreciate a good mud bath; animals that have access to water and a place to wallow are likely to be a lot more active than those in dry exhibits.

HORN SHARPENING *After years of sharpening, the rhino's horns will wear down and then grow back.*

HORN SHARPENING

Rhinos rub their horns back and forth on trees or rocks to sharpen them. Horns rubbed on rocks tend to wear down

faster than those sharpened on wood, but the horns do regenerate, especially when the animals are younger.

SLEEPING

Rhinos do most of their feeding in the early morning and evening, spending the sultry mid-portion of the day resting in the shade or in a wallow. They catch their Z's either lying down or standing on all fours, with head hanging and ears still moving of their own accord. Several times during their slumber, rhinos will bolt up and briskly trot about, as if keeping themselves on their toes. In addition to their own vigilance, rhinos have a partner in crime detection—the birds that tribal people call "the

DUNG SCRAPING   *Rhinos scrape their hind legs through the communal dung pile, tracking the strong scent wherever they walk.*

HUMANS AND RHINOCEROSES

Perhaps because they are so durable, we like to believe that rhinoceroses are indestructible as well. The truth is that rhinos as a species are quite fragile. Because each female mates only once during her estrus and has young only once every 3 or 4 years, it takes a while to replace a rhino that has died. In the days before human intervention, this was never a problem because rhinos lived long lives and their death rate was low. Today, however, human hunters are bleeding some rhino populations for their horns, drought is threatening others, and loss of habitat hangs heavy over all.

As a result, black rhinoceros survey counts are like countdowns, ticking ever lower as you read this. Even in game preserves, where survey aircraft and armed guards protect rhinos, poachers are taking bigger and bigger chances. The monetary rewards for those who succeed are extravagant.

Ground-up rhino horns are used medicinally in China, South Korea, Taiwan, and Thailand, and as an aphrodisiac in northern India. Since raw horns can command $20,000, the ground-up powder is surely among the world's most expensive drugs. Ironically, there is no pharmacological basis for the powder's supposed powers. The horn is composed of keratin, a perfectly ordinary protein that is found in common substances, including our fingernails.

In the 1980s, a surge in the demand for rhino horn came from North Yemen, where horns are used for making handles of special jambia daggers. The daggers are given to young Yemen men when they reach puberty as a symbol of their manhood and devotion to the Muslim faith. For years, only an elite group of Yemenis could afford the rhino-horn daggers, which can cost as much as $30,000 (1988 prices). In 1970, when the civil war ended, Yemenis began to cross the border and make high salaries in nearby oil-rich Saudi Arabia. As personal income increased fivefold, the demand for rhino-horn daggers increased along with it, making rhino hunting well worth the risk to some poachers.

In June of 1991, biologists reported that only 3,400 black rhinos were left in Africa, down from 2 million at the turn of the century. It's hard to say for sure how many rhinos are left today, because a constant barrage of poaching diminishes their numbers daily. No matter how inspiring, the concept of the indestructible rhino is only wishful thinking. As long as this powerhouse of an animal is square within the cross hairs of modern greed, even the fastest gallop cannot help it escape.

rhino's police." These tickbirds eat insects on the animal's back as it sleeps, calling raucously and waking the animal whenever they see something or someone approaching.

In areas where rhinos are constantly harassed by humans, they reverse their schedule so that most of their activity is at night. If the harassment were to lessen, researchers suspect that the rhinos' schedules would revert to a normal, day-active state. Thankfully, rhinos in zoos feel comfortable enough to feed during the day and sleep 8 to 9 hours a night. If you're ever in the zoo after hours, stand quietly near them and you might hear something familiar—snoring.

FRIENDLY BEHAVIOR

Since rhinos are not social creatures, instances of friendly behavior are rare. You may, however, witness a **greeting ceremony** between two mothers, each with a calf in tow. The adults will first touch noses, reassuring each other that no harm is intended, and then the calves will do the same.

CONFLICT BEHAVIOR

Scent Marking

It helps to remember that the rhino's goal is to avoid conflict, and, if possible, to avoid meeting other rhinos at all. To this end, males frequently mark their feeding range with long-lasting scents: dung, urine, and even flakes of their own skin. In the wild, all male and female rhinos in an area will **defecate in the same pile**, each animal adding its contribution, then **scraping** through the pile with its hind feet, as if spreading the scent message around. As it walks away, it probably carries the scent on its hoofs, laying down an odoriferous path. **Rubbing** against bushes and rocks may also leave behind scented flakes of skin and dried mud.

The males also scent by **spraying urine** back through their hind legs, aiming at bushes, clumps of grass, tree stumps, and other conspicuous points along trails or range borders. They pump the urine 9 to 12 feet behind them, producing a fine aerosol that perfumes a large expanse and lasts for days. The male's spray seems to contain status information useful to other travelers in the area. Subordinate males coming across this urine, for in-

URINE SPRAYING *To broadcast their personal scent far and wide, rhinos hose down vegetation with a 12-foot stream of urine. Males spray to keep other rhinos away, while females use their urine to attract eligible males.*

THREAT CHARGE *The threat charge rarely ends in a goring; it is a bluff display that prompts the opponent to "back down or else."*

stance, know exactly who they are dealing with and usually go the opposite way. A female will also spray urine, but her mission differs somewhat from that of males. She leaves her chemical calling card only when she is in heat, as a way to inform males that she is ready to mate.

In the zoo, sanitation procedures prohibit the buildup of manure in the displays. Nevertheless, rhinos still defecate in the same spot each day, and though it is removed, the odor remains. Because zoo-keepers know how important the home range is to the rhino, they take special precautions when introducing new animals to the exhibit. They usually try to stage the introduction in a neutral area, so the pair can make their peace on common ground.

Threat

Sometimes, despite their precautions, rhinos cross paths and have to settle their differences on the spot. Their negotiations climb a predictable curve, from bluff threat to actual goring. The Hollywood image of the charging, stampeding rhino tells only half the story and misses much of its subtlety.

It's true that a rhino, if provoked, will warn the perpetrator with a **threat charge.** It will lower its head, roll its eyes, prick up its ears, raise its tail, curl its upper lip, and let out a screaming groan as it charges. It's a frightening display that does exactly what it's designed to do: it prevents a real attack from taking place. Once treated to a show of the rhino's willingness to fight, most adversaries are more than happy to turn around and find a quieter corner of the woodland. In zoos, where space is limited, keepers must provide a substitute for this "break and run" escape. They often install visual shields such as hills, rocks, and trees so a less dominant animal can simply hide.

Before they work up to a charge, however, rhinos usually try a more subtle threat gesture. When two males are still some distance away, they will turn their heads from side to side as if to **display their weapon**—its size, their size, and their willingness to fight if need be. They may also **spray urine** and **break branches** as a way of showing dominance. When they meet in closer quarters, like at a water hole or a wallow, the resident male may warn the newcomer by **horning**—repeatedly jerking his head and horn into the air. He may not even bother to stand; horning offhandedly from a prone position is usually enough to let the newcomer know that he should wait his turn.

When females and males meet, the interaction is altogether different. If the female is in heat, courtship may begin. If not, they both emit a puffing **snort** when they see each other. The bull approaches the female with a fake **two-step charge**; he takes short, mincing steps, snorting occasionally, shaking his head from side to side, and jerking his horn into the air. (This display may also be directed toward zookeepers or visitors who get too close.) If the female approaches or attacks him, he will run away, then circle right back to her with the same light, prancing step. The pair may repeat these testing displays for hours until one of the animals finally leaves.

Fighting

Occasionally in the wild, a rhino meets a male rival that ignores his warnings completely. The resident may charge, and the challenger may call his bluff, thrusting the two into all-out warfare. After a series of **lowered-head charges**, the combatants **jab upward** or **club** each other with their horns, sometimes inflicting serious, gaping wounds. If the resident male loses and is exiled by the intruder, he will go to an adjacent range and try to overthrow that resident, who will in turn try to unseat a third. In some areas, researchers have noticed that wounded and dead males begin to show up all of a sudden once this domino effect starts. Before long, however, the displacing reaches its natural end, and things settle back to normal.

SEXUAL BEHAVIOR

To mate successfully, bulls and cows must overcome their natural animosity toward one another. This process takes time, and until they are comfortable, you are likely to

Lip curl

Horn jousting

COURTSHIP *To find out whether a female is ready to mate, the male sniffs her urine, then curls his upper lip over his nostrils to trap the odor and analyze it. Rhinos gently spar with their horns as a prelude to mating.*

see many aggressive elements popping up in their courtship displays. Mating may take place during any month of the year, but field studies in Africa have pointed to possible rutting periods—one in March and April and another in July—when sexual activity seems more commonplace.

Courtship

An estrous female signals her availability by **spraying** short bursts of hormone-laden urine. The male is magnetically drawn to the scent, and to get a good read on the hormones, he performs a **lip curl** (or flehmen) to trap the scent in his nostrils for thorough investigation. Once he ascertains that the female is in heat, he will follow her and solicit her attention until she agrees to stand still for mating.

He moves toward her with a short, **stiff-legged gait**, dragging his hind limbs behind him on the ground as if scraping through dung. At the same time, he may perform symbolic aggressive displays, such as **wiping his lowered horns** on the ground or against a bush or **darting** forward and backward, spraying urine. This war dance seems to help the rhino raise his courage to attempt a mounting, which, given the power and attitude of the female, can be a dangerous undertaking.

In response to his early solicitations, the female is likely to attack, forcing him to scurry off. Ever persistent, he **circles** back repeatedly until she relaxes somewhat. They may then stand face-to-face and **joust** gently with their horns, or he may **prod** her gently between the legs or on the belly. These exchanges may go on for several hours before the male gets close enough to **lay his head on her back**. When she lets him do this without attacking, you'll know they are about to mate.

Copulation

Bracing with his head, the male lifts himself onto the female's back and gets in position for copulating. He merely mounts at first, however, holding this position for 10 minutes or so and then repeating it up to 20 times before actually copulating.

Copulation may last anywhere from a few minutes to $1^{1}/_{2}$ hours, during which time the male ejaculates every 1 to 2 minutes. This lengthy mating ritual is what first prompted people to use powdered rhino horn as an aphrodisiac. Scientists claim there's no connection, however, between what's in the rhino's horn and how long they can perform. Nor, for that matter, is there any evidence that what works for rhinos would work for humans. The myth lives on, however, and rhinos continue to lose their lives because of it.

PARENTING BEHAVIOR

Birth

Like their wild counterparts, pregnant rhinos in the zoo prefer to be alone when they give birth. The mother delivers a single, tiny calf, weighing a mere 4% of her own weight. Within an hour after the birth, the little armored tank is standing on

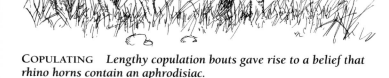

COPULATING   *Lengthy copulation bouts gave rise to a belief that rhino horns contain an aphrodisiac.*

wobbly legs and searching for its mother's teats. The mother and calf remain in protective seclusion for the first couple of weeks, getting to know each other's scents and voices and forging a long-lasting bond.

Caring for the Calf

The mother and calf remain remarkably close for 3 to 5 years until the female has her next calf. The calf follows behind the mother most of the time, traveling with ease through the vegetation she tramps down. When danger threatens, the mother **swings broadside to shield** her calf. On the rare occasions when a cow loses sight of her calf, she **mews** in a high-pitched voice that brings the calf running. In the same way, when the calf is in distress, it lets out a bellowing **squeal** that calls the cow to the rescue.

Weaning

Even though the calf is weaned at 1 to 2 years, it travels with the mother during the 15 months of her next pregnancy and right up until the birth of the new calf. At this point, the mother begins to brush off her first calf with threats until it finally wanders off in search of another connection. It may join forces with a fellow calf or with another adult female who is with or without young. Eventually, the young rhino leaves this company as well, beginning its life as a solitary adult.

BLACK RHINO BEHAVIORS TO LOOK FOR AT THE ZOO OR IN THE WILD

BASIC BEHAVIORS

LOCOMOTION		FEEDING	DRINKING	HORN
☐ fast walking	☐ trotting	☐ browsing	☐ MUD WALLOWING	SHARPENING
	☐ galloping	☐ dung feeding		☐ SLEEPING

SOCIAL BEHAVIORS

FRIENDLY BEHAVIOR
☐ greeting ceremony
CONFLICT BEHAVIOR
   Scent Marking
☐ communal defecating
☐ scraping
☐ rubbing against vegetation

☐ urine spraying
Threat
☐ threat charge
☐ weapon display
☐ urine spraying
☐ branch breaking
☐ horning
☐ snorting
☐ two-step charge

Fighting
☐ lowered-head charge
☐ jabbing upward
☐ clubbing sideways
SEXUAL BEHAVIOR
   Courtship
☐ urine spraying
☐ lip curl
☐ stiff-legged gait

☐ horn wiping
☐ darting back and forth
☐ circling
☐ horn jousting
☐ prodding
☐ laying head on back
☐ *Copulation*

PARENTING BEHAVIOR
☐ *Birth*
   Caring for the Calf
☐ shielding the calf
☐ mewing
☐ squealing
☐ *Weaning*

STRETCHING   *Before and after activity, giraffes stretch to get the gases in their blood flowing again.*

GIRAFFE

Humans have always loved to spin yarns about giants, both those we fear and the gentle, misunderstood giants that we love. Perhaps that's why the tallest land mammals on earth so easily capture our imagination and our heart. Giraffes are gigantic the moment they are born, measuring 6 feet tall and weighing 150 wobbly pounds. Male calves grow 3 inches taller each month until, as adults, they may reach 17 feet and weigh a ton. Their heart alone weighs 27 pounds and pumps a bathtubful of blood every 3 minutes. Their head is 2½ feet long, their neck 6 feet high, yet the combined 550-pound package is held up by only seven neck vertebrae, each at least 10 times longer than ours. Their tongue stretches 18 inches end to end, and their hoofs (which are really toenails) are the size of dinner plates. When wielded, these plates are capable of crushing a lion's skull—a capability that giraffes, by the way, rarely use.

Which brings us to another amazing thing about giraffes: as big and as powerful as they are, they are basically peaceful animals. Isak Dinesen (author of *Out of Africa*) once described a moving giraffe herd as "a family of rare, long-stemmed, speckled, gigantic flowers slowly advancing." Their mammoth height is designed not to help them overwhelm prey or predator, but rather to allow them to be connoisseurs of a certain layer of vegetation in their habitat. Giraffes are browsers, which means that they eat the twigs, buds, leaves, and branches of woody shrubs and trees. Thanks to their extension-ladder legs and neck, they are able to browse as high as 20 feet above the ground, earning them an uncontested place at the savanna table.

Luckily for the habitat, these big eaters don't take all their food from one source, but instead move from tree to tree as they feed. This movable feast approach may have arisen because of the unique reception giraffes receive at their favorite tree, the acacia (*Acacia drepanolobium*). Through the wonders of natural selection, acacias seem to have struck a deal with a certain species of stinging ants (*Crematogaster* spp) that lives in the black hollow galls on

VITAL STATS

ORDER: Artiodactyla

FAMILY: Giraffidae

SCIENTIFIC NAME: *Giraffa camelopardalis*

HABITAT: Savannas and open woodlands

SIZE: Height to horn tip: Male, 15–17 ft   Female, 13–15 ft

WEIGHT: Male, 1,765–4,255 lb   Female, 1,213–2,601 lb   Average, 1,750 lb

MAXIMUM AGE: 26 years in wild, 36 years in captivity

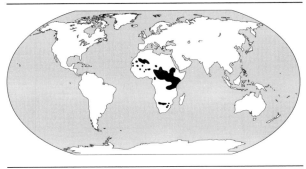

their branches. Besides providing shelter, the acacia secretes nectar from the leaf stem that nourishes the ant colony. When a giraffe feeds on an acacia, the resident ants swarm quickly onto the animal's face and neck. The giraffe will tolerate their stings for only so long before moving on, saving the tree from excessive pruning. This arrangement has led biologists to suggest that the nectar served up by the tree may be a kind of "protection money" paid to the ant for its giraffe-repelling services.

BASIC GIRAFFE BEHAVIORS TO WATCH FOR

LOCOMOTION

At first glance, the giraffe looks like a collection of mismatched animal parts. Its head looks small atop the gigantic stalk of a neck; its front legs are taller than its back legs; and its torso is not long enough to allow the front legs to move in a diagonal stride without hitting the back legs. Despite their ungainliness, all the parts work beautifully together. The giraffe **paces** when it walks by swinging both right legs forward, then both left legs. When **galloping**, it brings both hind legs forward and places them on the outside of the front legs. Pitching its neck backward and forward for momentum, the galloping giraffe moves like a giant rocking horse, swallowing 10 feet with each stride and clocking up to 36 miles per hour. By throwing its neck forward at just the right moment, the gangly giraffe can even **leap** fences as high as 6 feet. Most of the time, however, there's no need for great speed, and the proud, elegant giraffe simply strolls.

FEEDING

Giraffes are picky eaters, mainly interested in the leaves and shoots of trees and shrubs, which they supplement with vines, flowers, seedpods, and even fruits in season. They **browse** for 10 to 12 hours a day, concentrating most of their activity in the 3-hour periods after sunrise and sunset. Their long legs take them to where the highest quality vegetation grows: the savanna woodlands during the wet season and the streamside woodlands during drought. The reward for this flexibility is that they always have enough to eat and can therefore breed whenever they want.

Giraffes are the pruning shears of the savanna garden. They nip the undersides of tall tree canopies, shaping the crowns into spreading, flat-bottomed umbrellas. They prune very large bushes into hourglass shapes by eating into the midsection, leaving the branches above and below untouched. They trim smaller bushes from the top, nibbling them to a permanently dwarfed 6 feet or so.

Giraffes use their 18-inch tongue to lasso leaf bundles, and their lobed teeth to strip the branches clean. A good supply of saliva and tough grooves on the roof of the mouth allow the giraffe to swallow

BROWSING *His long legs, long neck, and 18-inch tongue allow the male to reach as high as 20 feet into the canopy. The female's reach is somewhat shorter.*

thorny branches with impunity. Since giraffes are ruminants (like cows), they can eat now and digest later. In fact, if you watch the giraffe's throat after mealtime, you'll often see a large bulge rise up; this is the undigested food coming back up for a second chewing. Also, listen for belching, which is how ruminants release the byproducts of digestion: methane gas and carbon dioxide.

DRINKING

Here's a problem for the giraffe. Its front legs are so long that its neck, phenomenally lengthy by most mammals' standards, is too short to reach the ground. To get a drink, the normally dignified giraffe has to **splay** its legs, dip its head down, and then hitch back up with a lurching motion. Or if it prefers, it can **kneel** down to get water. Either way, a drinking giraffe is vulnerable, especially around a water hole thick with salivating predators. To keep its unguarded moments to a minimum, the giraffe drinks quickly, furtively looking around whenever it can.

One bright spot in this drinking dilemma is the giraffe's gravity-defying circulation system. Think about what happens to your blood pressure when you bend down and right yourself suddenly. For the giraffe, which has to lower its head 17 feet, "a little dizzy" would not even begin to describe the effect. With its head dipping this far below its heart, the sudden increase in blood pressure could send the blood slamming into the brain were it not for the giraffe's intelligent vascular gatekeeping system. As the giraffe lifts and lowers its head, highly elastic valved arteries leading into and out of the brain contract or expand as needed to control blood flow.

BODY CARE

One advantage to having a 6-foot neck is that you can put your teeth just about anywhere you need to scratch. Giraffes **groom** themselves from tip to tail by biting and licking. To relieve a stubborn itch, they **rub** their neck on a tree trunk, their head on a branch, or their belly over the top of a thornbush. When flying insects swarm, they clear the air for yards around with a **switch** of their 6-foot, hair-tipped tail.

Certain grooming maneuvers take a team effort. The tickbirds that live with the giraffe—the buffalo weaver and the red- and yellow-billed oxpeckers—make their living by tugging out ticks and blood-sucking flies from the skin of their host. Even though this pecking sometimes keeps wounds from healing properly, the giraffes don't shoo the birds away. In addition to their grooming service, tickbirds also warn giraffes by shrilly calling at the first sign of predators or humans.

SLEEPING

After feeding in the late afternoon, giraffes retire to an open area with good visibility for predator patrol. They feed here until midnight, then warily lie down

DRINKING *The giraffe must splay its long front legs in order to reach the water. Because this position is difficult to recoup from, an imbibing giraffe is a vulnerable one.*

and ruminate for 2 or 3 hours, catching sleep in bits and pieces. You can tell when a giraffe is sleeping deeply because it lies down with its forelegs tucked beneath it, its neck bent way back, and its head resting on an outstretched hind leg or on the ground. It sleeps like this for only 3 to 4 minutes at a time in the wild. All together, adult giraffes catch only 20 minutes of deep sleep a night, calves about an hour. In the predator-free zoo environment, giraffes may luxuriate in deep sleep a little longer.

You can also catch giraffes taking a siesta during the hottest mid-portion of the day. For this lighter sleep, giraffes either stand or lie down with their neck erect, ears twitching, and eyes alternately open and closed.

SLEEPING   *Deep sleep is short-lived, totaling only about 20 minutes a night.*

How Giraffes Interact

FRIENDLY BEHAVIOR

Giraffes hang out together in loose, 2- to 10-member groups that have no obvious leader and may consist of all males, all females, or any combination of ages and sexes. Each group's roster is constantly changing as giraffes cross paths within their overlapping, 40- to 50-square-mile home range. Even groups that are miles apart are together in a sense, since their up-periscope eyes allow them to see one another above the trees and all the way to the horizon. These visual connections allow giraffes to keep an "all points bulletin" out for lions and other predators. If one group panics and flees, other groups within sight will also heed the warning.

Touching

To reinforce the bonds that allow them to coexist peacefully, giraffes frequently use behaviors that you can easily see at the zoo: nosing, licking, and rubbing. **Nosing** is when one giraffe touches the other lightly and briefly with its nose, presumably smelling it. Newborns perform special nosing ceremonies when they meet: they join noses and sometimes tongues, put their heads down, and then jump apart. These ceremonies create and cement bonds between them. Though **licking** lasts a little longer than nosing, it's still a brief affair. Favorite licking locations are the trunk, neck, mane, or horns, and sometimes even the eyelids. Also watch for giraffes **rubbing** their heads, legs, or entire bodies against one another. Researchers find that giraffes exchange these niceties more often with certain individuals, which may mean that giraffes have friends.

Communal Defense

The giraffe has keen eyes perched atop its watchtower neck. When a giraffe spots a human or a predator, it stares at it fixedly with nostrils flaring and ears spread stiffly. This **alarm posture** prompts all the other giraffes in the group to stare the same way, usually turning their bodies at an angle to the disturbance and switching their tails from side to side. The giraffe closest to the intruder allows it to come within 20 to 30 yards before giving a sharp **snort** and a lurch that sends the whole herd galloping away. Giraffes are not the only ones watching one another for signs of alarm; if giraffes vacate the premises, elands, hartebeests, gazelles, and other savanna animals will often follow suit.

CONFLICT BEHAVIOR

For the most part, giraffe society is peaceful. Male giraffes do their serious jockeying for position when they are young, and from then on, an occasional threat is all that's needed to reaffirm the hierarchy. If an unknown giraffe wanders into a group, however, the sparks immediately begin to fly. These encounters, called necking matches, have various levels of intensity, from the gentle and almost affectionate to the extremely aggressive. Before getting to this stage, however, giraffes perform simple threat gestures, usually just by repositioning the neck.

Threat

You can easily tell the dominant bull by his proud bearing. He walks with his **neck erect** and his head and chin held horizontally and pointed forward. (For comparison, it helps to notice that a standing, calm giraffe holds its neck at an angle of about 55 degrees from horizontal. When the giraffe is resting or chewing its cud, its head may sink a little lower.) When the dominant bull wants to be where a subordinate is standing, he simply walks toward it, prompting the lesser giraffe to bow out gracefully. For a sterner warning, the dominant giraffe will lower his neck to a flat, **horizontal** position and extend his head all the way forward. For a final touch, he may erect his neck again, tilt his chin down, and **point his horns** at his opponent.

Submission

Rather than fight, a submissive giraffe can either **retreat** or erect its neck and point its **nose upward,** a move that removes its horns (weapons) from sight. If the intruding giraffe doesn't flash this "I give up" symbol, a necking match may follow.

Low-Intensity Necking

To prepare for necking, two male giraffes stand side by side, shoulder to shoulder, facing in the same direction. They erect themselves proudly,

Threat

Submission

CONFLICT BEHAVIOR *The bull on the left exerts his dominance by flattening his neck in a horizontal position, while the lesser bull, wanting to avoid a fight, lifts his horns up and out of commission.*

Low-intensity necking

High-intensity sparring

NECKING CONTESTS *Standing shoulder to shoulder, the giraffes lean into each other, rubbing and entwining their necks together. As the action heats up, they wind their necks out to the side, slam them together, then jab at each other with bony horns. For more serious battles, opponents stand head to tail and direct their sledgehammer blows to the flanks and loins.*

chests inflated and legs spread slightly to brace themselves. They then test each other's strength by **leaning** flank against flank so hard that if one giraffe suddenly stepped away, the other would probably crash to the ground. Finally, one giraffe, usually the larger, lowers his head, curves it out a short distance to the side, and then swings it back in, gently sliding it against his opponent's neck. The giraffes **rub their necks** over and under for a while, like snakes wrapping and unwrapping around each other. This leisurely ritual may continue for up to 20 minutes, interrupted by grazing breaks, mountings, or by one of the giraffes simply wandering away. Occasionally, if the mood turns aggressive, neck rubbing may escalate to actual sparring.

High-Intensity Sparring

The less violent of the two sparring positions is the **head-to-head** stance, in which the giraffes stand shoulder to shoulder, facing in the same direction. They brace their legs in a firm straddle and curve their necks out to the side and slightly back as if winding up for the blow. On the return stroke, the two **necks slap** together with a sound that you can hear up to 50 yards away. The contest has a rhythmic quality; as one giraffe leans in to strike, the other evades the blow by withdrawing his neck, then leans in himself. The neck slaps you'll hear, although loud, are not especially damaging. The real kicker comes at the end of the blow when one giraffe **jabs its horns** on the underside of his opponent's neck or shoulders. After a few years, males wear all the hair off their horns this way. They also develop extra knobs of bony material on their skull that both protect and add weight, making their head an even more formidable weapon.

Giraffes really put their horns to the test in the more energetic **head-to-tail** sparring, in which the giraffes stand side by side but face in opposite directions. Bending their necks back as far as they can reach, the giraffes strike at each other's flanks, loins, and thighs with unrestrained blows. These battles seem to provide genital stimulation, since one or both opponents commonly display erections during the contest. The scuffle continues until one of the giraffes finally feels inferior and retreats a couple of steps. The victor may then **mount** him, and sometimes a third male will come and mount both the contestants. Once scores are settled, giraffes don't seem to hold grudges; moments after a sledgehammer fight, you'll see the two warriors grazing side by side or gently rubbing necks together.

Older bulls generally live peacefully within the group until their dominance hi-

HUMANS AND GIRAFFES

A giraffe was once floated 2,000 miles down the Nile River to become the star inhabitant of one of the world's first zoos, established by Queen Hatshepsut of Egypt some 3,500 years ago. Since that time, people's fascination with giraffes has never waned. In 1827, the people of Paris saw their first live giraffe paraded down the street as part of a city celebration. The crowds were so enamored of the animal that they nearly rioted to get a good look. The visit set off a giraffe mania in Paris, prompting women to wear their hair combed straight up in a "giraffe coiffure," a trend that lasted for almost a year.

Cave paintings of giraffe hunts in South Africa prove that Parisians weren't the first people to appreciate the grace and beauty of the giraffe in motion. In addition to meat, native Africans used the giraffe's skin for shields, sandals, bullwhips, and drums, and its tendons for stringed musical instruments, thread, and bowstrings. The Europeans' arrival led to the killing of many more giraffes. Today, although laws protect giraffes in national parks, poaching still continues, presumably for meat as well as for the trinket market. This magnificent animal is often slaughtered just for its tail; the long black hairs are used to make bracelets, fly swatters, and good-luck ornaments.

Even more threatening to giraffe populations is the perception that giraffes must move aside to make room for domestic livestock. In reality, there need not be a contest. The browsing giraffe currently poses little threat to herders because it uses vegetation that domestic species neglect. In fact, plans have been drawn up that would allow giraffes to become an integral part of the pastoral scene in Africa. If part of the land were maintained for giraffe habitat, and hunts were conducted in a sustained-yield fashion, experts predict that free-roaming giraffes could provide one-third of the meat requirements of a pastoral family. This would reduce people's dependence on sheep and goats, the animals responsible for much of the overgrazing and erosion problems in Africa.

Today, the amazement that swept the crowds of Paris a century and a half ago is rekindled each time a person sees their first real giraffe at the zoo. They are so large, so elegant, and so peaceful! People's affection quickly turns to outrage when they learn that giraffes are still being mistreated in some parts of their wild habitat. Giraffes may be bigger than we are, but we're somehow able to comprehend their vulnerability and the need to keep the world safe for giants like these.

erarchy breaks down or a new male enters the area. Most newcomers wander by during breeding season, when they are cruising for estrous females. As soon as a nomad arrives, the dominant male will ask him to "step outside," and the winner will gain the right to mate with the estrous female.

SEXUAL BEHAVIOR

Male-Male Pairing

The necking contests in all-male groups have a decidedly sexual flavor, as evidenced by the fact that males typically get erections during the contests and one often mounts the other as a finale. Homosexual behavior is common in many species, but it is poorly understood. Behaviorists have traditionally attributed male-male mountings to shows of dominance or to a lack of access to females. Giraffe researchers,

Testing

Lip curl

COURTSHIP *The male nudges the female until she releases a stream of scent-laden urine. He samples the stream, then curls his upper lip over his nostrils to closely analyze the scent for signs of estrus.*

however, are quick to point out that giraffe males will mount each other even when surrounded by females. Perhaps, say some, the mountings introduce males to sexual behavior and whet their appetites, a practice that may ultimately facilitate mating between the sexes.

No matter how sharp their appetite, however, males must work their way up to the top of the ranking order before they can mate with estrous females. Subordinate bulls may consort with a female early in her receptive period, but by the time she is ready to mate, only the dominant bull in the area courts her.

Courtship

Bulls wander from group to group **testing** females to see if they are in estrus. Since the proof is in the urine, the bull has to nudge the female's underside or lick her tail until she releases a sample. He catches the urine in his mouth and then lifts his lip, exposing his teeth in a **lip curl**. This gesture seals his nostrils, trapping odors in a cavity where smell-sensitive Jacobson's organs can thoroughly analyze the fluid for telltale hormones.

Copulation

Early in her estrus, the female may ignore the male's advances, walking away as he trails closely behind. When fully ready, she stands still and the male mounts, resting his chest on her rump and putting his forelegs on either side of her back.

PARENTING BEHAVIOR

Birth

You can tell when a female giraffe is about to give birth by her rising anxiety. She paces nervously and barely eats for hours. Finally, you can see the calf's forefeet protruding from her birth canal, covered by jellylike pads that protect both mother and newborn. With each contraction, the mother spreads her hind legs and extends her head and neck, exposing more and more of the calf. Between contractions, she repeatedly gets up and lies down as if she can't make up her mind. Anywhere from 20 minutes to several agonizing hours later, her final push sends the calf falling as much as 6 feet to the ground. Births may take longer in zoos, since mothers need not hurry to get their calves out before predators interfere.

Caring for the Calf

In the first few days after the birth, the cow spends time **licking** and **nosing** her calf, both to clean it and to memorize its unique smell. This odor signature will come in handy in the following months, when mother and offspring must locate each other in the midst of a herd. Sound also helps, so mothers and calves keep in touch with **mooing** calls. A calf in distress may also **bleat** to get its mother's attention.

Communal Care

Newborns join calf groups, called kindergartens, when they are 1 to 2 weeks old. The calves are already neighbors because most giraffe mothers go to the same general area for birthing. These traditional sites are somewhat removed from high traffic areas of the savanna.

The calves stay together under the supervision of perhaps one adult, allowing the other mothers to range far afield to find the high-quality foods they need to produce rich milk. In the meantime, the calves spend their first month playing, stopping only twice a day when their mother returns to suckle them. Being in a group helps them stay alert to predators such as hyenas, lions, leopards, and African wild dogs. After the first month, they don't play as much, acquiring the calm, aloof look that giraffes are known for. After 5 or 6 months, the calves begin to follow their mother (or another female) on feeding expeditions. After 6 to 18 months, they become independent and begin to travel in groups, usually segregated by sex.

CARING FOR THE CALF   *While licking and nosing her calf, the mother becomes intimately familiar with its unique smell and skin pattern, enabling her to keep track of her calf in a herd.*

NECK RUBBING *In the mellowest form of conflict behavior, giraffes slowly rub and wrap their necks together.*

GIRAFFE BEHAVIORS TO LOOK FOR AT THE ZOO OR IN THE WILD

BASIC BEHAVIORS

LOCOMOTION
- ☐ pacing
- ☐ galloping
- ☐ leaping

FEEDING
- ☐ browsing

DRINKING
- ☐ splayed legs
- ☐ kneeling

BODY CARE
- ☐ grooming
- ☐ rubbing

- ☐ tail switching
- ☐ SLEEPING

SOCIAL BEHAVIORS

FRIENDLY BEHAVIOR
Tactile Contact
- ☐ nosing
- ☐ licking
- ☐ rubbing

Communal Defense
- ☐ alarm posture
- ☐ snorting

CONFLICT BEHAVIOR
Threat
- ☐ erect neck
- ☐ horizontal neck
- ☐ horn threat

Submission
- ☐ retreat
- ☐ nose-up posture

Low-Intensity Necking
- ☐ leaning
- ☐ neck rubbing

High-Intensity Sparring
- ☐ head-to-head
- ☐ neck slapping
- ☐ horn jabbing

- ☐ head-to-tail
- ☐ mounting

SEXUAL BEHAVIOR
- ☐ *Male-Male Pairing*
Courtship
- ☐ testing
- ☐ lip curl
- ☐ *Copulation*

PARENTING BEHAVIOR
- ☐ *Birth*
Caring for the Calf
- ☐ licking
- ☐ nosing
- ☐ mooing
- ☐ bleating
- ☐ *Communal Care*

RUNNING *The male (left) and female ostrich use blistering speed to avoid predators.*

OSTRICH

Somewhere along the line, ostriches became cartoon characters in the popular imagination—giant, cowardly birds with a tutu of feathers around their midriff and their head stuck in the sand. In reality, ostriches are fierce and prudent birds, well adapted to the vagaries of the African savanna. With lions, leopards, and cheetahs lurking behind each saltbush, they have learned to keep an eye peeled and an ear cocked at all times.

The head-in-the-sand myth was probably started by some early ostrich observer who saw the bird stretched out flat for its midday snooze. To keep a low profile when sleeping or incubating its eggs, the ostrich stretches its neck along the ground and folds its legs beneath its body, leaving only the round hump of its torso sticking up above the plain. From a distance, this hump looks surprisingly like one of the rounded saltbushes that grow on the savanna, and the short, fluffy tail looks just like the bush's shadow. Capping the illusion is the sand-colored head, which conveniently disappears against its background.

When they stand, however, 8-foot-tall ostriches make exceptional targets for keen-sighted predators. Their best defense is a good offense, namely, the safety-in-numbers strategy that protects family groups and herds at water holes. Thanks to their panoramic eyes (the size of tennis balls) and sensitive ears, ostriches don't miss much from here to the horizon. Caution is their rule, and news of the slightest disturbance will spread like wildfire, causing the entire herd to take flight.

Not literal flight, of course. Ostriches, like emus, kiwis, cassowaries, and penguins, are flightless birds, living examples of a unique detour in bird evolution. Rather than lifting their wings for an aerial escape, ostriches depend on their long, powerful legs and tough toes to spirit them away. When running is out of the question, a 300-pound ostrich will stay and kick, packing 500 pounds of oomph per square inch. Once you've seen an angry territorial male strike out at an opponent, you'll never think of ostriches as silly again.

VITAL STATS

ORDER: Struthioniformes

FAMILY: Struthionidae

SCIENTIFIC NAME: *Struthio camelus*

HABITAT: Savannas to woodlands

SIZE: Length, 6 ft   Height, 8¼ ft

WEIGHT: Maximum 345 lb

MAXIMUM AGE: 70 years

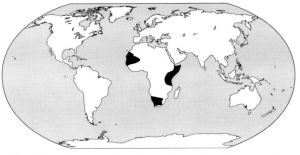

Basic Ostrich Behaviors to Watch For

LOCOMOTION

Ostriches can **run** at 50 miles per hour (faster than a horse) in brief stints, taking amazing 15-foot strides. By raising one wing and lowering another, they can tack right or left or spin completely around. Unlike most birds, ostriches don't have interlocking barbs that zip their feathers together for flight; instead, each loose plume remains open to the wind, presenting less of an obstacle as the ostrich runs.

Ostriches are also tireless walkers, a trait that comes in handy when they face long treks to water during the dry period of the year. They **walk** not on the soles of their feet, but on two toes, reinforced with a lug pad that wears well on the gritty savanna trails. An elongated foot forms the lower half of each leg, and the knob that looks to us like a knee is actually an ankle.

FEEDING

Because ostriches can cover large areas and **peck** with precision, they are able to select the highest-quality leaves, shoots, flowers, and seeds from the sparse savanna menu. After accumulating several pecks'-worth in their gullet, they gulp down a large ball of food. As you watch them feeding at the zoo, look for this bulge working its way down their long neck.

Feeding presents the ever-wary ostrich with a dilemma: every time it dips down to dine, it becomes vulnerable to stalking predators. In a compromise measure, the ostrich alternates pecking with up-periscope **perusing**, making for rather nervous-looking mealtimes.

PREENING

The ostrich uses its long neck and broad, flat bill to groom its fluffy feathers from stem to stern. Once one bird starts to preen, others catch the urge, just as they do with yawning.

OSTRICH ALERT!

When danger rears its head on the plain, ostriches are liable to disperse suddenly in all directions. The dominant male tries to keep his various mates together by zigzagging toward and away from them, throwing his wings high, and performing elaborate flash displays to attract their attention. He may try to get behind his mates and corral them in a certain direction, then suddenly speed away again, uttering notes of fear and threat.

Dominant female ostriches (major hens) will also pull rank when faced with a potentially dangerous situation. For instance, when a group of wary ostriches approaches a water hole, they naturally stop dead in their tracks, as if waiting for one brave soul to be the first to make an appearance onshore. If none of the birds volunteer, the major hen may choose a young or low-ranking ostrich as a sacrifice. She kicks and rams the bird from behind until it steps forward, springing the trap if a predator is lurking nearby.

DUST BATHING

Like many birds, ostriches enjoy a good roll in the dust, perhaps to discourage ticks and other parasites or to sop up excess oil in their feathers. Several ostriches may indulge together, waving their wings up and down to create a swirling dust storm.

PANTING

An animal as large as an ostrich can be hard pressed to find shade on the savanna. Like dogs, they pant to cool themselves, evaporating moisture from their mouth tissues and releasing heat in the process. Adult ostriches relieve their chicks by spreading their wings and inviting them to stand in the only shade in town.

SLEEPING

Ostriches sleep out in the open, keeping a herdful of ears tuned to surprise attacks. With their bodies prone, they either hold their necks up in an S shape or lay them flat on the ground. Every bird in the herd yawns, beds down, sleeps, wakes up, yawns again, and becomes active at the same time. Thanks to synchronized movements like these, ostriches can present a united front to their enemies.

YAWNING AND STRETCHING

Yawning and stretching are good ways to improve the flow of blood and oxygen through the body. This flow tends to stagnate a bit when an animal has been sleeping or standing still for a while. When the brain "realizes" it's not getting enough blood, nor the oxygen that comes with it, it calls for a yawn—the quickest way to vacuum in fresh oxygen.

Besides oxygen, the body also sometimes craves carbon dioxide, a gas that helps stimulate breathing. When the breathing center in the brain becomes sluggish, it calls for a stretch. Stretching contracts the muscles, releasing carbon dioxide to jump-start the breathing process. Both yawning and stretching usually occur just before or after sleeping, and they usually appear in tandem.

Chances are you'll see lots of yawning and stretching at the zoo, since life here is more relaxed than it is on the veldt, and ostriches can afford to be more sedentary. You're most likely to see ostriches yawning and stretching in communal rest areas or while waiting their turn at the water hole (or trough). They'll also yawn and stretch while guarding the nest, settling in for a night's sleep, or when woken suddenly by a disturbance.

Once one ostrich yawns, it won't be long until the rest of the herd (and probably the

STRETCHING AND SLEEPING   *The flat-out sleeping posture may have spawned the myth that ostriches bury their head in the sand.*

zoogoer watching them) follow suit. While this is not a deliberate attempt to communicate, it does serve a social function by helping to synchronize the group's activities. Low-ranking individuals know that if a higher-ranked ostrich is calm enough to yawn, the coast must be clear. At night, this relaxes tensions and puts the birds in the mood for sleep. Without this group "wind down," the nervous ostriches would scatter at each and every sound, losing the sleep they need to perform as a group the next day. When morning comes, more yawns will ripple through the herd, sending oxygen coursing to their brains and muscles, readying them for anything from a long trek to a hasty retreat.

HOW OSTRICHES INTERACT

FRIENDLY BEHAVIOR

Flocking

Ostriches flock together to stack the odds against predators. The more ostriches on lookout duty, the more likely it is that a lion or cheetah will trip the alarm before getting too close. Even if a sly intruder does manage to slip by the sentinels, ostriches are still better off in a herd. A herd saturates the predator with choices, so that each bird is less likely to be the one caught.

Ostriches flock in even larger groups when they need access to a precious resource, such as water in the dry season. Gatherings at water holes quench more than just thirst, however. After their exhausting journey, the birds can rest, preen, fall asleep, and even socialize within the protective sphere of the group. Strangers become

HUMANS AND OSTRICHES

The ancients coveted the loose, feathery plumes of the ostrich long before socialites in the Western world began wearing them in their hats. The symmetrical plumes were considered a symbol of justice in Egypt, and Romans used the feathers for the "Mohawk" decoration on soldiers' helmets. Tribal Africans used ostrich feathers in their ceremonial costumes, and by the 1500s European women were wearing elaborate plumed hats to their own ceremonies. When American fashion designers resurrected ostrich-plume hats in the 1800s, plumes became South Africa's fourth largest export after gold, diamonds, and wool. The market for these plumes suddenly evaporated in 1914, however, when open-topped automobiles came into vogue. It seems that the delicate hats proved impractical when traveling at the dazzling speed of 20 miles per hour, and in their stead, tight-fitting bonnets became the rage.

Nevertheless, the damage had already been done. Ostriches had been hunted to extinction in their former habitats in Asia and the Middle East. Today, only five of the nine original ostrich species still exist. The African ostrich is bred on ranches in Africa and the United States now, providing most of the ostrich feathers, as well as leather goods, meat, and eggs for market. As a result, overhunting is no longer the greatest threat to the wild ostrich; destruction of its habitat is. At the turn of the century there were 750,000 ostriches in the Cape Province alone. By 1984, there were only 150,000 in the entire world.

acquainted, families adopt stray immatures, and males and females forge bonds that lead to mating. The birds show an amazing amount of tolerance around a water hole, yet if you watch closely, you'll notice that some birds are clearly dominant over others. This ranking system lends order to flock life and keeps the birds from wasting their energy in squabbles.

CONFLICT BEHAVIOR

Threat

The golden rule of water-hole conduct is to recognize the dominant birds and treat them accordingly. Luckily, this is not hard to do, since dominant males frequently perform displays that even nonostriches (like you) will recognize. Just look for the black and white bird (females are brown and gray) that is standing very erect with his **tail pointing up**, addressing passing birds with hisses, snorts, boos, and other **threat calls**. The next-highest-ranking male will hold his tail just horizontal or slightly above horizontal, while the rank and file hold theirs in a subordinate, drooping position.

THREAT AND SUBMISSION   *The ostrich in the foreground fluffs its plumage, raises its tail, vocalizes, and gapes as if to say, "I'm ready to fight." The lesser bird drops its neck into a ∪ shape, lowers its tail, and closes its bill to signal, "I surrender!"*

Ostriches use a gaping gesture called the **open-mouth threat** when approached too closely by other ostriches, birds that fly too low, large reptiles or mammals, or even zookeepers. To distinguish a threat gape from a yawn, pay careful attention to the throat skin. In the threat mode, the skin under the beak is perfectly taut, unlike the loose, dangling skin of a yawn. Threatening ostriches will also vocalize, and as with a Geiger counter, you can tell by its calls when the ostrich is getting "hot." At the mild end of the scale, you'll hear a hissing note, escalating to a loud, harsh snorting, then finally to two-syllable calls.

Submission

Subordinate ostriches can squeak out of a sticky situation by demonstrating their submission to the high-ranking male. They hold their tail down, lower their body somewhat, and dip their neck way down into a ∪-shape.

Fighting

If the hint is not taken, two ostriches may engage in a spectacular **kicking** bout. Standing face-to-face with wings spread slightly for balance, they strike out with powerful legs, aiming for each other's tender underbelly. Though brief, these fights are all that's needed to establish a pecking order.

SEXUAL BEHAVIOR

Male ostriches are in the enviable position of being in demand when it comes time to mate. One reason is that they live more dangerously than females. Their gaudy black and white plumage makes them conspicuous at nesting time, not only to females, but also to predators. During the 6 weeks that the parents are incubating the eggs, predators kill more males than females, presumably because the female's subtle plumage blends in better with the landscape. In addition, hens mature more quickly, so there are always more females than males eligible for mating. Finally, even if a male is eligible, he may not be lucky enough to have a territory of his own. With all these strikes against them, breeding males are usually outnumbered three to one by breeding females.

Pre-Courtship

KICKING   *The muscles that power the ostrich's speedy retreats can also pack a mighty punch.*

For breeding females looking to meet one of these rare males, a water hole is a likely hunting ground. (Females will perform the same solicitations at your zoo exhibit, however.) Anxious to mate, the females boldly **solicit** males by urinating, defecating, and raising their wings to show off their newly molted plumage. Not willing to share the stage, they vehemently peck, kick, and drive away other female suitors as well as immature birds that are unlucky enough to pass by.

In the meantime, the breeding males complete their molt and start to get stirred up. They attack one another with short, mad dashes, raised-wing displays, and snorts, hisses, and boo calls. Adorned with new black and white plumage, their bare head and legs are flushed with pink, and their penis and cloacal wall are red, in vivid contrast to the pale organs of nonbreeding males. With penis erect and glowing, the males will ceremonially urinate, defecate, hold high their flashing wings and tail, and parade their goods in a **penial display** before the admiring females. A flock of 40 or more fully dressed males whirling around like this is a striking sight.

When the sexual ceremonies are over, each breeding male will be paired in a permanent way with only one hen—the major hen—with which he will share a bond for several years. He will also form looser bonds with two or more minor hens that he will also impregnate that year. Interestingly, the major hen seems to have some say in which minor hens come on board that season. Watch as she exerts her dominance, letting some hens close to her male while chasing others away.

Establishing a Territory

Before the first rains come, the male, either alone or with his new mates, will go to a nesting territory away from the water hole. It may be one that he has used before or one that he has to wrest from another cock. If a male ostrich doesn't secure a piece of real estate for the nest, a female will not even look at him much less mate with him. Young cocks usually have to fight with older, experienced cocks for many years, getting plenty of kicks for their trouble, but no claim of their own.

Once a cock does win a claim, he works hard to defend his 1-square-mile estate from other aspiring males. The **territorial display** consists of repeated wing flicking and posturing with both wings raised. He also spends much of his time **booming** out a four-syllabled *boo boo boooh'hoo* call that says, "I rule the roost here." You'll hear this call most frequently during early morning hours and from late afternoon to midnight. The intruder that advances despite these warnings can count on being **chased**.

Courtship

The male breaks from this territorial defense just long enough to isolate and mate with each of his hens. The couple wanders around together with their wings curved out and downward in an **0-posture** and their heads lowered as if feeding. Their

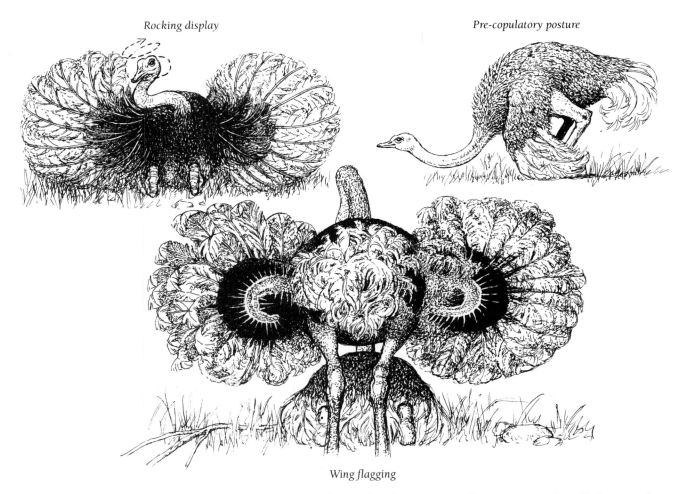

Rocking display

Pre-copulatory posture

Wing flagging

COURTSHIP   *The male stirs a duststorm with his wings while his head twists in a corkscrew pattern. Suitably impressed, the female presents herself for mating. The male runs over, spreads his wings high, and prepares to mount.*

minds are not really on food, however, as you can see by the rapid, superficial pecks at sand, pebbles, or grass. The point of this **symbolic feeding** is to synchronize their movements until they mirror each other's every action.

Once in tune, they retire to the symbolic nest site chosen by the male for a round of nest-site displays. To begin, the male raises his wings straight up (**flagging**), then beats his right and left wings alternately and rhythmically, flashing the stunning white feathers for added effect (**sweeping**). He then drops to the ground for his **rocking display**, in which he fluffs his tail and sweeps the ground with one wing after another until the dust whirls. All the while, he is singing his booming song and twisting his neck in a corkscrew shape.

The female, in the meantime, is still pretending to feed, walking in a very slow, measured gait around him. Finally, she stops and assumes her **pre-copulatory posture**: she lowers her head, flattens her neck into a shallow arc, droops her tail, and curves her wings out and downward in an 0 shape. This shows that she is ready to squat for mating.

Copulation

The male leaps from his make-believe nest and runs to her with his wings held very high and still. She drops to the ground and, as they mate, moves her head back and forth slightly, pecking and shifting grains of sand. The cock maintains his balance by beating his spread wings. After 30 to 40 seconds of flapping, he jumps off and they part.

PARENTING BEHAVIOR

Egg Laying

The male, with some help from the female, usually scrapes out a shallow, 10-foot-wide nest in a dried-up riverbed or other sandy spot. The major hen lays one egg every 2 days or so, for a total of about seven. The eggs seem huge—24 times the size of chicken eggs—but are small enough for the major hen to cover 20 of these globes with her warm body. Some will be hers, and some will belong to the 10 or more minor hens that also lay their eggs in her nest. The major hen and the cock do all the incubating for 6 long weeks, while the minor hens walk away scott-free.

Scientists were puzzled for years by this seemingly altruistic behavior. Why would the major hen invest all her energy to raise other hens' eggs? As they looked closer, they realized what a risky proposition nesting is for ostriches and how lucrative it is for predators. The wide-open nests draw aerial predators such as Egyptian vultures (which drop rocks to break the

ROLLING OUT THE EGGS *Up to 10 ostriches may lay their eggs in the same nest. Since the major hen can only incubate about 20 of the large globes, she rolls out the rest, careful to oust only eggs that are not her own. This circle of evicted eggs will satiate predators, thus protecting her eggs in the nest.*

eggs) as well as jackals, hyenas, lions, and other land predators. If all the eggs in the nest were those of the major pair, each devoured egg would be a direct loss. Since the eggs come from several different hens and the predator has a wide choice, the survival odds improve for the major hen's eggs.

To further improve her odds, the major hen reserves preferential treatment for her own eggs. Since she can sit on only 20 of the 40 or more eggs in the nest, she rolls the surplus out about 3 feet, where they will grow cold or be eaten. Amazingly enough, she somehow knows which eggs are hers. Though she's careful not to throw any of these out, she still winds up keeping and caring for some that are not her own. In this way, the minor hens get a chance to pass on their genes with a minimal energy investment. The male makes out any way you slice it. Since the major and some of the minor eggs share half his genes, he benefits from a good portion of the eggs that hatch.

Incubating the Eggs

The parents take turns sitting on the nest (the female by day and the male by night) to keep the eggs warm and safe from predators. Notoriously aggressive at this time, ostriches have been known to kill young lions with their strong kicks. Predators occasionally manage to get past their ironclad guard, however, and it's enough to keep them coming back for more.

Defending the Chicks

When the eggs finally hatch, the parents lead the chicks to secluded places where they can teach them about edible insects and plants without showing them off to predators. Should a predator appear, the ostriches launch into an elaborate **distraction display** to capture the predator's attention and lead it away. The male acts as the star decoy. Running from side to side and back and forth, he spreads and beats his wings rapidly, repeatedly calling *boo*. Every now and then, he drops dramatically to the ground, whirling his wings in the dust for a while before resuming his frenzied, zigzag run. As soon as the predator is occupied, the female calls the chicks and marshals them to safety.

Communal Care

Impeccably timed, the ostrich hatch usually coincides with the short rainy period (which varies from year to year and place to place), so a flush of tender new growth awaits the newborns. Later, when the chicks are older, they gather into **crèches** of 30 to 100 chicks, all supervised by 2 or 3 of the original parents. After communal incubation, this

DISTRACTION DISPLAY *The male makes a spectacle of himself to lure the predator away from the female and chicks.*

communal day-care seems a logical next step. For the caregivers, the surplus chicks act as a buffer; if a predator raids the crèche, it's more likely that one of the other chicks, not their offspring, will be the victim.

OSTRICH BEHAVIORS TO LOOK FOR AT THE ZOO OR IN THE WILD

BASIC BEHAVIORS

LOCOMOTION	FEEDING	☐ perusing for predators	☐ DUST BATHING	☐ SLEEPING
☐ running	☐ pecking	☐ PREENING	☐ PANTING	☐ YAWNING AND STRETCHING
☐ walking				

SOCIAL BEHAVIORS

FRIENDLY BEHAVIOR	☐ *Submission*	*Establishing a Territory*	☐ sweeping	☐ *Incubating the Eggs*
☐ *Flocking*	*Fighting*	☐ territorial displays	☐ rocking display	*Defending the Young*
CONFLICT BEHAVIOR	☐ kicking	☐ chasing	☐ pre-copulatory posture	☐ distraction display
Threat	**SEXUAL BEHAVIOR**	☐ booming calls	☐ *Copulation*	*Communal Care*
☐ tail pointing up	*Pre-Courtship*	*Courtship*	**PARENTING BEHAVIOR**	☐ crèching
☐ threat calls	☐ female solicitation	☐ symbolic feeding	☐ *Egg Laying*	
☐ open-mouth threat	☐ penial displays	☐ flagging		

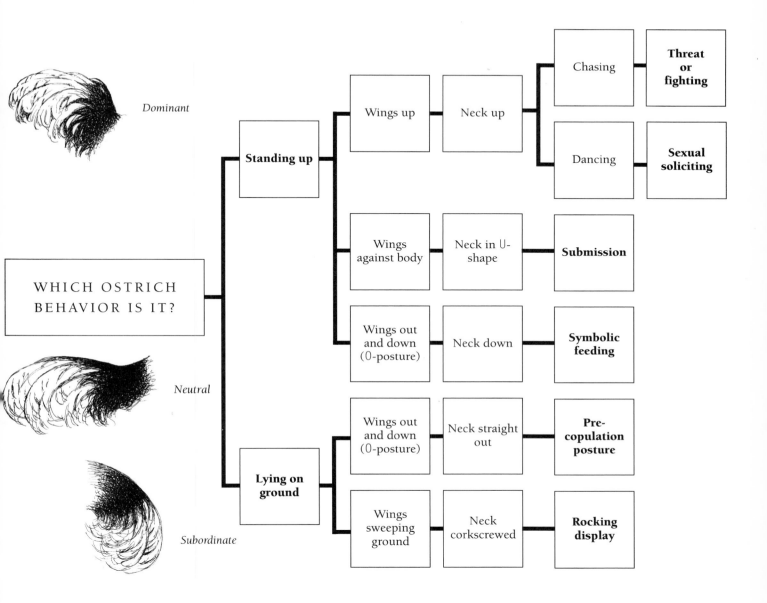

WHICH OSTRICH BEHAVIOR IS IT?

Dominant

Neutral

Subordinate

Standing up
- Wings up → Neck up
  - Chasing → **Threat or fighting**
  - Dancing → **Sexual soliciting**
- Wings against body → Neck in U-shape → **Submission**
- Wings out and down (0-posture) → Neck down → **Symbolic feeding**

Lying on ground
- Wings out and down (0-posture) → Neck straight out → **Pre-copulation posture**
- Wings sweeping ground → Neck corkscrewed → **Rocking display**

STRETCHING   *Easing one side of the body at a time, the flamingo recirculates gases in its blood.*

GREATER FLAMINGO

I f flamingos make you think of piña coladas and palm trees, you're not alone. The leggy pink bird with the upside-down smile appears in a thousand vacation brochures, usually pictured on a sky-blue pond, framed by blossoms and exotic greenery. Biologically speaking, something is definitely wrong with this picture.

In real life, flamingos inhabit some of the most *un*inhabitable wetlands on earth. They head for evaporating alkali or soda lakes and salty lagoons—remote, untracked wetlands surrounded by featureless flats that are too harsh to support grasses, much less blossoms. The waters are thick with blue-green algae and teeming hordes of crustaceans and other edibles. Underneath, the muddy floor is oozing with billions of protozoans, annelid worms, and insect larvae that feed on a nutritious mat of algae, diatoms, and bacteria. How does paradise sound so far?

Think of it from the flamingo's perspective. Algae-coated mud flats are a virtual smorgasbord to a bird that dredges and filters food through its bill. The lack of trees and shrubs means there's plenty of room to breed the way they like—in colonies of hundreds or thousands of birds. Finally, since most soda-encrusted lagoons are far from land predators, sunbathers, and water skiers, flamingos can raise their families in relative peace.

So where did the rumor about flamingos in paradise get started? It may have begun in 1931, when a flock of flamingos was imported from Cuba to Miami to grace the grounds of the Hialeah Race Course. The birds stayed for a day and

VITAL STATS

ORDER: Ciconiiformes

FAMILY: Phoenicopteridae

SCIENTIFIC NAME: *Phoenicopterus ruber ruber*

HABITAT: Shallow salt or soda lagoons and lakes

SIZE: Height, 41–61 in   Length, 49–57 in
Wingspan, 55–65 in

WEIGHT: Male, 8 lb   Female, 6.5 lb

MAXIMUM AGE: 44 years in captivity

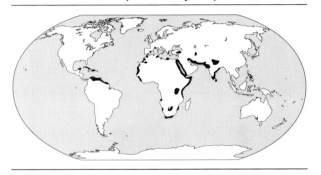

then promptly flew off, giving rise to repeated reports of wild flamingos in the resort areas of Florida's eastern coast. In reality, those birds were no more native than the snowbirds who catch the red-eye express from Yonkers when the mercury falls. Now you know the truth.

BASIC GREATER FLAMINGO BEHAVIORS TO WATCH FOR

LOCOMOTION

It's rare that you'll see a flamingo go it alone; they're usually amassed nearly shoulder to shoulder in highly gregarious groups. On the ground they **walk** sedately, moving in unison away from any perceived danger. Their stiffer, more choreographed walk is called **marching**, a goose-step affair that is actually a courtship display. Flamingos lose their normal decorum and **run** when threatened, especially when they are without their flight feathers during the molt. In the wild, flamingos also run before taking off, flapping their wings to build up speed so they can hoist themselves aloft.

Flying against a bright blue sky, these huge pink, crimson, and black birds are something of a spectacle. Each bird stretches out in the shape of a cross, with its long neck leading the way, legs trailing behind, and 5 feet of wings forming the crossbar. Flocks of these pink crosses fly in formation, creating a V, a diagonal streak, or a loose line undulating like a child's streamer across the sky. By the time the chorus of harsh honks filters down to earth, it sounds like the croaking of a thousand frogs. **Landing** is an ordeal because of the tremendous momentum flamingos build up in flight; like DC-10s, they circle until they spot a landing strip that will allow them to run several yards before coming to a stop.

At the zoo, where you won't see flamingos flying or landing, you may catch them **swimming**, something they are surprisingly good at, thanks to their webbed feet.

FEEDING

Flamingos are the baleen whales of the bird world. This refers not only to their size (which *is* pretty spectacular), but also to the way they pump water through their bills, straining out tiny food particles, in much the same way that gray and blue baleen whales do. This bill design really sets flamingos apart from other birds.

Most birds have sturdy upper bills that work against smaller, flexible lower bills. In flamingos, the design is reversed; the lower bill is deep and troughlike, and the thin upper bill fits like a lid on the top of a box. But here's the real switch: when flamingos feed, they turn their bill upside down, transforming the frown into a smile. This places the thin upper jaw (the box lid) on the bottom, where it can dig through the mud and work against the larger beak, *almost* like other birds.

A thick, spiny tongue fits in the large part of the bill, and the flamingo works this like a piston, sucking water in and pushing it out three or four times a sec-

FEEDING *The flamingo slurps a slurry of water and mud, then pumps it out again, straining out edible particles with its mouth filter.*

ond. Before the water exits, it squeezes past a filtering network of horny plates that line the edges of both the upper and lower bill. Any food suspended in the water gets caught in the screen and swallowed. Since greater flamingos have a larger mesh than other species of flamingos, they specialize in larger catches such as invertebrates, insects, crustaceans, mollusks, and annelids.

Watch for the various techniques that flamingos use when feeding:

- **Treading**: The flamingo places its bill in the mud and then slowly walks in a circle around it, stirring the mud as it goes. It slurps at the rising cloud of silt, catching small crustaceans before they settle back down to their mud homes. Look in the shallow end of your flamingo exhibit for signs of this kind of feeding: a small mound of mud with a circular runway around it.
- **Mud Dredging**: The flamingo merely walks forward, moving its filtering bill back and forth through the algae-coated mud, leaving a thin, meandering track.
- **Tipping Up**: In deeper water, where flamingo feet can't reach bottom, a flock of birds may float on the water's surface like giant pink swans with their heads submerged underwater. To find tipping-up flamingos, watch for pink and white feather dusters (their rumps) sticking out of the water.
- **Ground Feeding**: In zoos, flamingos will pick solid pieces of food such as grain off the ground, tossing them to the back of their throat to swallow them.

Flock members gabble to one another like geese as they feed, using the characteristic three-note nasal sound that inspired their South American name: *cho GO go.* Other listeners claim that the constant gabble sounds more like *eep-eep, cak-cak, eep-eep, cak-cak,* with females giving the higher pitched *eep-eep* and males the lower-pitched *cak-cak.* You can hear the gossiping of a large flock up to a mile or more away, so you may want to listen for them even before you get inside the zoo.

COMFORT BEHAVIOR

Some examples of flamingo comfort behaviors that you can see at the zoo are **preening**, **stretching**, and frequent **bathing**. You'll also see them **foot shaking** to remove excess water before entering the nest.

SLEEPING

Flamingos commonly rest in a one-legged pose, with their neck curved or folded along their back and their bill tucked into their feathers. In breeding colonies, sleeping birds pack together so tightly that one bird's loss of balance could set off a domino reaction. Since they have no formal bedtime, you may spot flamingos taking siestas throughout the day. If you see one sleeping, chances are the whole flock will be at it soon, since flamingos in a group normally synchronize their sleep periods. They are

SCRATCHING *A one-legged scratch takes a bit of balance.*

also notorious creatures of habit, so you're likely to see them sleeping just about the same time every day.

HOW GREATER FLAMINGOS INTERACT

FRIENDLY BEHAVIOR

Aerial photos of wild flamingo congregations are enough to boggle the mind. The tens of thousands of birds look like a slick of pink paint on a shimmering blue canvas, a slick that spreads and splits as the birds move. Although these jamborees are primarily associated with breeding, flamingos will also group together outside of breeding season to exploit abundant food or colonize new areas during a drought. Even though sex is not a part of the picture at these gatherings, the flocks perform many of the courtship displays described in the Sexual Behavior section. The flocks at your zoo are liable to engage in these off-season bonding ceremonies as well.

CONFLICT BEHAVIOR

Alert Behavior

You can tell from a distance when flamingos are disturbed, and so can other flamingos. In the **alarm posture**, they snap to attention, raise their necks straight up, and reach as if on tiptoe to peer over the heads of the other flock members. They look this way and that, moving their heads from side to side and proclaiming with a deep-chested **alarm honking**.

Threat

Aggressions can flare at any time, but they're especially common during breeding season, when tempers are hot and fuses are short. When a strange flamingo threatens a courting or incubating pair, the dominant bird (usually a male) will perform the intimidating **hooking threat**. This threat may precede or follow a fight, or, in a sexual context, you may see it before and after copulation. The riled bird stalks purposefully toward the opponent, its neck extended and inclined forward slightly, its shoulder and back feathers fluffed to look all the more imposing. It holds its bill sharply downward in a hook, aiming the point toward its own chest. If the intruder doesn't get the hint, the annoyed bird will usually launch into the neck-swaying threat.

In the **neck-swaying threat**, the flamingo lowers its head even further, stretches its neck almost straight out and moves it from side to side in sinuous, horizontal arcs. At the peak of its performance, it may rotate its bill, aiming the pointed end of the frown straight at the offending party. All the while it keeps its shoulder and back feathers strongly

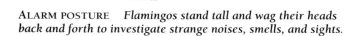

ALARM POSTURE *Flamingos stand tall and wag their heads back and forth to investigate strange noises, smells, and sights.*

erected, "like a giant chrysanthemum," said one researcher. The deep, grunting honk of alarm becomes louder, harsher, and sometimes higher pitched as the bird gets more aggressive. Just before a fight breaks out, the flamingo gapes open its beak slightly.

Fighting

If the intruder is still unconvinced, the two birds may begin to **spar**, exchanging neck-swaying threats, but just missing each other's bills. In a full-blown fight, the birds stand breast to breast and **peck** at each other. Before the injuries can get serious, however, the rival that's losing usually retreats.

Submission

A flamingo that wishes to walk by another without posing a threat will lower its head and sleek its feathers. In contrast to a puffed up, erect dominant, its plea for peace is loud and clear.

THREAT   *When a courting pair is disturbed, the dominant flamingo hooks to warn the intruder. If the intruder doesn't take the hint, the flamingo begins the more intense neck-swaying display.*

BITE THREAT AND SUBMISSION   *A nip from a stout flamingo bill can be painful, so warnings are usually heeded.*

SEXUAL BEHAVIOR

Much of the group behavior you'll see at flamingo gatherings has a sexual element, which helps the birds synchronize their libidos during breeding season (March through June). Synchronism is important because the flamingo's habitat is so changeable. If flamingos are not ready to mate when productivity and water levels in the salt pans are just right, they might miss a window of opportunity. Since zoo flocks are liable to break into these ceremonies before, during, and even after breeding season, it pays to look sharp at any time of year.

Courtship

HEAD FLAGGING

Head flagging, the most common sexual display, is similar to the alert posture but has an exaggerated quality that pegs it as a communication gesture. A few birds in the group begin by issuing loud, two-syllable honks and flagging their heads from side to side in precise, horizontal arcs. Other birds soon join in, flagging faster and faster as the excitement builds. After a minute or two of head flagging, the birds usually perform a wing salute.

WING SALUTE

Suddenly, each bird stops honking and head flagging and places its bill straight up in the air, uttering a series of short, low-pitched grunts. With a quick flourish, it cocks its tail, raises its shoulder feathers, and quickly open its wings to the sides, revealing a splash of black in a field of pink. After about 3 or 4 seconds of showing off, each bird abruptly snaps shut its wings and proceeds to a twist-preen display.

TWIST PREEN

This display is derived from grooming movements, but because it's so stiffly executed, you'll know it's a display and not just an itch being scratched. Each bird twists its neck and bill back to one side and briefly flashes its wing downward, exposing the black primary feathers. It may then go through the motions of preening for 1 to 2 seconds, although it's not really cleaning its feathers.

INVERTED WING SALUTE

You'll commonly see this maneuver in sequence with head flagging and wing salutes. With its neck extended, each bird tilts its body forward and down until its tail is cocked higher than its shoulders. It then lowers its neck, flashes its wings open, and holds them behind its back for a moment. The bend in the wing is pointed downward, and the black primaries are tipped skyward. From the front you see a quick flash of red and hear a soft, nasal sigh. In a second or two, each bird is erect again, breaking to feed or preen or perhaps continuing with more head flagging.

GROUP DISPLAYS   *Foreground, clockwise from lower left: Flamingos begin their display by wagging their heads side to side, honking harshly. Next comes a grunting wing salute, a bottoms-up inverted wing salute, or a ritualized preening display. Background: At any moment, the entourage may march forward in a group tango, then just as suddenly stop and trail their bills as if feeding.*

WING-LEG STRETCH

Like the twist preen, this display resembles comfort stretching, but is stiffer than the real thing. As they stretch the leg and wing on each side outward and to the rear, you'll hear a brief call that is best described as a satisfied sigh.

MARCHING

Count yourself lucky if you're able to see the most unusual of group encounters: the march. Look for it at the peak of courtship season, when the birds are whipped up into an almost hypnotic frenzy. The march begins mysteriously when hundreds or even thousands of birds freeze, stand fully erect, and walk off together with a quick goose-step gait. As if cued by an invisible conductor, they'll suddenly turn and walk just as quickly in another direction and then another. It looks for all the world like a group tango set to music beyond the range (and realm) of our hearing.

FALSE FEEDING

To add spice to an already strange and beautiful display, hundreds or thousands of marching flamingos will suddenly slow down, bend their necks forward in unison, and trail their bills in the water. Though they do not actually feed, they make chomping movements with their bills for several seconds. When they straighten up again, they pick up the pace of their walking to about two steps per second (fast-forward, flamingo-style).

Pair Formation

If you visit regularly, you'll notice that some birds mate with the same partner year after year, while others lead a more promiscuous lifestyle. The group ceremonies give the birds a chance to either reconnect with old flames or start new relationships. You can recognize pairs by their subtle shows of solidarity. For instance, partners will often sleep and feed together, come to each other's aid, or pretend to feed while walking quickly side by side.

Copulation

Courtship and pair formation displays go on in the group for weeks or months, but most copulating occurs during nest building and just before egg laying. The pair moves away from the group, and the male (the larger of the two birds) begins following the female with a **hooking** posture. In this instance, the hooking posture is no longer a threat, but a sign of his increasing sexual interest. Before copulating, the female bends her neck in a **false-feeding** posture. The male, in close pursuit, extends his neck diagonally, trying to touch her with his bill or part of his breast. Eventually, the female stands still and spreads her wings, signaling, "OK." The male jumps at the chance, and the two birds **copulate** by placing their cloacas together for a few seconds. Afterward, they may call together with necks stretched toward the ground in a kind of **triumph ceremony**.

PARENTING BEHAVIOR

Nest Building

Once flamingos choose a nest site, they have a hard time taking no for an answer if other birds have already occupied the site. They usually attack, vehemently and persistently, until they drive the property owners away. If no nest exists, they start shaping billfuls of mud into a squat tower that will eventually be 6 to 18 inches high, 17 to 20 inches wide at the base, with a depression in the top to cradle the single egg. The tower raises the egg so it won't be hard-boiled on the hot ground or swamped by an unexpected flood. Nest building is an ongoing project for the flamingos; repairs and embellishments continue even after the egg is laid.

Not all flamingos follow the nuclear family model. Male-male courtship is common, and occasionally you'll see a female and two males begin a family. Interestingly, the male pairs go to the trouble of building nests even though they will harbor no chicks.

Caring for the Eggs

Both sexes take turns **incubating** the egg for up to 32 days. They relax their wings as they sit to form a makeshift roof over the nest; any rain that does trickle in is quickly

HUMANS AND FLAMINGOS

Thank goodness for the sun! If it weren't for the fact that sunlight quickly fades the beautiful blush-tones of plucked flamingo feathers, the fashion trade might have hounded flamingos the same way it did ostriches. Another saving grace for these glamorous birds was that their habitat was remote, unforgiving, and at least until now, unappealing to humans. In recent years, salt and soda extraction operations have begun to crop up in traditional flamingo breeding areas, and their lagoons are being dredged to create harbors for the fishing industry. Both developments bring more people to the area, with the inevitable alterations and "improvements" that are likely to discourage flamingo breeding.

Even a slight increase in human pressure is disruptive to flamingos. A low-flying aircraft, for instance, can cause a world of damage in a colony by starting a stampede that tramples both eggs and young. Feral hogs and dogs introduced by nearby human communities can also unleash a menace that the flamingos are not adapted to deal with. These exotic birds are very picky, and if conditions are not just right for breeding, they won't build nests. After several chick-less seasons, their overall numbers will start to dwindle.

Some human activity may be compatible with flamingo breeding, however. Ironically enough, flamingos have taken to feeding in certain salt lagoons that were actually *created* by the salt and soda extraction industry. As long as villages don't build up around the lagoons, and some parts are set aside for breeding, there may be a chance that the birds and the miners can work cooperatively. Since the greatest threat to flamingos is the loss of habitat, wherever we can create more breeding ground without too much human presence, so much the better!

FEEDING THE CHICK *The parents produce a nutritious red liquid in their throat glands, then dribble it into the chick's beak.*

sucked up by their bills. Both birds are extremely protective at this time and quite willing to fight, especially when birds from other colonies dare to intrude.

Hatching time is a critical imprinting period for flamingos. You'll see the adults with their heads right against the hatching egg, broadcasting a high-pitched **contact call**, *kurruck-kurruck-kurruck,* right through the shell. This way, the chick learns its parents' calls so that once it hatches, it will be able to recognize and find them in the crush of other flamingos in the colony. Locking in identities is essential since the young flamingo is completely dependent on its parents to provide food for the first several weeks. The chick's bill won't even begin to turn down for 14 days, and even then, its specialized filter feeder is not quite ready to use.

Feeding the Chick

Instead of bringing food morsels back to the nest, both male and female flamingos drool a blood-red **secretion** from their bills into the chick's. This extremely rich fluid comes not from their stomachs, but from glands in their throats and upper digestive tracts. Each drop contains 15% fat, 1% red blood cells, and large amounts of canthaxanthin, a pigment that turns flamingo feathers pink. Because this pigment stays in the chick's liver at first, its feathers are gray, not pink. As the chick matures, each new set of feathers will contain more of the rosy pigment.

Adult flamingos acquire additional carotenoid pigments (similar to the pigments of carrots) in their wild diet of insects, crustaceans, and algae. Zoo dieticians are also careful to choose foods with plenty of carotenoids to keep flamingos "in the pink." If the birds lose their color, zookeepers have found, they also lose interest in breeding.

Communal Care

After about a week at the nest, the still flightless chicks begin to explore on their own, returning to their parents only for meals. The young band together into **crèches** of hundreds or even thousands of birds, chaperoned by only a few baby-sitting adults that may or may not have young of their own. When the parents return from food gathering, they issue the dinner call and the young come running.

In the zoo, where flocks are smaller and parents are rarely out of sight, feeding is more convenient. The gangly chicks simply **beg for food** by calling and pecking at the parents' neck feathers. Sometimes you can spot persistent older chicks still lobbying for a free meal. In a gesture that seems to glue the adult to the spot, the chick lowers its neck to the ground, shimmying as far as it can under the adult's breast. Once in place, it stretches its neck upward and calls loudly and frequently until it gets its way.

GREATER FLAMINGO BEHAVIORS TO LOOK FOR AT THE ZOO OR IN THE WILD

BASIC BEHAVIORS

LOCOMOTION
- [ ] walking
- [ ] marching
- [ ] running
- [ ] flying
- [ ] landing
- [ ] swimming

FEEDING
- [ ] treading
- [ ] mud dredging
- [ ] tipping up
- [ ] ground feeding

COMFORT BEHAVIOR
- [ ] preening
- [ ] stretching
- [ ] bathing
- [ ] foot shaking
- [ ] SLEEPING

SOCIAL BEHAVIORS

FRIENDLY BEHAVIOR
(See sexual)

CONFLICT BEHAVIOR

Alert Behavior
- [ ] alarm posture
- [ ] alarm honking

Threat
- [ ] hooking threat
- [ ] neck-swaying threat

Fighting
- [ ] sparring
- [ ] pecking
- [ ] *Submission*

SEXUAL BEHAVIOR

Courtship
- [ ] head flagging
- [ ] wing salute

- [ ] twist preen
- [ ] inverted wing salute
- [ ] wing-leg stretch
- [ ] marching
- [ ] false feeding
- [ ] *Pair Formation*

Copulation
- [ ] hooking
- [ ] false feeding
- [ ] copulating
- [ ] triumph ceremony

PARENTING BEHAVIOR
- [ ] *Nest Building*

Caring for the Eggs
- [ ] incubating
- [ ] contact call

Feeding the Chicks
- [ ] dribbling secretion

Communal Care
- [ ] crèching
- [ ] begging for food

FIGHTING   *In the final stage of battle, crocs bite the base of each other's tail.*

NILE CROCODILE

While the dinosaurs came and went in a heated rush, their crocodilian relatives kept going strong. Modern-day crocodiles, alligators, and their kin look and act much as they did 100 million years ago. That original design, as effective today as it was in the Cretaceous Period, has helped make them the longest-running reptile show on earth.

One of the cornerstones of their success is the very thing that frustrates some zoo visitors. Crocodiles seem downright lethargic; they are nearly motionless for a good portion of the day, and unless you know the inside story, you might think they're boring. Far from it. Crocodiles that are silent and supposedly inactive are actually busy keeping themselves alive. A basking crocodile is soaking up warmth from the sun to power its metabolism, and a hiding crocodile is all eyes and ears, sleuthing after prey. Besides, lying still as a log helps the croc add an element of surprise to its explosive attacks.

By laying low until they need to move, crocs are able to save energy, which means they can live on less food and put their energy into growing, interacting, and raising young instead of hunting all the time. Growing to a large size enables crocs to help themselves to anything from frogs to water buffaloes and grants them immunity from most predation as well. Likewise, their social energies are well-spent, since good communication helps crocs divide their habitat and live in close quarters without too much turmoil. Finally, the energy spent on reproduction is what sep-

VITAL STATS

ORDER: Crocodilia

FAMILY: Crocodilidae

SCIENTIFIC NAME: *Crocodylus niloticus*

HABITAT: Shorelines of rivers, lakes, and coastal areas

SIZE: Female, 8–10 ft   Male, 10–12 plus ft, have been known to reach 16 ft

WEIGHT: Maximum ever recorded, 2,200 lb

MAXIMUM AGE: More than 50 years

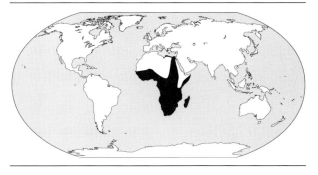

HIGH WALKING *When leaving the water and walking overland, the croc straightens its legs and lifts its belly off the surface.*

arates crocodiles from lizards, snakes, turtles, and other reptiles. While their cousins lay their eggs and run, all crocodilians spend months protecting their eggs, helping them hatch, and getting their young off to a good start. Defending their offspring against the multitude of hungry predators at the water's edge is a full-time job—a job that, like most of their activities, crocodiles do best lying down.

Although the following write-ups refer to Nile crocodiles, other members of the crocodilian clan (other species of crocodiles, alligators, gharials, and caimans) behave in much the same way.

BASIC NILE CROCODILE BEHAVIORS TO WATCH FOR

LOCOMOTION

If vigilance were a behavior category, crocodiles would be among the most active animals on your zoo tour. Keeping watch in the water means **floating** with just their nostrils and eyes above the surface or **cruising** with their thick tail sweeping from side to side. While **swimming**, they lash their tail for power and press their legs flush against their body to create a fishlike figure. One partly webbed foot extended like a rudder can help them turn on a dime.

When crocs need to really *move,* they can **lunge** forward at incredible speeds or even **leap** out of the water like big fish. These speedy assaults usually signal a territorial chase or a rendezvous with tasty prey. After a long, patient stalk, the crocodile seems to erupt into action—tail flailing, jaws snapping, and head thrashing its prey to and fro.

On land, the croc has three modes of moving—the high walk, the gallop, and the belly run. In the **high walk**, the croc straightens its legs to lift its belly off the ground. The hind limbs, being slightly longer than the forelimbs, carry most of the weight and hoist the rear of the croc higher than its shoulders. Watch for this diagonal stride (similar to that of four-footed mammals) when crocs are hauling themselves out of the water or moving overland.

When time is of the essence, the croc will **gallop**, bringing its hind limbs under its belly as it pushes back with its forelimbs. Past a certain speed, the gallop degenerates into a **belly run**. Crashing to the ground on its polished belly scales, the croc writhes from side to side, rowing as fast as it can with its limbs. Watch for the belly run whenever the croc is traveling down steep banks.

FEEDING

Food goes a long way for the slow, methodical crocodile. They survive on little more than 50 meals a year in the wild, although, admittedly, some of the meals are on the heavy side. Fully mature Nile crocodiles are quite capable of taking down an antelope, zebra, warthog, or even the gigantic Cape water buffalo! Dining on giants like these is something crocodiles literally have to grow into. Juveniles feast on small frogs, fish, and even insects, but as they grow, their capacity for large prey also grows.

A credit to their flexibility is that no matter how large they become, crocs never lose their taste for smaller prey, nor their ability to snap up a small frog or fish in their path.

Lying in the shallows with its jaws agape, the **still hunting** crocodile stays incognito except for a bump of nostrils and eyes above the water. It may even disguise these by floating next to a log or other object. From this position, the croc can smell, see, hear, and probably feel the vibrations of passing fish with its sensitive jaws. With a trigger-quick chomp, the "log" comes to life, snagging the wriggling fish, tossing it in the air, and repositioning it for easy swallowing. Crocs may also pursue their prey by **swimming** after them.

Besides playing the underwater field, crocs also stay tuned to the lineup of mammals that come to the shoreline to drink. When a croc spots a likely meal, it submerges without so much as a ripple and begins to submarine toward shore. Flexing every dense muscle in its body, it breaks from the water and **lunges** up the bank toward the terrified prey. Digging its hind legs into the mud to lever itself for-

LUNGING   *From floating log to exploding torpedo, the croc mobilizes in the wink of an eye.*

ward, the croc can advance several body lengths with astonishing speed. If the bank is steep, it can vault straight up nearly 5 feet and even hook its head over the top of the bank to make it up and over.

The crocodile's enormous jaws hold an impressive arsenal of 70 razor-edged teeth, each replaced as many as 45 times over a lifetime. When these serrated pincers bear down on the prey's leg, they not only pierce the skin, but also crush the bones. If the crocodile can reach the prey's muzzle, it snaps a vise grip on it and then holds the animal under until it drowns. When the carcass is large, the croc sinks its teeth into one spot and then whirls in the water until it twists off a bite-sized chunk. Often, it makes a series of deep bites first and then tears along the dotted line.

Before sinking its teeth into land prey, a croc will sometimes swing its bony head like a sledgehammer to **strike** and stun the animal. There are also various **tail hunting** tactics. In one, the croc uses its tail to bend reeds down, thus spilling the contents of weaverbird nests along the riverbank. In another, the croc swims parallel to the shoreline, creating a riffle with the scales at the top of its tail. The small fish in the shallows move ahead of the disturbance, and the croc neatly curves its head toward shore to trap them. Crocodiles are also quite willing to **scavenge** carrion and may even allow other crocodiles to share the feast.

BREATHING

Because crocodiles still have the lungs of a landlubber, they must surface periodically to get some oxygen. Conveniently, their nostrils and eyes are located on top of their head, allowing them to breathe and keep watch while exposing only a few bumps

GAPE *An overheated crocodile opens wide to let air circulate over its mouth tissues. As the moisture evaporates, heat is released, and the croc cools down.*

above the surface. Crocodiles can up-periscope discreetly in placid water, but when the wind churns up whitecaps, they have to hold their whole head out of the water to breathe. Rather than blow their cover, crocodiles usually shy away from rough water.

When they go under, they squeeze their nostrils shut with strong muscles to keep water out of their lungs. A watertight flap at the entrance to their throat allows them to open their mouth to snag fish without drowning.

WARMING UP AND COOLING DOWN

Temperature control is a common theme in all reptile behavior. Since reptiles can't generate enough body heat metabolically, they must soak up heat from the sun, warm water, or maybe even a heat lamp in an indoor exhibit. Their days revolve around rising and falling air temperatures. In the morning, when chilled and groggy after a long night, they emerge from the water to **bask** in direct sunlight. They also seek this free warmth after meals to raise their metabolism for digestion.

As the sun ascends and becomes more intense, their internal thermostat warns them that they are getting too warm. To cool off, they lift their top jaw way back in a **gape** that exposes the moist inner lining of their mouth to the air. As the moisture in these membranes evaporates, the body gives up heat. When it gets too hot for even this trick, crocodiles **seek shade** or **return to the water**.

At the end of the day, crocs crawl back out onto land to catch the last rays of the setting sun. As night closes in, they slide back into the water, which holds on to the heat of the day longer than the air does. Eventually, even the water cools off, and by the time the sun rises again, crocodiles are more than ready for a good long sunbath.

As you can see, temperature considerations can mean the difference between suffering and surviving for crocodiles, both in the wild and at the zoo. Crocodiles that have no way to warm up in their exhibit will refuse to eat. Or if they do eat on a cold stomach, the food could rot there instead of being digested. All in all, basking environments are something you may want to ask your zookeeper about, especially if the crocs at your zoo are showing signs of poor appetite.

HOW NILE CROCODILES INTERACT

At one time, crocodiles were dismissed as rather primitive, small-brained animals that rarely deviated from their "programmed" behaviors. That was before researchers started to explore the subtleties of their interactions. Crocs relate to one another according to their place in the dominance hierarchy, an order that is easiest to observe around breeding season or whenever crocs congregate around food sources. The unusual thing about this ranking is that it's flexible. When crocs really need one another's help, even warring factions of dominants and subordinates are sophisticated enough to "set aside their arms" and cooperate. As you watch, keep in mind that these social behaviors, so unusual among modern-day reptiles, may have been business-as-usual for the crocodile's closest relatives, the dinosaurs.

FRIENDLY BEHAVIOR

Cooperative Feeding

We humans know that there are some tasks, like folding the bedsheets, that are just easier with two people. In the same way, crocodiles find that cooperation pays off when they are eating small- or medium-size prey. By themselves, they can tear pieces from a large carcass by twirling in the water until the piece twists off. When the prey is smaller, however, it just twirls with them. When this happens, they may bring the problem over to a nearby crocodile who will oblige by steadying the carcass in return for some leftovers. Curiously, there is no squabbling at these shared feasts.

Cooperative Hunting

Cooperative hunting also calls for a cease-fire on aggression. At certain times of year, rivers flood their banks and spill into shallow pools in the floodplain. Crocs will sometimes form a semicircle at the entrance to these pools, creating a net of mouths that snag all the fish that enter. Again, there is no fighting over the prey, perhaps because leaving their stations would create a gap in the net.

CONFLICT BEHAVIOR

Threat

If you had weaponry like the crocodile's, you'd no doubt want to avoid conflict with a similarly equipped crocodile, especially if it was as large or larger than you. Dominant males are larger and more aggressive than subordinate males and twice as large as females. These top crocs control access to mates, nest sites, food, and living space. (There is a hierarchy among females as well, but ranking is obvious only in disputes over nesting sites.)

A dominant male is easy to spot. He advertises his size and status by **swimming boldly** with his head, back, and tail above the water's surface. When an intruder threatens to cross into his territory, the dominant croc may respond with a **head slap**. In this display, he rears up out of the water, gapes open his jaws, then slams them shut just as his lower jaw hits the water's surface. The impact is explosive; the loud popping and sharp smack are especially dramatic in the still of night. During the day, the sheer size of the displacement wave makes

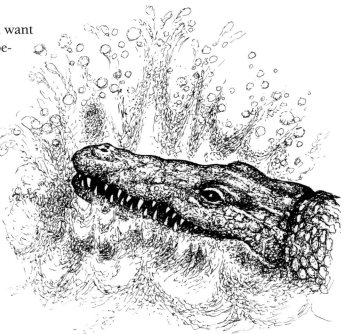

HEAD SLAPPING   *If you smacked the water with the flat of a large shovel, you'd begin to appreciate the startling sound made by an annoyed croc.*

a visual statement as well. Immediately after a head slap, watch for turbulence in the water as the male **exhales bubbles** through his mouth and nose and **thrashes his upraised tail** from side to side.

In another intimidation move, the male **chases** the intruder at high speed, jaws open, tail thrashing, and body planing like a motorboat. After the chase, the male may expose his entire body above the surface of the water in an **inflated posture** designed to exaggerate his size. If the subordinate doesn't retreat quickly enough, it may be **bitten** at the base of the tail—a taste of things to come.

Submission

Subordinate males or females that don't want to tussle with the male can wave the white flag by **lifting their snouts** out of the water. In stark contrast to the inflated dominant, they expose only their head, keeping the rest of their body **submerged**. As soon as they can, they **retreat** underwater and out of sight.

Fighting

Occasionally, two males will meet in a territorial contest or a scrimmage to decide which one is "king" of the waterway. As they lie face-to-face, they snort through their nostrils, causing small fountains of water to shoot into the air in what is called **narial geysering**. Finally, one male charges the other, and with toothy mouths gaping, they **lunge**, **growl**, and **snap** at each other for some time. The dominant finally achieves a **lock hold** on the snout or tail of his opponent. If the opponent manages to break away, the victor doggedly pursues him for another round of water-frothing fighting. The war ends when one crocodile swims away, sometimes snout lifting to indicate that he has lost the battle.

SEXUAL BEHAVIOR

In some parts of the Nile crocodile's range, the crocs stay home to mate and the males simply defend the home ranges that they maintain throughout the year. In Kenya's Lake Rudolf, however, groups of more than 200 breeding crocodiles gather away from their home range. When they arrive, the males hammer out their rank differences, and by the time it's all over, fewer than 15 dominants emerge and stake their territorial claims. These dominant males assume the right to mate with all the females in their area and will threaten and chase any subordinate males that try to intrude. The females show no particular allegiance to these top males, however, and if a subordinate can sneak in for a mating, he will. He has to be quick about it, however, since the superior will not hesitate to interrupt the tryst with head slaps and chases. Besides these outbursts of aggression, breeding season is also filled with expressions of mutual desire by both male and female. You're most likely to see this amorous behavior in the zoo during the late spring and early summer.

Pre-Courtship

To advertise himself to females, the male lifts his head and fills the air with a deep-throated **roar**. Females seem irresistibly drawn to this call and will either approach the male or allow him to approach them. As he grows near, the female **snout lifts** to show him that she means no harm.

For eons, crocodiles had been the top predators in the rivers and lakes of Africa. They lived in harmony with their environment, and their place at the top of the aquatic food chain was unchallenged by any reptile, fish, or mammal. That is, of course, until modern humans entered the picture. In less than a century, our hunger for crocodile skin and our desire to make the waters "safe" for people have come close to eliminating this reptilian success story. Some were killed for their skins, but most were killed for just being themselves.

As with most hatreds as deep as this, the root of crocodile hatred is fear. After all, crocodiles are large and powerful enough to eat the same mammals that we eat, and that makes us uncomfortable. When our livestock come lumbering down to the water to drink, crocodiles smell an easy meal. Likewise, when a woman comes to fill her jug by the water's edge, she also fits the profile of a prey species. She is a mammal who comes down to the same place at nearly the same time each day, and she is 14 times lighter than the largest crocodile and rather slow to react. As a result, back when crocodiles still lived to ripe old ages, some of the larger individuals began to exploit this prey source. For every person dragged into the river, dozens of crocodiles were killed to avenge the death.

Retribution was so mighty and so swift that large crocs were eventually eliminated from the more densely populated parts of Africa. The smaller crocodiles that were left were more likely to shy away from people than to eat them. Nevertheless, the human desire to kill crocodiles lived on, if not for revenge, then for money. In the 1950s and 1960s, crocodile hides became international objects of desire, and their populations were hunted to the bone. By the 1970s, wild skins were so hard to come by that people began raising their own crocodiles, and producer nations finally passed laws to protect their wild stocks. Today, the only hides that can be exported legally are those that bear an indelible mark from one of these ranching operations.

Despite the export ban on wild crocodile products, a well-paying market still exists for illegal or poached hides. In 1988 a crocodile handbag sold for $1,500 to $3,000, and shoes fetched anywhere from $600 to $800. As a result, poachers can make more in a night of poaching than they do in 6 months of honest labor. For today's crocodile hunters, armed with high-powered rifles, night scopes, and speedboats, it takes only a few nights to decimate entire populations. If they get caught, the fine is often less than their profit.

Some people feel that a boycott of all crocodile products may be the only way to shrivel demand. Others, acknowledging that boycotts are rarely complete, are looking to the crocodile ranching industry for help. In many parts of the world, an animal is considered worthy of protection only if it is valuable to humans. Crocodile ranchers depend on wild populations for the future genetic diversity of their own stocks. As long as ranchers are making money, they will have a vested interest in protecting wild crocs throughout the world.

Even if producer nations are able to eliminate poaching, the despairing news is that crocodiles may still have a hard time rebounding to their former abundance. Native habitats along rivers and lakes have changed dramatically. Where rain forests have been logged, for instance, silt now washes into rivers by the ton, causing shallower streams, fewer fish, and more destructive flooding of crocodile nests. The logged soil is also less able to hold nutrients, so excess nitrates and phosphates get washed in with every rain. Algal blooms thrive on these nutrients, taking precious oxygen away from fish that crocodiles eat. Besides silt, mine trailings and other pollutants are clogging and poisoning the waterways where crocodiles raise their young. Close to human settlements, boat traffic is forcing mating pairs into marginal habitats that may not be suitable for breeding.

In the end, whether it's with guns or silt or chemicals or propellers, it looks as if we're getting back at the "man-eating" crocodile after all. One can only wonder if the species that outlasted the dinosaurs can survive this onslaught as well.

COURTSHIP SPLASH DISPLAY *The courting male vibrates his trunk muscles, sending out low-frequency sound waves that cause little eruptions of water.*

Courtship

The male then responds with the courtship splash display and/or the fountain display. In the **courtship splash display**, the male arches his tail, neck, and head out of the water and begins to emit **subaudible vocalizations** that sound to us (at close range) like a roll of distant thunder. In clear water, you can actually see the croc's trunk muscles contracting as he creates these low-frequency sounds. The sound waves cause the shallow water across his back to bubble up and bounce off. He stirs the water up even more by **thrashing his tail** from side to side and opening and closing his jaws. He may also treat the female to the **fountain display**, in which he dips his snout under, swells his neck, and snorts air through his nostrils, spewing a font of water several feet into the air.

He then catches up to the female, and they **swim parallel** for a few minutes before he takes the lead and turns to face her. Next they **swim in circles**, with the female intermittently **snout lifting**, **groaning**, and **snapping her jaws**. The two may also occasionally submerge for a bout of **bubbling** or **snout rubbing**. Finally, the male may place a forelimb over the female's back and begin to **ride** her as a prelude to mounting.

Copulation

The male usually mounts the female in shallow water. Clasping her with his stout forearms, he tucks his tail and body beneath her so that their cloacas line up. You'll see their tails entwine and thrash together in the water for a minute or two, and then the pair will part.

PARENTING BEHAVIOR

Beginning life in an egg is a perilous fate. Mongooses, spotted hyenas, baboons, warthogs, bushpigs, storks, and monitor lizards consider crocodile eggs a delicacy, and even small hatchlings are appetizing to many creatures. Perhaps that's why crocodile mothers are so picky about choosing a nest site that can be carefully guarded.

Digging the Nest

When crocodiles shop for a nest site, they look for soft soil along riverbanks, in dry streambeds, or on islands. In captivity, dominant females usually snag the best sites in the exhibit. Though the nest itself should be in the open, it should have bushes and trees nearby that can provide shade for the female, who will hawkishly guard the nest for the

COURTSHIP *The female snout lifts in response to the male's alluring fountain display.*

next 3 months. The site also has to have the right thermal characteristics so that the eggs can be kept at the magic temperature. Nests that are too cold will produce only females, while nests that are too hot will produce predominantly females. Males occur only at midrange temperatures. No one seems to know exactly how crocodiles manage to select the right site and build the perfect incubator, but averaged over millions of years, a balanced mix of males and females has always been produced!

Maybe it's in the excavation. Using her sharp front claws, the female digs a hole up to 2 feet deep, then lays her eggs in layers, tamping down soil between each layer. On top she spreads additional soil, then presses it with her body to erase the signs of disturbance. Despite her efforts, some of the eggs won't make it to hatching and others will end up in the stomachs of predators shortly after hatching. To offset these losses, crocodiles lay an overabundance of eggs: 48 to 60 per clutch.

Guarding the Nest

The female keeps a close watch on her nest from the shade of nearby bushes or trees. Although the male hangs out nearby, he dares not approach the nest and leaves all the direct defense to the female. During the 2 to 3 months of incubation, she neither wanders away nor feeds and is quick to **growl**, **lunge**, or **charge** at anything that comes too close. Much to the predators' delight, however, even this armor has its chinks. Individual crocodiles differ in the thoroughness of their vigilance, and persistent predators can eventually slip by a female when she's off-duty.

Freeing the Young

If the eggs manage to survive predators or occasional floods, they begin to swell inside their muddy incubator. Each shell cracks into pieces, but the leathery inner layer of the egg remains intact. Luckily, the hatchlings are born with an escape tool—a deposit of hard lime on the snout called the egg tooth. The final barrier to freedom is the ceiling of packed-down soil, which by this time is usually baked to a cement-like hardness. Without the mother's help, most hatchlings would not be strong enough to break through, egg tooth or no egg tooth. To enlist her help, they begin a chorus of loud barks or *eeau* **cries** that you can plainly hear. The cries rally the female from her torpor, and with claws and jaws flying, she **digs** her offspring from their prison.

If you're lucky enough to witness such a release, you may worry at what comes next. The same female that fasted for 3 months to protect her young takes

FREEING THE YOUNG *The female breaks the eggshells in her mouth (gently!) and then totes the hatchlings to the water.*

some of the still unbroken eggs into her jaws! No, she's not eating them (though early observers assumed this); she's merely **rolling the eggs** gently between her tongue and palate to help the young escape their bonds. She releases them a few eggs at a time and then lays her gigantic, open mouth on the ground so that the whole gang of foot-long miniatures can crawl in. When her mouth is full, she **carries** the hatchlings down to the water's edge, opens her mouth, and swishes back and forth in the water to release them. In captivity, males sometimes help with this procedure.

The young hit the water swimming, heading instinctively for vegetative cover. If crocodile hatchlings have "learned" one thing through evolution, it's the need to be secretive in their predator-dense habitat. The baby crocs spend their first few weeks in nursery groups protected by a jungle of water plants and their watchful parents. They live off the internal yoke sac (about the size of a chicken yoke) that they are born with. When their free food runs out, they quell their hunger by snapping at beetles, jumping at dragonflies, and ambushing giant water bugs, frogs, and toads.

Communal Defense

You'll witness one more example of this reptile's unusual social network whenever hatchlings are in trouble. As soon as one tiny croc broadcasts its loud, long **distress call**, the other hatchlings in the area start crying too. This high-pitched chorus alerts not only their parents, but all adults within earshot, prompting them to **investigate** the complaint and swiftly threaten and possibly **attack** any intruders that they find.

The hatchlings stick with their parents for up to 12 weeks before finally dispersing. As they head out, their previous attitude toward adults reverses itself. Instead of drawing close to the larger crocs for safety, they now look for pools and streams that are not inhabited by adults and scurry away from anything larger than themselves. Comfortable only with one another, the juveniles may together dig burrows as much as 10 feet into a riverbank. For the next 5 years or so, the young crocs may retreat to the tunnels any time they're plagued by predators or extreme temperatures.

NILE CROCODILE BEHAVIORS TO LOOK FOR AT THE ZOO OR IN THE WILD

BASIC BEHAVIORS

LOCOMOTION
- ☐ floating
- ☐ cruising
- ☐ swimming
- ☐ lunging
- ☐ leaping
- ☐ high walking
- ☐ galloping
- ☐ belly running

FEEDING
- ☐ still hunting
- ☐ swimming
- ☐ lunging
- ☐ striking
- ☐ tail hunting
- ☐ scavenging
- ☐ **BREATHING**

WARMING UP AND COOLING DOWN
- ☐ basking
- ☐ gaping
- ☐ seeking shade
- ☐ returning to water

SOCIAL BEHAVIORS

FRIENDLY BEHAVIOR
- ☐ *Cooperative Feeding*
- ☐ *Cooperative Hunting*

CONFLICT BEHAVIOR
  Threat
- ☐ swimming boldly
- ☐ head slapping
- ☐ bubbling
- ☐ tail thrashing
- ☐ chasing
- ☐ inflated posture

- ☐ biting
  Submission
- ☐ snout lifting
- ☐ submerging
- ☐ retreating
  Fighting
- ☐ narial geysering
- ☐ lunging
- ☐ growling
- ☐ snapping
- ☐ locking hold

SEXUAL BEHAVIOR
  Pre-Courtship
- ☐ roaring

- ☐ snout lifting
  Courtship
- ☐ courtship splash display
- ☐ subaudible vocalizations
- ☐ tail thrashing
- ☐ fountain display
- ☐ parallel swimming
- ☐ swimming in circles
- ☐ snout lifting
- ☐ groaning
- ☐ jaw snapping

- ☐ bubbling
- ☐ snout rubbing
- ☐ riding
- ☐ *Copulation*

PARENTING BEHAVIOR
- ☐ *Digging the Nest Guarding the Nest*
- ☐ growling
- ☐ lunging
- ☐ charging
  Freeing the Young
- ☐ hatchlings crying

- ☐ digging the hatchlings out
- ☐ rolling the eggs in her mouth
- ☐ carrying hatchlings to the water
  Communal Defense
- ☐ distress calls
- ☐ investigating
- ☐ attacking

ASIAN FORESTS

FEEDING   *Pandas use all four paws to peel bamboo stalks and strip off the leaves.*

GIANT PANDA

The most adorable stuffed animal in the world weighs over 200 pounds, has massive jaws, scalpel-sharp claws, and might not take too well to being squeezed and dragged by one arm. Nevertheless, the panda's universal charm has made it the perfect "poster species"—the ambassador in the World Wildlife Fund's fight to save endangered animals.

Besides winning human hearts, how does the panda's coloring help it survive in the wild? Curiously, no one seems to know for sure. Camouflage is one guess, since the dark and light pattern complements the shadow and light of the bamboo forest. This theory starts to stall, however, when you consider that the panda has no natural enemies to hide from. Maybe the pattern accentuates social signals in some way, or helps pandas recognize one another from a distance so they can avoid socializing. Another theory suggests that the black absorbs heat while the white reflects it, helping pandas maintain an even temperature. The fact is, like many other aspects of animal biology, we don't really know why pandas are black and white.

The Chinese have their own legend to explain it. Long ago, when giant pandas were pure white, a little girl came across a panda and a leopard fighting. Knowing the panda would be killed, she tried to save it, but the leopard turned and killed her. The grief-stricken panda assembled all the pandas in the world to give the girl a proper funeral. As was the custom at panda funerals, they all wore black arm bands. The mourners were so sad, legend goes, that they wept, rubbed their eyes, and hugged themselves for comfort. In the process, they spread the black dye from the arm bands onto their eyes and arms and backs, and even onto their ears, which they covered to block out the sounds of wailing. (Seems as plausible as the warming and cooling explanation to me!)

Here's another mystery about pandas. Why have they evolved to be such specialized plant eaters when their teeth and digestive tract are designed for meat eating? On the surface, it looks like an evolutionary mistake. Compared

VITAL STATS

ORDER: Carnivora

FAMILY: Ursidae

SCIENTIFIC NAME: *Ailuropoda melanoleuca*

HABITAT: Mountainous bamboo forests

SIZE: Length, 4–5 ft   Shoulder height, 2.5 ft

WEIGHT: 165–360 lb

MAXIMUM AGE: Possibly 30 years in captivity

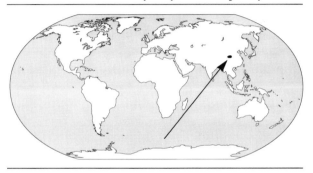

with other plant-eating animals, the panda's digestive tract is short, which means that food can't linger very long for nutrient extraction. Besides, the panda lacks the gut microorganisms that help break down cellulose, a tough component of plants. Most other herbivores have the right equipment and are able to digest 80% of the food they eat, while pandas net a paltry 17%.

Nevertheless, as you probably guessed, natural selection hasn't left the panda completely in the lurch. To make up for what they lack in digestive equipment, pandas have developed ways to shovel in massive amounts of food in record time with-

HUMANS AND GIANT PANDAS

Judging from inscriptions on Chinese tombs, pandas have fascinated people for well over 2,000 years. For eons before that, pandas survived despite their slow breeding rate (one young every 3 years of adult life) and their strict dependence on bamboo. They also had to deal with the cyclical nature of that chosen fare. Each year bamboo plants resprout from rootstocks, not seeds. At the end of the bamboo life cycle (which may be anywhere from 15 to 125 years, depending on species), the plants flower once, set seed, and then die. It takes 6 years for new plants to germinate, and in the meantime, pandas and other bamboo connoisseurs are simply out of luck. When China was still pristine and untrammeled, bamboo die-offs presented no problem. When one species of bamboo faded, pandas could simply wander to another part of the forest to find a species growing on a different schedule. That was when pandas had plenty of room to roam.

Today, a billion people live in China, competing for food, land, water, wood, and other resources. By contrast, there are perhaps 1,000 remaining pandas crowded into a rapidly diminishing habitat on the eastern edge of the Tibetan highlands. Farms and villages have hopelessly fragmented the last wild haunts of the giant panda, isolating them in small populations that have no way to reach one another. These populations are threatened by inbreeding and by the impending disaster of bamboo die-offs. In 1975 and 1976, pandas that were protected by the Chinese government faced such a die-off of the arrow bamboo. With no other species to choose from and no migration corridors to take them to other preserves, an important group of the world's pandas simply sat down with their heads in their hands and starved to death.

The Chinese government, justifiably proud of its national treasure, is working hard to save its remaining pandas. It has established 12 panda preserves and has taken steps to stem illegal poaching. The main markets for panda skins are in Hong Kong and Japan, where, in 1989, dealers could get as much as $100,000 for a single pelt. A villager who kills a panda may see $3,000—a veritable fortune to a Chinese peasant.

The World Wildlife Fund, which has adopted the panda as its mascot, is now working with Chinese scientists to devise better ways to protect pandas from the gun and the saw, as well as from the genetic time bomb of inbreeding. One way is to plant bamboo corridors between reserves so that the panda can travel to find new mates. Ultimately, however, we know that the human expansion into these areas must be slowed if these schemes are to work. In the meantime, the 100 or so pandas in captivity around the world are lavishly protected and admired. Zoos in the United States, China, Mexico, Spain, and Japan that have bred pandas are working hard to transfer what they have learned to other zoos.

out a single wasted motion. They can pluck, break, and peel a bamboo stem almost faster than the eye can see, and before they finish one stalk, they're loading in another. They rake the branches through their mouth until they collect a bunch of leaves, which they retrieve with one hand like an edible bouquet. Somehow, in the midst of this rapid-fire feeding, pandas manage to ingest only the most nutritious parts of the bamboo.

Their digestive process is equally speedy, allowing them to keep eating without getting stuffed. Unlike horses and cows, which take 24 hours to cycle through a dinner, pandas excrete waste in 5 to 13 hours, allowing them to chow down some more. It's not unusual for them to put away 40 pounds during 16 or more hours of feeding a day. On the other side of the equation, pandas conserve calories by taking it easy, avoiding social encounters, and keeping their ranges small and their pregnancies short. It also helps to have a large body, which allows them to burn calories more slowly and retain the heat longer than small animals can.

Alone Together If you're lucky enough to visit a zoo that has pandas, you'll notice something about how pandas choose to spend their energies. Despite the best-laid plans of matchmakers in the zoo world, panda mates spend most of their time ignoring each other. It gives you the feeling that they're alone together in their extravagant exhibit. As it turns out, this aloofness is nothing personal, but merely the way that pandas act in the wild. In the dense bamboo forests high in the mountains of China, mist and vegetation screen pandas from their neighbors, leaving each animal to its solitary life.

Group living, as a rule, is only an advantage if an animal needs help getting food or warding off enemies, or if it has lots to learn from other group members about the ins and outs of survival. In the panda's case, survival is straightforward; food is usually abundant, and there are no predators large enough to threaten it. As a result, pandas keep their distance from one another except when it comes to sex. Unfortunately, though, as anyone who followed the early breeding attempts of Hsing Hsing and Ling Ling at the National Zoo knows, even sex is not always a mutual meeting of the minds.

The trickiness associated with mating, it seems, is not just confined to zoos. Even in the wild, pandas aren't casual or prolific breeders. Compatibility is always an issue, especially since pandas have to compromise their normally antisocial natures in order to mate. Females can be picky and, because the world's captive community is so limited, finding a compatible mate can be like dating in a very small town. The match has to be well-timed to boot, since the panda's 2- to 3-day estrus period comes only once a year. Finally, even when cubs are born, they are so small, fragile, and dependent on the mother that survival is tenuous at best.

Whatever the reason for breeding failures, people in conservation circles feel rather desperate to come up with a solution. Given the fact that there are only 1,000 or so pandas left in the wild, the zoo population may someday be the world's only source of live cubs. If and when we find the secret to guaranteed breeding success, it will not be a moment too soon.

Basic Giant Panda Behaviors to Watch For

LOCOMOTION

One of the panda's most endearing traits is its roly-poly, pigeon-toed shuffle. When you think about it, an animal with few natural predators living by itself inside a well-stocked pantry of food really has no need to move fast. Besides, its **leisurely walk** is just right for conserving energy and making the most of a low-nutrition diet. Only when startled, by a loud noise somewhere in the zoo for instance, is it liable to **trot** off rapidly. It can also **climb** trees to scout out the area, escape from intruders, or even to snooze. To ascend a tree trunk, it digs in its sharp claws and pulls itself up like an old-fashioned pole climber. The smaller the diameter, the faster the climb. One unusual form of locomotion that you might see is a few steps of **backward walking**, usually practiced by females during courting season.

FEEDING

Pandas are living, breathing bamboo-processing plants. Although they don't quite have the digestive tract for it, they do have some other specialized gear, such as claws for hooking stems and elongated wrist bones for holding them steady, powerful jaws full of cell-crushing molars, a horny esophagus, and a gizzardlike stomach. Their slow metabolism (a function of large body size) is also a boon, allowing them to subsist on a nutrient-poor diet.

Pandas spend most of their active hours collecting, preparing, and finally devouring food, using all four paws to full capacity. It doesn't matter what position they're in—sitting upright, flat on their back, or lolling on their side—pandas just keep on **stalk peeling** and **leaf stripping**. And if it's not bamboo (which it is 99% of the time, even at the zoo), it may be wildflowers, vines, wild grasses, honey, or even a little meat.

All that food has to go somewhere. Check out the grounds at your zoo exhibit. No matter how quickly the keepers rake up, it's hard to stay ahead of the 48 pounds of waste produced per animal every 24 hours.

GROOMING

Another solo activity you're likely to see is fur grooming. Pandas use their forefeet to **rub** their head and face from the nape of the neck, over the ears, and down to the muzzle in a movement that's similar to face washing in rodents. They also **scratch** themselves with all four limbs and occasionally lick their fur. To scratch the places that paws won't reach, pandas stand or squat against a vertical surface and rub themselves up and down against it. Or they may **roll** on their back or belly along the ground, writhing in obvious pleasure.

SCRATCHING *Pandas use their sharp claws to gingerly scratch an itch.*

Because the panda's anal glands secrete scent, some of these grooming moves may also have a marking function, as you'll see in the How Giant Pandas Interact section. In addition to spreading their own scent around, pandas seem to like picking up strange ones. Pandas will often dig up a piece of sod, for instance, and rub it all over their body. Or they'll choose rolling sites that are mostly wet earth, and then rub and roll until they've blackened their fur.

It's not that pandas are averse to being clean. In fact, they will gladly take to water if it's available, cooling their body as they **bathe.** Stand back when they exit; like your favorite dog, pandas like to shake out all the water in their short, dense fur. You'll see females rolling, writhing, and bathing most compulsively when they are in heat. Their skin and their mood seem to be equally irritated at this time, and scratching their itch becomes a preoccupation.

You'll notice that pandas strike some unusual poses to get in a good rub, roll, or scratch. The playful quality of their movements may have something to do with the fact that these giants didn't have to worry about predators, food shortages, or fights for much of the time they were evolving. This "don't worry, be happy" theme led to the relaxed look that characterizes modern panda movement.

SLEEPING

In the wild, pandas nap for 2 to 4 hours at a time between feeding bouts, snoozing on their side, back, or belly, either sprawled or curled up. In zoos, their feeding bouts come at two specific times, and the pandas spend the rest of the day resting. Even when they're sleeping, pandas manage to look adorable. Incredibly flexible, they are able to bend their well-padded body in an endless variety of poses. One of the biggest crowd-pleasers is the legs-propped-up-on-a-tree-and-the-forearm-draped-over-the-eyes pose.

HOW GIANT PANDAS INTERACT

Visual signals are a rather moot point in panda society; their round faces are inexpressive, their tails are stubs, they have no crest or mane to erect, and their ears aren't flexible enough to cock forward or flatten. The reason pandas never developed these visual accessories is obvious when you look at their lives in the wild. Pandas live in dense, fog-enshrouded stands of bamboo where they can't see each other well. When they do see each other, they usually hightail it in opposite directions.

Much of their communication, therefore, is accomplished through scent markings left like graffiti throughout their habitat. When they do want to meet, usually for mating, they follow the scent marks to find one another. Once they're face-to-face, pandas switch to sound, relying on an extremely detailed vocal language to express all shades of mood, from the amorous to the angry (see the Listening to Giant Pandas chart, p. 201).

Silence is another telling medium. Pandas that are playing or simply being friendly, without a sexual or aggressive agenda, do not vocalize at all. This sound/no sound rule can help you interpret many of the panda behaviors that you see at the zoo.

FRIENDLY BEHAVIOR

Social Play

You're most likely to see play behaviors in spring (March to May) as the pandas break down their social barriers in preparation for breeding season. One panda will invite another to play by rolling itself up into a ball and **somersaulting**. The pair may also **mount** each other in a nonsexual, nonaggressive way or **stand over** and on top of each other. From here, they may collapse into a **wrestling** match, in which the partners sit facing each other, grabbing, pushing, clawing, and swatting with their forearms and hind limbs. The matches may include head tossing, lunging, or biting, but it's all performed in slow motion, with none of the vigor or intensity of real fights. **Paw-swatting** contests are most apt to occur when the opponents are at different heights, especially when the female is on a raised surface looking down at the male.

CONFLICT BEHAVIOR

Scent Marking

PAW SWATTING *This roughhouse move can be a playful gesture or an aggressive threat.*

The secret to keeping peace in the bamboo forest is to mark your territory with your scent so other pandas know you're there. Pandas rub secretions from their anal region onto posts, tree trunks, exhibit walls, or on the ground, usually along paths that they habitually tread. Depending on who reads the mark, the scents may either separate pandas or help bring them together. Outside of breeding season, one sniff of a strange panda is usually enough to send intruders ambling away. In season, however, a female's scent mark advertises her sexual readiness and draws males to her. She is most apt to accept a male whose scent she recognizes, perhaps because she's been smelling his scent marks in her range all year long.

There are several characteristic postures that will tell you that a panda is marking

SCENT-MARKING POSTURES

1. **Squat:** Squats down on its hind legs and rubs its rump on a low object with a back-and-forth or circular motion.

2. **On all fours:** Backs up to a vertical surface such as a post or wall, raises its tail, and rubs.

3. **Leg cock:** Lifts its leg, braces with its hind paw or whole leg, and twists its body to bring its anal gland in contact with the surface.

4. **Head stand:** Here's a comical one; the panda backs against a vertical surface, flips onto its forepaws, leaning its raised legs against the surface as it marks.

5. **Body rub:** Uses its entire body to rub the scent in thoroughly.

Leg cock

Head stand

Squat

SCENT MARKING   *Pandas use a variety of postures to spread their personal scent throughout their feeding area.*

its exhibit. Pandas also get into these positions when marking with urine or with a combination of urine and anal gland secretion. As they mark, pandas may weave and bob their head with their mouth open. They leave a dark spot of secretion that is so thick and sticky that it would have to be scraped to be removed. As a final touch, they may leave a visual flag in the form of **peeled bark** or **claw marks** on tree trunks that will guide other pandas to the scent mark.

Aggressive Threat

In the wild, panda conflicts usually occur around breeding time, when three or four males may vie for the right to mate with an estrous female. Occasionally, males and females get into tussles with one another, too. In the zoo, exhibit-mates may get into fights over food, water, or simply which animal gets the best lounging site. Most conflicts begin with threat sequences, which can sometimes escalate to all-out fights if they're not resolved.

Often, no more than a **hard stare** is needed to settle out of court. The annoyed panda raises its head and stares directly, while a panda that is more submissive will turn its head away, lower its body, or run.

When two pandas of equal confidence meet, they threaten each another by tensely circling around, performing a **lateral body display**, and jockeying for posi-

UPRIGHT POSTURE *Defensive pandas rear up on their hind legs, trying to avoid rather than accelerate a fight.*

tion. As the tension mounts, the pandas may **paw swat** at each other or, crouching low to the ground, they may **push** each other with the full force of their body. Listen for throaty, ominous **growling** that lets you know this is not a play fight.

Sometimes, however, the pairing is not equal. In a stand-off between a male and a female, the female usually takes the offensive by **moaning** and **barking**, which prompts the male to either leave or defend himself.

Defensive Threat

If he stays, he rears up in a defensive **upright posture**, either standing on his hind legs or simply raising his upper torso to a sitting position. Rearing up is designed not to accelerate a fight, but to discourage it. Using his head and forelimbs, he may try to **push** his partner off balance or simply threaten her into withdrawing.

Fighting

If one of the pandas doesn't withdraw, the other may suddenly lunge, drape its body over the partner, and begin to **bite**. It concentrates its bites on the black body parts: the saddle, ears, and limbs. Fights are especially common when males and females meet in nonsexual encounters. The female often mounts or stands over the male, making stabbing motions with her head and biting his back or side. She shakes her head vigorously as she bites, tightening her toothy grip until the male finally withdraws with a squeal.

SEXUAL BEHAVIOR

Pre-Courtship

Estrus is a once-a-year event in the life of a female panda, and her peak of receptivity is a scant 2 or 3 days, usually in the spring. Zookeepers, eager to encourage breeding, have gotten good at recognizing the signs of a female that is approaching her peak. You can look for them too.

One to two weeks before estrus, the female's nipples and genitals begin to swell and redden beneath all that fur. She becomes impossibly **restless**, constantly changing position, **rubbing** and **rolling** on the ground, and **bathing** in water more frequently. At the same time, she begins bleating, chirping, and grunting more, eating less, and vigorously **scent marking** by rubbing her anal glands on logs and stones. This is the only time during the year that a female regularly marks scent posts. As the magic window of opportunity draws near, watch for **backward walking**, in which she backs up a few steps, tossing or shaking her head while bleating.

Courtship

As courtship heats up, the female becomes less aggressive and more determined to mate. She allows the male to approach, and he begins to follow her. She even solicits his attentions by backing up to him with a **tail-up posture**: rump elevated, tail raised and held horizontally, and forequarters lowered so there is a depression in her back. She may even tuck her head under her chest or brace it against a wall or tree while elevating her hindquarters. If he doesn't respond, she may roll onto her back, squirming and writhing, and reach gently up to him with her forepaws. The courted male may **sniff** and manipulate her tail region, again getting a snoutful of scent that perhaps prepares him for mating as well.

Copulation

During copulation, the male stands almost upright behind the female, raising his head, bleating, and making mouthing gestures reminiscent of the typical carnivore neck bite. He mounts and dismounts constantly before actual intromission. One research group counted 48 mounts in 3 hours, with only one 4-minute intromission during that time. After only a couple of days of intense mating, the couple separates to resume their solitary lives.

PARENTING BEHAVIOR

Caring for the Cub

On top of all the other peculiarities of panda breeding, their young are born weighing only 4 ounces, which, sizewise, would be comparable to a 2.5-ounce human baby that could fit on a tablespoon. Alarmingly small. Parenting, therefore, is serious business that occupies the female full-time for at least 8 months, and sometimes for up to 2 years until she is occupied with her next offspring.

The cub is born in September in a secluded birth den, usually an ancient tree hollow or a

COURTSHIP   *The female solicits the male for mating.*

LISTENING TO GIANT PANDAS	
SOUNDS	**WHAT THEY MEAN**
bleat	sexual
chirp	sexual
huff	conflict
snort	conflict
growl	conflict
roar	conflict
chomp	conflict
bark	conflict
moan	conflict
jaw clapping	conflict
yip	conflict
honk	fear or distress
squeal	fear or distress

natural rock crevice that the female has lined with bamboo twigs and grasses. She must hold the tiny newborn constantly in the crook of her arm, or, when moving, she **carries** it in her mouth. The cub's loud, squeaky voice seems to insist that she be careful. The cub **nurses** up to 14 times a day, creating a tight schedule for the female that still must feed herself megadoses of bamboo to stay alive. Now she's eating for two, providing the milk that will fatten her cub from a frail 4 ounces to the 90 pounds it will be a year later.

Survival Realities

The female's solicitous nature goes only so far, however. If twins are born (60% of all births), the mother will deliberately reject or ignore the weaker cub and pay more attention to the stronger one. In the wild, this ensures that at least one cub will survive because it will have the best care possible. In captivity, however, where zookeepers are holding their breath over each panda birth, researchers are investigating ways to hand-rear these weaker cubs so they can save as many giant pandas as possible.

CARING FOR THE CUB   *Tiny panda cubs enjoy an unusual style of transit.*

GIANT PANDA BEHAVIORS TO LOOK FOR AT THE ZOO OR IN THE WILD

BASIC BEHAVIORS

LOCOMOTION
- [ ] leisurely walk
- [ ] trotting
- [ ] climbing
- [ ] backward walking

FEEDING
- [ ] stalk peeling
- [ ] leaf stripping

GROOMING
- [ ] rubbing
- [ ] scratching

- [ ] rolling
- [ ] bathing
- [ ] SLEEPING

SOCIAL BEHAVIORS

FRIENDLY BEHAVIOR
Social Play
- [ ] somersaulting
- [ ] mounting
- [ ] standing over
- [ ] wrestling
- [ ] paw swatting

CONFLICT BEHAVIOR
Scent Marking
- [ ] squatting
- [ ] on all fours
- [ ] leg cock
- [ ] head stand
- [ ] body rub
- [ ] bark peeling

- [ ] clawing
Aggressive Threat
- [ ] hard stare
- [ ] lateral body display
- [ ] paw swatting
- [ ] pushing
- [ ] growling
- [ ] moaning
- [ ] barking

Defensive Threat
- [ ] upright posture
- [ ] pushing
Fighting
- [ ] biting

SEXUAL BEHAVIOR
Pre-Courtship
- [ ] restlessness
- [ ] rubbing
- [ ] rolling
- [ ] bathing

- [ ] scent marking
- [ ] backward walking
Courtship
- [ ] tail-up posture
- [ ] sniffing
- [ ] *Copulation*

PARENTING BEHAVIOR
Caring for the Cubs
- [ ] carrying
- [ ] nursing

SOUNDING THE ALARM *The first peacock that senses danger lets loose with a* kok-kok-kok-kok *alarm call.*

PEACOCK

One morning, a sudden screaming—half-human, half-baboon—erupted outside my office window. I knew that the neighborhood was normally deserted at this time of day, with kids away at school and parents at work. Gingerly, I peeked out to see the source of the commotion: two nonchalant female peacocks and one heavily costumed male strutting his heart out. Courtship in my backyard!

A call to the Como Park Zoo confirmed my suspicions. It seems that most zoos, like the one near my former home, let their peafowl run free. Zoos can do this because peacocks are rather sedentary birds; they remain where they are as long as they have a dependable source of food, water, and a good roosting tree. But every now and then, as I can attest, a disruption in their home range or the onset of breeding season will cause them to take off for a few days.

I couldn't help but notice how very wary my visitors were; as they fed on my shrubs, they constantly cocked their heads from side to side and way up high, inspecting the branches of my giant basswood and scanning the sky above. In their wild haunts, peacocks must constantly be alert for hawk eagles, jackals, martens, tigers, leopards, and civets. They choose their feeding spots carefully, preferring an open area where they have a clear view of anything coming. They also stay within sight or earshot of one another, issuing group alarm calls at the first hint of danger. Alerted peacocks rarely stay to see what the trouble is; even though they are as large as turkeys, they are able to either run for cover or fly up at a sharp angle, gaining height quickly even in close quarters. If they must defend themselves, they use sharp spurs on their legs to disarm an attacker the way fighting cocks do.

Something works against all these defensive tactics, however, making every adult male peacock vulnerable. No matter how sensitive his eyes and ears, or how quick his response, the male carries with him a heavy, awkward train of 150 feathers that trails up to 5 feet behind his body. This trailer not only gives predators a big target to aim for, but also keeps

VITAL STATS

ORDER: Galliformes

FAMILY: Phasianidae

SCIENTIFIC NAME: *Pavo cristatus* (Indian Peafowl)
Pavo muticus (Green Peafowl)

HABITAT: Clearings with scattered trees or shrubs

SIZE: Male length, 6–8 ft, including train
Female, 2–3 ft

WEIGHT: Male, 9–13 lb   Female, 6–9 lb

MAXIMUM AGE: 20 years

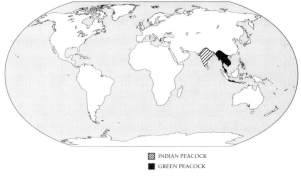

▨ INDIAN PEACOCK
■ GREEN PEACOCK

the peacock from escaping quickly. Confounded by excess baggage, he has to land frequently during flights, often bogging down so badly in vegetation that he can't get going again. How could natural selection have overlooked such a glaring flaw in design?

Behaviorists have been looking for a method in the madness of the peacock's train for years. There must be something that redeems its obvious drawbacks, otherwise males wouldn't go to the metabolic trouble to grow one. As it turns out, the train's strong point is the same one that justifies many of the designs in the animal kingdom: it helps the peacock pass on his genes. When properly "dressed," a male can attract a harem of three to five peahens. If the showy display he performs is convincing enough, he will edge out competing peacocks and win the right to sire the harem's offspring, thereby multiplying his genes in the population.

But just what is it about the peacock's fan that puts the female in the mood for mating: The iridescence? The graceful shape? The spots that look like eyes? The truth is, what wows humans may not impress female peafowl at all. Instead, the sheer size of the fan may be the irresistible feature because of what it says about the bird that carries it. Namely, if a bird with a large fan has survived to breeding age despite his unwieldy handicap, he must be both strong and wily. In the same way, females may value colorful plumage not for its beauty, but because the sheen shows that the bird is free of parasites. Females that can spot these superior traits are rewarded by passing on their genes to offspring that, like their "handicapped" father, are more likely to survive and reproduce.

When you think about handicaps as indicators of superior genes, the unwieldy rack of the elk also makes sense. As you browse at the zoo, you'll probably spot other strange and showy body parts that have evolved in the animal kingdom. In many

HUMANS AND PEACOCKS

People have been keeping peacocks in captivity for more than 4,000 years. In the old days, you had to be some kind of ruler or religious figure to own one. A pair of peacocks in ancient Greece, for instance, cost the equivalent of $3,500. (In today's market, $50 will buy a pair.) In southern Asia, peacocks had their own royalty status, and people worshiped them for their ability to ward off snakes, tigers, and bad luck.

To this day, sacred law forbids Hindus to harm the bird. As a result, wild Indian peafowl have taken to congregating near settlements and shrines in the Indian hills where they are guaranteed handouts and immunity from persecution. They roost by the hundreds, painting the trees with stained-glass colors and filling the air with their ringing cries. Unless religious doctrine changes, these semitame birds will forever enjoy protection in their native country of India.

The green peafowl do not enjoy the same sanctuary, however, which may be why they are wilder and more prone to aggression than their Indian peafowl relatives. Though feistier, green peacocks are also in a far more vulnerable state. Nearly hunted out of their native habitats, captive breeders are working to rebuild green peacock numbers so they can one day reintroduce them to the wild. The problem is where to release them. As more and more jungles are cleared for agriculture and housing, there are fewer and fewer acres of safe habitat left. For the time being, zoos are the only places where these peacocks can enjoy a unique freedom—the freedom to visit someone else's backyard without fear.

case, you'll find that animals display these beauties or oddities not for vanity's sake, but because it's the quickest way to say, "Look what a good mate I would be."

LOCOMOTION

Though peacocks can take off quickly and angle up sharply, these flights are short-lived, especially for the male. The female, having no train, can **fly** a continuous series of flights lasting several hundred yards each, while the encumbered male gets bogged down long before her. When the bird lifts and fans its train, it presents a virtual sail to the wind. This explains, in part, the reeling and **strutting** you see when peacocks perform. Though it looks to us like they are strutting their stuff, the dancers are probably just trying to stay on their feet.

FEEDING AND DRINKING

After preening, a peacock's first order of business for the day is to eat and drink. Morning is spent searching the stream edge, clearings, and forest borders for grasses, grains, buds, flowers, leaves, vegetables, and berries. They use their large, clawed toes to break into termite mounds and their industrial-strength beak to find grubs and worms beneath the sun-baked soil. Frogs, mice, small lizards, and snakes (including an occasional young cobra) are also on the peacock's wild menu. To jazz up the prepared diet they receive in captivity, peacocks will stalk the zoo grounds, snapping up local residents that fly, burrow, or slither by. Like chickens, they wash down their meals by sucking water into their bill and then raising their head to swallow.

FEATHER CARE

Without their gorgeous feathers, peacocks would not only be cold, but also wet, mate-less, and earthbound. It's no wonder that both males and females spend nearly an hour a day caring for these feathery assets. **Preening** begins before dawn, when they are stuck in their roosting tree, waiting for the night's dew to dry before their feathers are flight-worthy.

As they wait, they stretch, shake their wings, and clean each feather, using their long neck to reach out-of-the-way places. Special muscles lift feathers away from the skin so the bird can smooth and straighten them with its beak. Before preening, the bird dips its beak into an oil-producing gland at the base of its tail. It **spreads this oil** on each feather to waterproof it and make it a better insulator.

DRINKING *Peacocks, like chickens, dip to drink, then raise their head to let water trickle down their throat.*

Preening

Dust bathing

FEATHER CARE *Peacocks comb and oil their feathers each day to keep them in strutting shape. Dust may be used to absorb excess oil and discourage parasites.*

Evidently, this oil is also a source of vitamin D, which the birds need to avoid a bone disease called rickets. The warmth of the sun activates the vitamin D, which the bird then absorbs through its skin and swallows while preening.

While a little oil is beneficial, too much can actually destroy waterproofing and attract oil-eating parasites. To discourage these freeloaders, peacocks dust their feathers to absorb some of the oil. At the zoo, you'll often see several peacocks **dust bathing** together—hollowing out cavities in the dry earth and using their wings to stir the dust into a storm.

You may wonder why birds as fastidious as peacocks look so ragged during a few weeks each summer. Chances are you've caught them during their annual **molt**, when they are replacing their old feathers with a new set. Though the new feathers appear by late summer, it will take 7 months before they are fully grown—just in time for spring courtship, when every peacock wants to look its best.

WARMING UP AND COOLING DOWN

Indian peafowl live in hot, dry habitats in India, but surprisingly, they are also able to withstand subzero temperatures. No matter how cold it gets, their metabolism keeps their internal furnace stoked to 106 degrees Fahrenheit. To hold in this heat, the bird **fluffs its feathers**, trapping an insulating layer of dead air against its body. When the mercury rises and the bird needs to cool down, it does just the opposite; it **sleeks its feathers** against its body, pushing out the air so heat can escape. Evaporation also releases heat, but since a peacock can't sweat, it **opens its beak**, letting air dry the moisture from its lungs. Though the peacock's legs are featherless, they contain little more than bones, tough tendons, and a few rudimentary nerves and veins, leaving them immune to both heat and cold.

SLEEPING

Peacocks don't need privacy for a good night's sleep. In fact, they don't mind being seen, as long as they're able to spot whatever is watching them before it gets too close. Their favorite vantage points are tall trees with smooth, branchless trunks that foil climbing predators. Roosting on the lowest, barest branches, the peacocks face into the wind and nestle together for warmth. When frost bites, they tuck their bills into their feathers to keep heat in. During the day, look for peacocks resting on fences, posts, or other perches that are high enough to keep the birds' trains off the ground.

FRIENDLY BEHAVIOR

Peacocks are essentially social birds that prefer to travel and sleep in large flocks for safety's sake. The only times you'll see them alone are (1) in the spring, when males begin to defend territories in which they will display and eventually mate with a harem of three to five females, and (2) when females sequester themselves for nesting and brooding. Once the young are large enough, the family will reunite and eventually join with other peacocks in a large flock.

Communal Roosting

Each evening at dusk, peacocks from all corners of the zoo will seek one another out, meeting at their favorite roosting trees. This nightly roundup routine is predictable and interesting to watch. In each group, the hens wait on the ground while the male peacock **scrutinizes** the roost site for possible danger. He struts about, cocks his head this way and that, and stops to listen. When satisfied that all is well, he takes a short run, beats his wings, and heavily ascends to the sleeping limb—a signal to the waiting hens that the coast is clear. They ascend after him (with less difficulty), and after a chorus of penetrating *he-on* **contact calls**, all settle down for the night.

Communal Defense

Peacocks are the night watch guards of the zoo. When I used to stroll past my neighborhood zoo at night, the sound of my footsteps near their roost tree would set off the security siren—the *kok-kok-kok-kok* **alarm call** of roused peacocks. In the wild, this group watchfulness allows peacocks to feed with their heads down or to sleep in relative peace, knowing that any member of their group that detects a disturbance will notify them all.

Social Play

Another form of comradeship you may see is the **tag game**, in which opponents—usually young peacocks—chase each other around a bush. For some reason unknown to us, they always travel counterclockwise, and the game always ends abruptly, sending all the birds scattering in different directions.

CONFLICT BEHAVIOR

Zoos very often have two species of peafowl on display. The Indian peafowl (also called a blue peacock) has a fanlike crest, white facial skin, and primarily blue plumage. The green peafowl has a tall, columnar crest, blue or yellow facial skin, and more green in its plumage. The green peacocks,

ROOSTING *The flock that sleeps together survives together, keeping a collective eye out for predators.*

which are wilder and warier than their Indian counterparts, have incredibly short fuses. In fact, if you run across a green peacock displaying on the walkway, you may want to give him plenty of room. (If it's two males facing off, load your camera and wait for the fireworks!) The Indian peacock, by contrast, is more likely to use simple threat displays, stopping short of actual fighting most of the time. The best time to see these displays, as well as the occasional fight, is at the start of spring breeding season, when the males are establishing courting territories.

Threat

Though the fence around his territory is invisible, the male peacock knows his borders and will threaten anything that approaches his domain. His first means of intimidation is to spread his gorgeous fan and perform the same **strutting display** that he uses in courtship. During nesting season, females also get defensive and are quick to erect their short tail in the same threatening way. If the display does not convince the rival to retreat, the resident may **rush** at it with **screams** of outrage. If the bird still doesn't yield, the two may **circle** each other with erect trains, maneuvering into battle positions.

As they jockey for position, watch how carefully each bird guards its back forty. One of the disadvantages of such a lush fan is that the wearer cannot see behind itself. If it's not careful, a well-timed peck on the rear could end its performance.

Fighting

A peacock fight is a tornado of flying feathers. The fighters parry and **slash** with their legs, each trying to sink its spurs into the other's flesh. The birds flap their wings, sometimes rising as much as 10 feet off the ground, furiously **kicking**, **pecking**, and struggling to avoid being slashed. The bird with the most injuries usually backs off. The best it can do is run away, trailing its torn feathers and hoping the victor won't chase it for a second round.

SEXUAL BEHAVIOR

For the polygamous male peacock, breeding success depends on casting a spell over as many females as he can. At the height of spring courtship, this Don Juan of the bird world is always ready; the minute a female comes anywhere close, he unfolds his fan and begins his act.

Courtship

In the **strutting display**, the male peacock uses his stiff tail feathers to lift the long plumes of his colorful train. He waits until the female gets close and then spreads the plumes outward in semicircular

FIGHTING *When elaborate threat displays don't do the trick, peacocks will engage in furious kicking bouts. Sharp spurs on their legs can leave deep slash wounds.*

splendor, madly beating his small wings behind the fan. He struts this way and that, as if showing off his best angles, and then shatters the silence with a loud *ming-ao-a* **scream**. (Jungle dwellers claim this call is the peacock's way of saying, "There will be rain." And it's true; peacocks do breed at the start of the summer monsoons.) To add percussion to the scream, he **shivers** violently, causing the stiff tail quills to rattle like rain on dry leaves. Going for an even more impressive effect, the male may spread his fan and then back up to the female, turning at the last moment to stun her with his shimmering array of gold, green, blue, bronze, and purple.

No matter how hard the male tries, the female seems indifferent at first. Instead of looking up at the display, she pecks and pecks at the ground in a gesture of **symbolic feeding**. This playing hard-to-get spurs the male on to even more energetic displays. For the male, persistence is the key to winning the female's attention and successfully passing on his genes. A drawn-out performance proves that he has stamina and may also give the female time to physiologically ripen for breeding.

Copulation

Finally, after as much as an hour of displaying, the male's persistence pays off and the female approaches him. Now it's his turn to play hard-to-get. Each time she gets close, he **turns his back** to her so she must race around him to see his fan. After several of these coquettish exchanges, the male begins pecking at the ground in the

Courtship   *The male whirls to show off his opulent train, while the female, seemingly bored with the show, stoops to feed.*

symbolic feeding display. Finally, the female reaches the spot where he is pecking and **crouches** so the birds can mate.

The Hopefuls

Once a male has attracted his harem, he can stop calling. Thanks to the hopeful younger males, however, the springtime air is still full of raucous peacock calls. Not that it does them any good. Immature males have a weaker voice, a less opulent train, and less overall strength than the adults, and most females have no interest in mating with them. Even if a female did express interest, the harem master would no doubt intervene before things went too far. This doesn't stop the randy males from trying, however. You can hear their calls day and night, even from a distance of 2 miles. In fact, Washington, DC, residents who live near the National Zoo have lodged formal complaints during breeding season, claiming that the anxious swains keep them awake.

Wooing Zoogoers

If you suddenly get the feeling that a peacock is aiming its alluring display at you, you may be right. Naturalists have noted that peacocks display longer and more often in front of humans than before hens of their own species, probably as a way of challenging us. You may also see them challenging their *own* reflection in the shiny side of a car.

THE ILLUSIONIST

We are not the first culture to wonder about the origins of the peacock's beautiful eyespots. The Romans believed that the peacock's eyespots originally belonged to the hundred-eyed giant named Argus. Hera, queen of the gods, had hired Argus to spy on her husband, Zeus, who was spending too much of his time spying on the lovely Io. One day, the unlucky Argus closed all one hundred eyes and fell asleep on the job. Infuriated, Hera plucked out Argus's eyes and hurled them at the peacock's tail, where they have remained forevermore.

Science's less illustrious explanation is no less mysterious. Though the plumes look multicolored, they are really brown. Each feather is made up of layers of keratin, a material found in human skin, hair, and fingernails. The microthin (1/10,000 of an inch), transparent layers have grooves and ridges that catch and splinter light. Scattered through the layers is a brown pigment called melanin. When the light hits this jumble of melanin and textured keratin, it reflects back an iridescent array of copper, bronze, gold, blue-green, and violet. Like the colors in a soap bubble, the colors in a peacock's fan are pure illusion.

Another misconception is that the peacock's train is composed of very long tail feathers. In reality, the tail feathers are the short, stiff quills that stand up behind the fan to hold it erect. The 150 "feathers" in the fan are actually protective extensions, or coverts, that cover the quills. Because the coverts are different lengths, the "eyes" at the end of each covert appear to be scattered here and there. If you're looking for actual wing feathers, you can find them quivering behind the spread fan. Unlike their bewitching neighbors, they show only their true color—a plain, serviceable brown.

PARENTING BEHAVIOR

Nesting

These highly visible birds have an uncanny way of making themselves invisible when it comes time for nesting. They look for tall clumps of grass, deep forest cover, stands of bamboo, shrubbery, or even marshlands in which to hide their nest. In the wild, they dig a shallow hole, 2 inches deep and 12 inches across, then line it with leaves. At the zoo, where the birds feel adequately protected from predators, they may not go to all this trouble. The female may simply lay her eggs here and there, then roll them together in a convenient hollow for brooding.

Incubating and Brooding

Peacocks lay three to eight eggs each, sometimes mixing their eggs in with those of other peahens in a communal nest. The various mothers take turns sitting on the eggs to keep them warm. These incubating birds develop a special brooding patch on their breast, an area where the plumage is thin and the skin temperature is higher than normal. Try to watch as the female tucks the eggs against this warm patch and wraps her remaining breast feathers around their exposed sides. Once they hatch, she will brood the chicks in the same way, keeping them warm and dry in all kinds of weather.

Feeding and Defending the Chicks

At first the female must leave the nest every so often to hunt for high-protein chick food such as grubs and insects. Upon her return, the chicks beg by chirping softly and pecking at her bill. Before too long, they are large enough to accompany her on feeding excursions, learning by her example which foods are tastiest. This can be a nervous time for the peahen. She keeps a hawkish eye out for intruders and thinks nothing of rushing at animals ten times her size—screaming and lunging with her fan fully erected. If an irate female ever confronts you in this way, be sure to watch for frightened chicks sheltering beneath her tail.

DEFENDING THE CHICKS   *A peahen can be a fierce adversary, especially when her chicks are in danger.*

PEACOCK BEHAVIORS TO LOOK FOR AT THE ZOO OR IN THE WILD

BASIC BEHAVIORS

LOCOMOTION
- [ ] flying
- [ ] strutting

FEEDING AND DRINKING

FEATHER CARE
- [ ] preening

- [ ] oiling
- [ ] dust bathing
- [ ] molting

WARMING UP AND COOLING DOWN
- [ ] fluffing feathers
- [ ] sleeking feathers

- [ ] opening beak

SLEEPING

SOCIAL BEHAVIORS

FRIENDLY BEHAVIOR

Communal Roosting
- [ ] scrutinizing
- [ ] contact calls

Communal Defense
- [ ] alarm call

Social Play
- [ ] tag game

CONFLICT BEHAVIOR

Threat
- [ ] strutting display
- [ ] rushing
- [ ] screaming
- [ ] circling

Fighting
- [ ] slashing

- [ ] kicking
- [ ] pecking

SEXUAL BEHAVIOR

Courtship
- [ ] strutting display
- [ ] screaming
- [ ] shivering
- [ ] symbolic feeding

Copulation
- [ ] male turns his back
- [ ] symbolic feeding
- [ ] crouching

PARENTING BEHAVIOR
- [ ] *Nesting*
- [ ] *Incubating and Brooding*
- [ ] *Feeding and Defending the Chicks*

WALKING UPRIGHT *Monitors occasionally take a few steps on their hind legs.*

KOMODO MONITOR

Although few zoos are lucky enough to have Komodo monitors, it's worth making a pilgrimage to those that do. It's not just because there are only 5,000 of these dragons left in the wild or because their home islands in Indonesia are on the brink of being exploited for their resources. Rarity and vulnerability aside, you should seek out these monitors because they are unlike any reptile you've ever seen or imagined.

Picture yourself walking the wooded trails of the small island of Komodo. You suddenly catch the pungent odor of blood in the air, a metallic, gamy smell. As you round the bend, you nearly step into a circle of enormous lizards feasting on a deer carcass. As long as alligators, they are a flat-footed 200 pounds of scaly muscle. As they slash at the deer flesh, you catch a glimpse of their teeth, jam-packed and as surgically sharp as a shark's. The long claws on their stubby toes steady the meat while their heads shake violently to free large chunks. Several movable joints allow their jaws to open wide enough to accommodate the whole hindquarters of a small deer, bones and all!

If you were to hide in the bushes and watch this feast for a while, you'd see even more unusual behavior. You'd see large monitors lording over the choicest pieces of the kill while small monitors perform careful, choreographed dance steps on the outskirts. You'd see males lavishing females with attention while sniping at one another with deadly accuracy. These dramas abound around carcasses because mealtimes are the only times when the normally

VITAL STATS

ORDER: Squamata

FAMILY: Varanidae

SCIENTIFIC NAME: *Varanus komodoensis*

HABITAT: Open savanna

SIZE: Male length, 10 ft   Female length, 6.5 ft

WEIGHT: Up to 200 lb

MAXIMUM AGE: More than 25 years

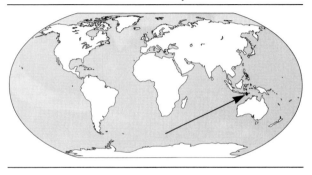

solitary monitors see one another. They tend to all their social business here; establishing rank, courting, and even copulating take place around the dinner table!

Afterward, each party returns to its piece of the island to bask, rest, hunt, and dig a shelter. Monitors lay down an impressive trail of personal scents to claim a place and isolate themselves. By keeping other monitors out of their personal hunting range, they're assured of having enough prey to satisfy their prodigious appetites (with enough left over to share). This kind of rationing of the habitat makes good ecological sense, especially for a species that lives within the finite borders of islands and can't simply move on if its food supply runs out. It's a lesson that we, who are essentially island-bound on our own planet, might do well to learn.

BASIC KOMODO MONITOR BEHAVIORS TO WATCH FOR

LOCOMOTION

High atop their stubby, bowed-out legs, **running** komodos can keep up an 8-mile-per-hour pace for a third of a mile. Their powerful tail switches back and forth for balance, and their hind legs swing in rhythm like a double-bladed kayak paddle. Witnesses say the beat of their feet against the earth sounds like the *rat-tat-tat* of a muffled machine gun. At slower speeds, they can stand on their hind legs and **walk upright** for a while. **Climbing** is no problem for young monitors, which prefer to spend their early years hunting for geckos and keeping out of harm's way up in the trees. Their talonlike claws help them get a leg up on rough, fissured bark. Monitors are also decent **swimmers** and **divers** and have been found at depths of 6 1/2 feet.

FEEDING

Monitors will eat just about any kind of meat they can get their jaws on. They'll gladly hunt down live animals or sniff out dead ones, traveling as far as 6 1/2 miles from their home range to find the source of a tempting odor. The size of their quarry doesn't seem to phase them either. Once they've set their sights, whether it's a grasshopper or a 1,300-pound water buffalo, monitors are tenacious adversaries.

As they grow, they test their mettle on many different types of prey. The young feed mainly on insects and geckos, while medium-size monitors go for rats and birds. By the time they're adults, monitors are capable of taking on animals larger than themselves, such as deer, horses, and buffalo.

Monitors do most of their serious live-prey hunting in the late morning, **lying in wait** along game trails. They assume a characteristic **alert posture**, keeping their tail down, but raising the front part of their body and lifting their head to see over the grasses. Besides keeping an eye out, they may occasionally cock their head and lick the air to test for

ALERT POSTURE *In the waiting game called hunting, Komodos keep an eye peeled for movement and a tongue poised to catch the slightest scent in the air.*

scents. If unsuccessful along the game trail, they move to the thickets where they can sneak up on deer in their daybeds.

Monitors will also **stalk** the enormous water buffalo, an animal 20 times more massive than they are. Creeping up slowly and silently from behind, the dwarfed predator turns surprise to its advantage, **lunging** and **biting** at a hind leg before the beast knows what hit it. The lizard holds on for dear life, riding the leg of the thrashing buffalo and eventually severing the Achilles' tendon with its serrated teeth, bringing the huge bovine thundering to the ground. Moments later, the monitor has gutted the carcass and is eating with gusto. These lizards pack away astonishingly large pieces, an adaptation that permits them to eat quickly, gulping down as much as they can while their competitors do the same. The monitor's high salivation rate undoubtedly helps these large, ill-fitting parcels slide down a little easier. All that's left at a feeding site are some wisps of hair and the spilled stomach and intestinal contents.

HUMANS AND KOMODO MONITORS

If you find yourself strangely repulsed or frightened by the idea of a lizard attacking a deer, you are not alone. Many of us have a fear of reptiles, perhaps inherited from our forebears who lived more directly in the path of poisonous snakes. Medieval storytellers fanned our fears by conjuring images of fire-breathing dragons that look surprisingly like a fully alert Komodo monitor. Could it be that these stories still play to our unconscious, biasing our attitudes toward these scaly predators? Whatever the reason, genetic or cultural, monitors hold a fearful fascination for us.

The villagers who live on the Lesser Sundra Islands in Indonesia where Komodo monitors roam have reason to be cautious, however. The milking goats that they keep tethered to their houses are a great temptation to the keen-nosed monitors. Sensing easy prey, the lizards will attack these goats in broad daylight, even when surrounded by barking dogs and screaming villagers. Some monitors are shot during these scenes, but that is not, surprisingly, the greatest threat to their continued existence. The real pall hanging over the Komodo monitor (and many other species) is the fact that their islands (especially the island of Komodo) are rich in oil and minerals. It is only a matter of time before exploration and drilling begin to whittle away the lizard's natural habitat. A Komodo Island without lizards, however, would be a very different place.

There are a number of ways that Komodo monitors keep the island ecology on track. When monitors hunt, for instance, they tend to pick on the slow, weak, or sick members of the population, thereby "weeding out" and strengthening the genetic character of the prey population.

Populations of predators on the island are also indebted to the monitor for providing them with nearly free meals. It seems that wounded animals that manage to get away from monitors invariably become ill with an infection caused by the monitor's bite. This makes them more susceptible to a second attack, either by monitors or by other predators.

Finally, monitors are important simply because they are the only carrion feeders on Komodo Island. Cleansing the island of dead animals is an ecological service that is vital to Komodo's health. If the monitor were to lapse into extinction on Komodo, the few monitors that zoos now exhibit would suddenly be considered more than bizarre and fascinating. They would be worth their weight in ecological gold.

One animal's kill often becomes a communal table. Over the next few days, monitors from near and far gravitate toward the scent and converge on the carcass. As many as 17 may feed in a day, though not all at the same time. It's a tense scene for many reasons, not the least of which is the fact that small monitors are in danger of becoming prey themselves—taken by the claws and jaws of the larger members of their own species.

DRINKING

Monitors drink the way snakes do: they immerse their head up to the eyes and suck up water. They then raise their head and let gravity take the water from throat to stomach. Monitors will also drink when completely submerged.

DISGORGING PELLETS

Though monitors can swallow almost all of their prey, they can't quite digest it all. Instead, they cough up pellets composed of hair, hoofs, teeth, and undigested portions of bones. When a monitor senses a pellet coming up, it lifts itself off the ground on all four legs, arches its neck, expands its throat, and opens wide its mouth. Finally, it sucks in its belly, shakes its head violently from side to side, and coughs, sending the pellet whizzing quite a distance through the air. Be on the lookout for these large projectiles of unwanted material (much larger and longer than fecal pellets) on the floor of your zoo's exhibit.

WARMING UP *These cold-blooded lizards spread out to soak up the sun after a chilly night.*

WARMING UP AND COOLING DOWN

Because they can't regulate their own body temperature the way mammals can, reptiles must use external warming and cooling devices such as sun and shade. Their daily cycles revolve around these temperature-regulating chores. Awake before dawn, monitors are usually not active until after 9 A.M., taking a full 3 hours of sunbathing to warm up to speed. To expose a maximum amount of body surface to the rays, the groggy lizard will lie spread-eagle, belly against the ground. In cloudy weather, it may have to lie out even longer. **Basking** is also a first priority after a big meal; without the heat needed for digestion, food could rot in their stomach and kill them.

Later in the afternoon, the lizards **seek shade** to avoid the equally dangerous results of overheating. With their upper body and tail lifted off the hot ground, they **pant** with open mouth, drawing cool air over their moist tissues to speed evaporation. When temperatures dip at night, they seek the cover of thickets or, more frequently, they **dig their own bur-**

rows. These retreats are usually moist as well as mild, insulating the lizards from shifts in climate.

DIGGING A SHELTER

As the long, sharp claws of the Komodo monitor tear into the sunbaked soil, its powerful front legs shoot a loose plume backward from the hole. Monitors have been known to dig 30 feet into a bank, but most stop long before that. Since they curl up into a U with their head and tail at the entrance, they usually dig only 4- to 5-foot burrows. Their favorite burrowing spots are in clay banks along creeks and in open hillsides at the junction of a valley forest and a higher savanna. If they don't dig their own, monitors can borrow the holes made by rodents, civet cats, wild boars, porcupines, or obliging zoo exhibit designers. They also take refuge in natural cavities, overhanging vegetation, hollow trees, and rock crevices. The air inside these getaways is usually more humid than the air outside, helping to keep their body moist during the dry season.

SCRAPING OUT A FORM

A form is a resting spot on the ground where monitors wait for prey, digest their meals, and bask in the sun. Before settling down, they carefully clear the vegetation and leaf litter away. Try looking for these tamped-down areas in the Komodo exhibit at your zoo, or watch for the monitors scratching and rearranging the landscape into new forms.

ROLLING IN SMELLY STUFF

Another delightful monitor habit is practiced exclusively by young lizards. Instead of eating the intestinal contents at a kill, the small fry roll and writhe and press their bodies into it, presumably rubbing the scent deep into their scales. Researchers have a couple of notions as to why young lizards would want so desperately to smell of prey guts. It may be that they are masking their own scent so as not to be cannibalized by the adults. The scent may also help the lizard reorient when something has chased it off a kill; it can simply follow its own scent trail back to the carrion.

CONFLICT BEHAVIOR

Reptiles would make great poker players. Cool and collected, they seem to possess an ancient wisdom picked up in the age of dinosaurs. Nevertheless, if you know what to look for, you can tell when a monitor is stressed out, usually because of some threat to its safety. For smaller monitors, the most immediate threat is the mere presence of a larger monitor that may attack or even cannibalize it. For this reason, a small monitor's first line of defense is to keep to itself.

Scent Marking

Most of the time, smaller monitors can avoid contact with larger monitors by reading the scent markings left around the resident's territory. The scents come packaged in fecal pellets left at the intersections of trails and other conspicuous places. When a

How Komodo Monitors Interact

small monitor encounters a fecal pellet, it checks it out thoroughly, circling around it and tasting it with its tongue for as long as 10 minutes. The information encapsulated in the pellet may include the sex, age, and breeding condition of the depositor, as well as how long ago the animal left its mark. A fresh pellet may tell the stranger, "Resident nearby, this area is closed." An older pellet may signal, "Proceed with caution," while a very old mark may mean, "Go ahead."

Appeasement

If a monitor wants to partake of a kill, however, it must shrug off its reluctance and stand shoulder to shoulder with other monitors, even ones much larger than itself. Like dogs and wolves, monitors have a clear-cut ranking system (usually based on size), and small monitors are therefore well-versed in how to appease the larger lizards. Before daring to come near, they traverse the outskirts of the circle with a stately **ritualized walk** that shows that they mean no harm and know their place. With a slow, stiff-legged, stereotyped gait, they throw their body from side to side with exaggerated undulations. They compress their torso laterally, arch their spinal

RITUALIZED WALK   *At a communal feast, smaller monitors defer to their elders by walking slowly and stiffly outside the circle, showing that they know their place and will wait their turn.*

column, and hold their tail up off the ground and straight out behind them. With neck arched and throat inflated, they lower or even cock their head as if tipping their hat to the dominant. Because it's important to hide weaponry, they keep their mouth closed and dare not hiss.

These appeasement gestures communicate humbleness and the promise that the smaller monitors will wait until the larger one has had its fill of the choicest meat. When the subordinates finally enter the circle, they are nervously attentive to the dominant's wishes. The slightest move will cause them to bow out of the way and let the dominant pass. Monitors may perform these same appeasement gestures when confronted by humans. Watch for them when the keeper approaches their exhibit.

When Komodo monitors are attacked or handled roughly, they may become so afraid that they **disgorge** their stomach or intestinal contents. This happens most often in the few days after a big meal, when they need a way to slim down before fighting or fleeing.

Tail bowing

Gape

THREAT   *A dominant monitor inflates its profile and keeps its tail bowed and ready to strike. A simple gape posts an ominous warning to would-be enemies.*

Threat

Monitors, like other animals, showcase their weapons when they want to threaten an enemy. A threat display usually consists of **gaping** (opening the mouth to expose the rack of teeth), and **tail bowing** (arching the tail back from the offender, as if "winding up" for a strike). A threatening monitor will also stand **broadside** to its opponent, expand and arch its neck, raise its back, and jack up its body—all to appear as large and imposing as possible. To heighten the effect, it may quiver and **lash** its powerful tail from side to side, spout a bit of foam at the mouth, and let loose a malicious-sounding **hiss**. All the while it keeps its eyes riveted on its rival, peering from underneath bony projections on its lowered head.

Attacking

If the opponent doesn't retreat or respond with appeasement behavior, the dominant monitor must defend its social standing with **tail lashing**, **lunging**, or **biting**. You can tell the dominant because it lowers its head and curves its neck in an S shape. Occasionally, fighting monitors will get up on their hind legs (**upright stance**) and

UPRIGHT STANCE   *With forearms locked around each other, fighting monitors stagger back and forth, trying to bite.*

grab each other's forearms, staggering back and forth and trying to bite each other. These bites can produce severe wounds, capable of killing either of the participants. Fights end when one of the monitors symbolically **mounts** or **scratches** the other, as if in a courtship situation, or when the loser simply retreats.

SEXUAL BEHAVIOR

Monitors do more than fight and grovel at feeding aggregations; they also meet and woo their mates. To soften the aggressiveness that monitors feel toward strangers, they perform courtship displays at every meeting even when it's not breeding season. This way, by the time mating season dawns, males and females may be comfortable enough to let down their guard for a few moments. This slow building of trust is important, especially in organisms that are not only capable of injuring, but also of cannibalizing, one another.

Courtship

Courtship begins when the male places his tongue on certain areas of the female's body: the point where the hind limbs join the body, the area just in front of the ears, and the side of the face. Researchers believe hormones in these areas may tell the male whether the female is ready for mating. Besides **tongue touching**, the male may also **nudge** the female with his snout and **rub** his chin on the sides and top of her body and neck. In turn, she may threaten him, or she may run, causing the male to **follow** close behind. When he catches up, he may **bite her neck** and **scratch** her with his claws to quiet her.

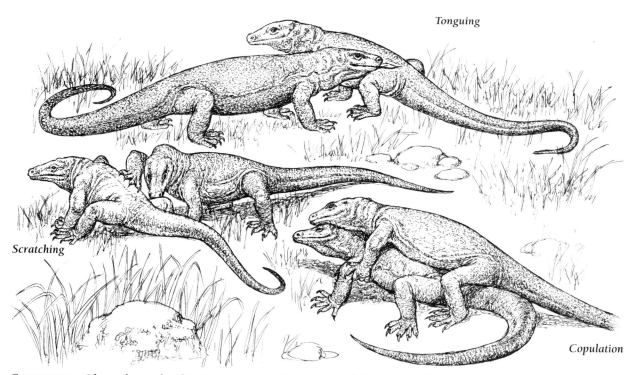

Tonguing

Scratching

Copulation

COURTSHIP *The male touches his tongue to scent-laden parts of the female's skin. He chases her and then scratches her back to quiet her. Since monitors could easily kill one another, the final mounting is a leap of faith.*

Copulation

Even after months of building trust, the act of copulation is still an iffy proposition. Because female monitors are more apt than males to use their teeth and tails in an attack, the male must clearly signal his intentions to avoid triggering an aggressive response. When it looks as if it's finally safe, he mounts quickly, biting the female's neck to anchor himself in place so when he tucks his tail beneath her, their cloacas will line up.

PARENTING BEHAVIOR

Egg Laying

Zookeepers report that female monitors become restless and begin breathing faster the day before laying their eggs. In zoos, you'll see them lay their eggs on the ground; with each delivery, they straighten their legs, lift their tail, and quiver. In the wild, they usually place their eggs in an underground nest, perhaps to protect them from the attentions of predators, including other monitors.

Life of a Young Lizard

Once hatched, the young monitors are literally on their own. They remain in trees most of their first year, feeding on geckos and avoiding larger monitors. Not until they are 3 feet long do they descend to start scavenging and eventually hunting. The hatchlings are instinctively nervous around large monitors and are quick to perform life-saving appeasement gestures.

CLIMBING   *Monitors stay in the branches during most of their first year to avoid dangerous run-ins with older monitors.*

KOMODO MONITOR BEHAVIORS TO LOOK FOR AT THE ZOO OR IN THE WILD

BASIC BEHAVIORS

LOCOMOTION
- ☐ running
- ☐ walking upright
- ☐ climbing
- ☐ swimming

- ☐ diving

FEEDING
- ☐ lying in wait
- ☐ alert posture
- ☐ stalking

- ☐ lunging
- ☐ biting
- ☐ DRINKING
- ☐ DISGORGING PELLETS

- ☐ WARMING UP AND COOLING DOWN
- ☐ basking
- ☐ seeking shade
- ☐ panting
- ☐ digging a burrow

- ☐ DIGGING A SHELTER
- ☐ SCRAPING OUT A FORM
- ☐ ROLLING IN SMELLY STUFF

SOCIAL BEHAVIORS

CONFLICT BEHAVIOR
- ☐ *Scent Marking Appeasement*
- ☐ ritualized walking
- ☐ disgorging

Threat
- ☐ gaping
- ☐ tail bowing
- ☐ broadside display
- ☐ tail lashing
- ☐ hissing

Attacking
- ☐ tail lashing
- ☐ lunging
- ☐ biting
- ☐ upright stance
- ☐ mounting
- ☐ scratching

SEXUAL BEHAVIOR
Courtship
- ☐ tongue touching
- ☐ nudging
- ☐ rubbing
- ☐ following

- ☐ neck biting
- ☐ scratching
- ☐ *Copulation*

PARENTING BEHAVIOR
- ☐ *Egg Laying*

WARM OCEANS

WALKING ON WATER *In the wild, as well as in shows, dolphins will build up steam and "walk" on their flukes.*

BOTTLENOSE DOLPHIN

What is it that draws us so hypnotically to dolphins? Is it their Mona Lisa smile? Their athletic grace? Their penchant for keeping sailors company and rescuing human swimmers? Whatever the reason, we're not the first generation to have studied and been smitten by dolphins. Ancient peoples revered these sea mammals as intelligent beings that could bring their messages to the gods. Even today, people believe that dolphins are the "mind in the waters," possessing an intelligence and spirituality that equal—or even surpass—our own. Proponents point to evidence such as brain size (larger than ours), the ability of captive dolphins to learn dozens of commands (complete with syntax), their ability to mimic sounds, and their fluency in "dolphinese," a complicated repertoire of whistles, clicks, squawks, and clacks.

Critics of the dolphin intelligence theory argue that the dolphin brain is not really all that big, if you consider the size of their body. Besides, they say, what looks to us like dolphins reaching out to humans may be nothing more than dolphins exhibiting natural behaviors that have nothing to do with us. When dolphins tag along next to our boats, for instance, they may be seeking no more than a fast ride on a bow wave, similar to the one they take in a whale's wake. In the same way, lifting drowning humans to the surface may come not from "kindness," but from the dolphin's automatic tendency to buoy up the bodies of ailing dolphin comrades. What about dolphins that drive schools of fish into anglers' nets or onto their shores? This "cooperation"

VITAL STATS

ORDER: Cetacea

FAMILY: Delphinidae

SCIENTIFIC NAME: *Tursiops truncatus*

HABITAT: Temperate and tropical oceans, mainly coastal waters, but also in bays and lagoons; sometimes ascends large rivers

SIZE: Length, 7–13 ft

WEIGHT: 330–440 lb   Maximum 1,430 lb

MAXIMUM AGE: 42 years in captivity

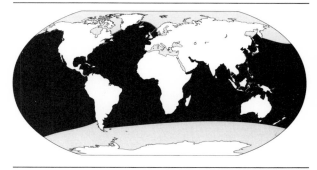

may be nothing more than the dolphin's inborn tendency to herd prey against shoals or beaches, where they're easier to catch. Actually, say skeptics, it's humans who have adjusted their routines to take advantage of the dolphin's natural fishing prowess, not the other way around.

Nevertheless, some things are hard for the critics to explain. Take the beach in western Australia where a community of dolphins nearly strands itself daily for the pleasure of greeting humans. Granted, the humans throw in fish, but more surprisingly, the dolphins often ignore the offerings and instead bring their own gifts—fish and seaweed—to give to the swimmers. Closer to home, a restaurant in Florida boasts a resident dolphin that has lived in the restaurant's enclosure for 15 years even though it can easily get back to sea, where it can see other dolphins swimming free.

Contact with humans *At Monkey Mia Beach on the west coast of Australia, a herd of dolphins regularly swims into the shallows to greet swimmers.*

In a world where so many wild animals flee from us, dolphins charm us because they seem to be more attracted to us than afraid of us. From a survival standpoint, dolphins never needed to evolve a fear of humans because we were not a part of their high-seas habitat. Today, dolphins that get used to swimming with divers or trainers exhibit gregarious behavior, much as they do with members of their own species.

Many evolutionary moons ago, dolphins became gregarious as a way of protecting and providing for themselves. Predators such as sharks and killer whales found it harder to attack dolphins that traveled in a tightly clustered group. Dolphins that pooled their senses could also scout out a broader area and find more fish than those working alone. Once a herd member sighted a school, they could all help crowd the fish into the shallows or encircle them for easy eating. Group living proved to be such a boon to dolphins that natural selection favored any behavior that fostered harmony. Communication is the cornerstone of a society, and as you'll see by watching and listening to the dolphins at your zoo, these skills are finely honed in dolphin culture.

The sociable dolphin has as many as 40 different sounds that it uses to communicate. While some scientists are scrambling to break the code of dolphinese, others are hard at work trying to teach dolphins to mimic human sounds. The quest to break the language barrier between humans and dolphins has lofty goals. In their wildest dreams, some researchers imagine a day when we can communicate with dolphins and perhaps learn from them about the history of the earth, the wonders of ocean depths, and so on. It's certainly something to ponder as you listen to zoo dolphins bark, squawk, and whistle!

LOCOMOTION

Dolphins evolved from four-legged animals that used to live along the edge of the sea. As they spent more time offshore, they evolved ever more seaworthy adaptations. Their body became streamlined—narrowing at the front and back like a submarine—and their skin became smooth and rubbery, with no hiding places for drag-causing turbulence. Their forelimbs shortened until only hands protruded from the body, and the skin around their five fingers merged to form flippers. The hind limbs completely disappeared, except for small vestige bones deep under the belly blubber. In their place, powerful tail flukes developed as the primary propelling force in **swimming**.

Here's how all the parts work together. When the dolphin lifts its tail, a swirling vortex forms at the trailing edge of the flukes, and the low pressure draws water over the dolphin's head and back, pushing the body both forward and downward. This sweeping flow of water prevents turbulence from forming along the dolphin's body, thereby cheating the laws of physics and allowing the dolphin to move through the water faster than a body normally would. To further "grease" their passage, dolphins produce copious strings of mucus from their eyes and shed a layer of oily skin cells that may cut down on drag. Jetting along at 18 miles per hour, dolphins use their flippers like hydroplanes for balance.

Porpoising

Formation swimming

Calf care

SWIMMING FORMATIONS *When traveling, dolphins swim in a stacked pack and arc out from the water at the same time to take a breath. Adults circle the young to form a safe playpen.*

Blessed with grace as well as speed, dolphins are natural gymnasts. They are quite capable of performing most of the behaviors you see in aquarium shows without being taught. Next time you are watching dolphins between shows or in the wild, look for **somersaulting**, **backflipping**, **breaching**, or even **"walking"** atop the water on their tail flukes. Some of these behaviors are part of exuberant play, while others are simply business as usual. For instance, **porpoising**, or synchronized leaping, is often seen in the wild as pods of dolphins travel, swim, and leave the water to breathe together. **Formation swimming** is another wild characteristic.

For the most part, dolphins stick to within 160 feet of the surface, taking advantage of surface waves to help them execute their gymnastic feats. A joyous **leap**, which you'll no doubt see at the zoo, can take them up to 20 feet in the air.

BREATHING

At some point in the dolphin's evolutionary journey, its two nostrils fused into one and migrated to the top of its head to form a blowhole. When the dolphin dips beneath the surface, it closes this hole with powerful muscles that keep the air in and the water out. When it rises again, it opens the hole, exhales with a blast, then refills for the next dive. If need be, the dolphin can survive on this portable air supply for as long as 7 minutes, mainly because of adaptations that store oxygen in the blood and muscles as well as in the lungs. Also, because its heart rate slows down as it dives, a dolphin can stretch the oxygen even further. When it surfaces for a breath (four or five times a minute), its pulse speeds back up to pump oxygen throughout its body.

FEEDING

When it comes to dining, dolphins are easy to please, and this adaptability has enabled them to survive in oceans throughout the world. In addition to their favorites—fish and squid—they'll also eat eels, worms, and even hermit crabs, gulping down as much as 33 pounds a day in captivity!

In one style of wild hunting, dolphins work as a team to herd fish into a tight cluster and then **encircle** them, eating away at the edges and bottom of the school. In another style, dolphins speed along the edge of the beach, **flinging** fish shoreward with their bow wave. Or they will crowd fish toward shore or against a coral reef, **trapping** them in a constricted area so they will be easier to catch. Even flying fish can't elude a leaping dolphin. If a fish proves too frisky, a dolphin simply stuns it with its tail flukes, **whacking** it as much as 30 feet through the air and then scooping it calmly off the surface.

Dolphins have also learned to feast on "trash" fish that commercial trawlers throw overboard and on the fish that come to investigate actual garbage dumped from barges.

RESTING

Dolphins don't sleep the way we do, but they are able to rest in the water without drowning by closing down only one-half of their brain at a time. The other half controls movement and breathing and keeps one eye open and alert to predators. In total, each half rests for about 3 or 4 hours out of 24.

In captivity, dolphins do most of their resting at night, though they will also take siestas after feeding during the day. Watch for them resting with their trunks horizontal and near the surface of the water, usually facing the current in the tank. Every so often a resting animal will beat a few slow strokes of its tail to bring it to the surface for a breath.

CRAVING TOUCH

Dolphins have such an affinity for touch that some trainers in zoos use pats on the snout instead of food to reward their dolphins during a show. In fact, this craving for tactile stimulation may be part of what attracts dolphins to human divers and swimmers. Knowing this, zoo personnel often get in the tank with their dolphins as a way of training, caring for them, or simply keeping them entertained.

In the absence of a diver, the dolphins often find something in the tank to rub against, such as a ladder, a brush attached to the floor, or a water jet. Touch becomes even more important in communication, as you will see in the next section.

HOW BOTTLENOSE DOLPHINS INTERACT

In the wild, dolphins hang together most commonly in groups of 2 to 15 animals. Their primary social groups, called pods, consist of adult females and their offspring that travel together for days or even weeks at a time. Sometimes several pods will swim or even hunt together for a few minutes or hours before dispersing again. As the water gets deeper or more open, the size of these herds tends to increase. Within this fluid and flexible community, dolphins spend most of their time with members of their own sex and age. Adult males are often found roaming the periphery of their home range, while females stick to core areas.

Mothers and calves form some of the tightest bonds in the community, swimming inseparably for up to 5 years. Daughters return to their mother's pod to raise

HUMANS AND DOLPHINS

For centuries dolphins have treated humans with tolerance and even trust. Unfortunately, we haven't always returned the favor. Off the Japanese island of Iki, for instance, thousands of dolphins are slaughtered each year by people who claim they are losing some of their fishing catch to these "competitors." In other parts of the world, dolphins are routinely killed for human consumption or used for crab bait. In the United States, dolphins risk losing their lives in military operations. The Navy trains dolphin draftees to retrieve mines that are too dangerous for human divers, a program that has ignited the ire of dolphin admirers.

As the tuna industry has learned, dolphin ad-mirers can be quite a force to reckon with. Until recently, dolphins died by the thousands off the sides of tuna fishing boats. Because tuna and dolphins naturally swim together, dolphins were caught in the nets along with schools of tuna. Most drowned before the crews released them, and some crews didn't even attempt releases. A recent letter-writing campaign and boycott of tuna products put pressure on the tuna industries to change their fishing techniques, and now three United States tuna companies have a "dolphin-safe" label on their products. It's a testimony to what customers can do when they are truly outraged and organized.

their own offspring, often in the company of other females with calves the same age. It's thought that the grandmothers in these groups play a vital role as repositories of learned information, such as where to find the best food in various seasons.

Young sons join all-male groups when they leave their families, forming friendships that last as long as 15 years. Instead of returning home the way daughters do, they take up traveling as they mature, making the rounds from one school of females to another.

FRIENDLY BEHAVIOR

Tactile Contact

Touch plays a key role in forging and maintaining the social relationships that are central to dolphin survival. As they swim together, watch for affectionate **rubbing**, which removes not only social tensions, but also barnacles and other parasites from the dolphin's skin. One dolphin may rub its torso against the flippers of another or brush its neighbor with its flukes, lower jaw, or flanks. Both dolphins may go in for a full-length body rub or simply pat each other over and over. In the **pat-a-cake** maneuver, which some trainers have adapted for shows, two dolphins swim belly to belly, rubbing their flippers together. Also watch for **beak-genital propulsion**, in which one dolphin places its beak in the other's genital slit (both males and females have them) and then gently pushes its partner around the tank.

Play

Zookeepers say that you can't keep a good dolphin down. When there are no visitors or shows to occupy them, they entertain themselves with whatever happens to be in their tank—toys, cleaning equipment, trainers, or feathers and leaves that happen to blow in. The **fetch** game (retrieving and bringing back items to a human partner) is one of their favorite impromptu acts. Dolphins may play **keep away** with one another, using whatever is at hand. **Chase** games are also common, as are friendly, competitive **races**, which start suddenly, as if someone had shouted, "Ready, set, GO."

You won't see as many threats and fights in dolphin play as you do in the play of other animals, but you will see plenty of **sexual play**. In one novel game, males use their erect penis to push or tow objects around the tank. If you watch awhile, you're bound to see others.

BEAK-GENITAL PROPULSION   *In this friendly gesture, one dolphin places its beak inside the other's genital slit and pushes it around the tank. Courting dolphins also play this game.*

Keeping in Contact

Dolphin sounds fall into three basic types: clicks, whistles, and the miscellaneous category of quacks, squawks, blats, and barks. Clicks are used for echolocating, but when dolphins want to communicate, they use whistles and sounds from the miscellaneous category. Each dolphin may have an identifying **signature whistle**, and other dolphins can name that tune in less than half a second. They can also imitate one another's signature whistles, enabling them, perhaps, to call one another's "names" in the chaos of a large herd. When dolphins whistle in unison, they are a flawless choir, perfectly matching one another's changes in pitch and amplitude. Most of the time, you can tell when dolphins are whistling by looking for a steady stream of air bubbles coming up from their blowhole. This doesn't always hold true, however; if a whistler wants to communicate incognito, it can get the word out without letting any air escape.

Dolphins also communicate with **nonvocal sounds** such as body parts slapping the water, jaw claps, bubble emissions, and blasts. Many are associated with frustration or aggression, but sometimes these nonvocal sounds simply help dolphins keep track of one another. As a porpoising group exits the water to breathe, for instance, their simultaneous blasts of white noise help them "count heads" and stay together. A large bubble of air rising from a blowhole may signal inquisitiveness, much like a puzzled look on a person's face.

CONTACT CALLS *After the birth, the mother and calf exchange signature whistles that will help them stay in contact.*

Group Rescue

When dolphins are sick or in pain, they issue a two-part **distress whistle**. Sometimes, other dolphins will stop their own calling as if to listen. Within seconds, they locate the victim, rush beneath it, and **lift** it to the surface so it can breathe. Drowning swimmers who have been lifted to safety report that dolphinian heroes are careful, efficient, and persistent. Dolphins have been known to lift an ailing comrade for each and every breath for days on end and then provide intermittent support for weeks.

GROUP RESCUE *When a dolphin is having difficulty swimming, companions will often lift it to the surface so it can breathe.*

Group Alarm

When frightened by a strange object in their vicinity, such as a human, a boat, a shark, or even a ball in the zoo tank, dolphins will immediately **fall silent**. The sudden hush in the normally noisy tank is as powerful as a scream. The dolphins then **bunch together**, suspiciously eying the cause of concern while cruising past it. Eventually, if all seems well, the curious dolphins dart in to **investigate** with sonar, and then dart away. Finally, they grow brave enough to touch or even manipulate the object before handing down their verdict. If the object is harmless they resume normal whistling, but if they sense danger they broadcast an **alarm whistle**—a high-volume, low-frequency call that carries well over long distances.

CONFLICT BEHAVIOR

Rank and File

Researchers have noticed that a kind of pecking order develops in some captive bottlenose dolphin communities. In the most common scenario, the largest male is on the top rung above all other adult males, and beneath the males come the adult females, juvenile males, and finally the infants. In the same way, the largest, oldest female is usually dominant over younger, smaller females.

The dominant animal is easy to spot because it has the run of the tank; it may jump higher in the show, be more aggressive, or enjoy a certain immunity when it comes to being pestered by the other dolphins. Dominance relationships among dolphins are not carved in stone, however, so be on the lookout for fluctuations in status. The best way to detect changes is to watch who threatens whom.

Threat

One of the dolphin's most subtle threat gestures is a **hard stare**, which usually prompts the subordinate to make way. If the dominant animal needs to shake a stubborn competitor from a coveted position (right in front of the trainer's bucket of fish, for instance), it simply swims above and settles down on top of it. In extreme cases, one or several dolphins may **pin** a rabble-rouser to the floor of the tank.

In the more direct **aggressive threat**, a dolphin shows its dominance by opening its mouth (thus showing teeth) or by arching its back and holding its head downward. To drive home the point, the animal may clap its jaws together, lunge at its opponent, or strike with a fluke or a beak. The mark of an aggressive gesture is its suddenness, providing a harsh contrast to the normally slow, fluid grace of dol-

THREAT *Frustrated dolphins slap the water with their flukes or with their whole body.*

phin movement. Another way a dolphin can show displeasure is to **slap the water** with its tail fluke, flipper, or whole body, making a loud sound as well as a splash.

A good time to look for threat gestures is when a dolphin gets angry or impatient with its trainer. A tail-fluke slap, which may well drench the trainer and the audience, is the animal's way of stamping its feet. It might also give a raucous, pulsed "Bronx cheer" from its blowhole to go along with the slap. Or it may vent an explosive air blast from the blowhole, triggering a spate of blowing in surrounding animals as well.

Submission

To show submission and avoid any further attack, a dolphin closes its mouth (thus hiding its teeth), faces away, and offers its vulnerable flank as if to say, "You can ram me, but I don't intend to ram you." Not to tempt fate, the fearful dolphin usually hustles out of the dominant's way.

Fighting

If you look closely at the bodies of dolphins in the wild, you'll notice scars—telltale proof that things are not always friendly in dolphin society. Dolphins may fight over an object, space, food, or even a favorite spot in the formation. Usually the provoker is a young male, and the dominant male responds with **tooth raking**, **biting**, **ramming** with the lower jaw, or **striking** with the powerful fluke. Fights are especially common when a young male "cuts in," courting a female that is already being escorted by the dominant male.

SEXUAL BEHAVIOR

Dolphins have been described as pansexual, a term that captures the essence of their freewheeling, unstructured sexuality. They are sexual at any time of year, at any time of day or night, both heterosexually and homosexually. Their erotic attractions even go beyond species lines. Human divers, for instance, must frequently deal with the sexual attentions of dolphins; while in Sarasota, Florida, a dolphin reportedly mounted a sailboat! Much of this sexual attention appears to be purely social, however, and outside of a reproductive context. Sexual gestures are also a mainstay of dolphin play, practiced with gusto by dolphins of both sexes, from the very young to the very old.

Courtship

While you'll no doubt see sexual behavior at any time of year, the main breeding season is spring through fall. During this time, watch for an adult male escorting and paying close attention to a particular female. She may be issuing broad sexual hints by swimming on her side in front of him and **presenting** her genital slit. To further arouse his interest, the female may rest her head on the male's back and **rub** slowly so her dorsal fin caresses his belly. Look again, and you may see the bull taking the lead with these same courtship moves.

An even more common entreaty to sex is **display swimming**, in which one partner circles the other, sometimes corkscrewing or swimming upside down. The dis-

DISPLAY SWIMMING *Courting dolphins swim circles around each other, flipping upside down, corkscrewing, and leaping from the water.*

play often becomes a **chase** with the partners taking turns corkscrewing, leaping, and belly-flopping back into the water with loud splashes. These excited chases are punctuated by lazy time-outs in which the dolphins float motionlessly for a while as if catching their breath. The male may hang head-down in the water displaying an erection, until the cow, often joined by others, imitates his posture.

The courting pair **caress** each other frequently, using all parts of their bodies, but concentrating on genitals, flukes, and flippers. Sometimes they cruise around the tank with the tip of one dolphin's dorsal fin inside the other's genital slit. The male may swim beneath the female in a typical **S-shaped posture**, vocalizing and taking a sonar reading of her underbelly. Also watch for **jaw clapping**, **nuzzling**, **rubbing**, **gentle tooth raking**, and **yelping**. Don't be surprised if the amorous pair start **knocking heads** together, a move which, though it looks as if it would bring only headaches, may actually bring them closer to copulation.

Copulation

If the female is not yet ready to mate, she may present her back to the anxious male or even strike him with her flippers. When the time is right, she relaxes and lets the male swim beneath her. He pushes her gently to the surface then turns belly-up, pressing his underside to hers so that their genital areas align. He clasps her with his flippers, and after a few thrusts, the animals gracefully fall away from each other to rest.

PARENTING BEHAVIOR

Birth

Gestation lasts about 12 months, which puts the birth right in the middle of next year's breeding season. Because this is the time when the males are most aggressive, females move away from the group to give birth. Usually one helper—a nonbreeding female relative—goes with her and assists in the birth and rearing of the newborn. To replicate wild conditions, zoos separate expectant mothers from males, and some put experienced mothers in the tank with first-timers so they can learn the ropes.

You'll know when the female is in labor by her exaggerated arching and tail flexing; she may even "stand on her head" with her flukes out of the water. It takes anywhere from 20 minutes to several hours for the female to push her infant out into the world. With each contraction, milk squirts from the mammaries and more of the calf appears. The final push sends the newborn out into the tank, flukes first, in a cloud of blood. As the female whirls around, the umbilical cord is torn, and the newborn takes its first trip up to the surface to breathe.

After the birth, there is usually a cacophony of calls and whistles from all the dolphins in the tank, including the continual whistling between mother and calf. The two animals probably learn to recognize each other through these sounds. Later, if the calf wanders out of sight, the female need only whistle to bring it close again.

Nursing

Bouts of touching and rubbing also help to cement the bond between mother and calf. The calf nurses for about 18 months, bumping the mother's undersides whenever it's hungry. Her teats offer a bounty of nutrients; in the first year alone, the calf gains 165 pounds and grows 2 feet.

Defending the Calf

In an ocean full of sharks and killer whales, a newborn dolphin is a real appetite teaser. Even male dolphins, which are particularly aggressive at this time, can take potshots at a vulnerable calf. As a result, the female dolphin watches her offspring hawkishly, herding it away from strange objects in the water, keeping it from beneath leaping dolphins, and defending it from the raucous attacks of the males.

Because the newborn is neither a coordinated nor a strong swimmer, it spends most of its time hovering above or to one side of its mother, being sucked along by the low-drag pressure wave caused by her swimming. In this **echelon formation**, the calf's coloring blends so beautifully with the mother's that predators may think they're seeing one animal instead of two. If the calf strays too far from this safe harbor, the female immediately retrieves it.

Communal Care

Mothers do get a break from all this vigilance as the calf gets older, thanks to the communal nature of dolphin society. Closely related sitters will often watch after youngsters while the mothers are out feeding. The young can also play together in a playpen formed by two or three females circling around them. In the relative safety of these day-care nurseries, the young learn how to interact with their peers. After 18 months or so, the mothers wean the dolphin calves and encourage them to swim on their own. Even so, each calf will maintain a close relationship with its mother for up to 6 years, giving it plenty of time to learn the traditions of the herd.

CARING FOR THE CALF *When her calf is born weak or stillborn, the mother will push it to the surface to breathe.*

ECHELON FORMATION *The calf "hitchhikes" in the slipstream caused by the mother's movement.*

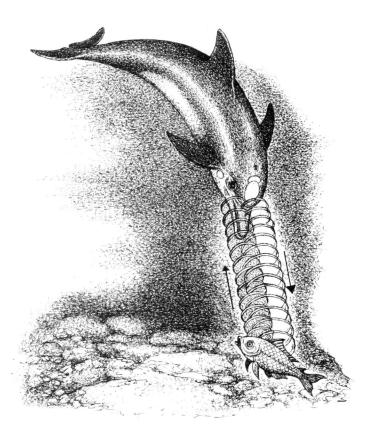

ECHOLOCATING   *To activate their sonar, dolphins focus clicks in the oil-filled melon atop their head. The sound waves bounce off the fish, return to the dolphin, then travel through the jaw bone to another oil deposit at the base of the jaw. These echoes give dolphins a three-dimensional picture of their prey.*

ECHOLOCATION: SEEING WHAT THEY HEAR

In 1946, the curator of the Miami Zoo proposed what was then an almost preposterous idea: maybe the dolphin uses its superior hearing to "see" objects in the water. Maybe all those clicks that the dolphin makes as it swims are returning to it in the form of echoes—a system that we ourselves use in sonar. Scientists have since proven that the curator was right. Because we search with only one frequency, however, our readings are rather rudimentary compared with those of dolphins. Dolphins send out a volley of frequencies, and the detailed echoes from those calls paint a three-dimensional picture so detailed that dolphins can tell a nickle from a dime even when the coins are buried in the mud. This sort of "sight" is also important when wild dolphins are tracking down prey or avoiding dangers, especially in murky water.

How do dolphins do it? Scientists believe that the calls originate in special air sacs in the dolphin's sinuses. The sounds are then focused in a sort of tunnel through the "melon," the bulbous reservoir of fats and oils atop the dolphin's head. The dolphin points and shoots a volley of sound waves at whatever it's trying to "see." The waves bounce off the target, and the returning echoes hit an acoustic window of thin bone in the dolphin's protruding lower jaw. The sound travels along the jawbones (think of the how those wraparound radios called bone phones worked) to another fat deposit at the base of the jaw, and then to the middle ear. Eventually, the sound is translated into a three-dimensional "picture" in the brain. All this happens in far less time than it takes to explain and with a precision that makes sonar engineers envious.

BOTTLENOSE DOLPHIN BEHAVIORS TO LOOK FOR AT THE ZOO OR IN THE WILD

BASIC BEHAVIORS

LOCOMOTION
- ☐ swimming
- ☐ somersaulting
- ☐ backflipping
- ☐ breaching
- ☐ "walking"
- ☐ porpoising
- ☐ formation swimming
- ☐ leaping
- ☐ **BREATHING**

FEEDING
- ☐ encircling
- ☐ flinging
- ☐ trapping
- ☐ whacking
- ☐ **RESTING**
- ☐ **SELF-STIMULATION**

SOCIAL BEHAVIORS

FRIENDLY BEHAVIOR
Tactile Contact
- ☐ rubbing
- ☐ pat-a-cake
- ☐ beak-genital propulsion

Play
- ☐ fetching
- ☐ keep away
- ☐ chasing
- ☐ racing
- ☐ sexual play

Keeping in Contact
- ☐ signature whistles
- ☐ nonvocal sounds

Group Rescue
- ☐ distress whistle
- ☐ group lift

Group Alarm
- ☐ falling silent
- ☐ clustering
- ☐ investigating
- ☐ alarm whistle

CONFLICT BEHAVIOR
Threat
- ☐ hard stare
- ☐ pinning
- ☐ aggressive threat
- ☐ slapping the water
- ☐ *Submission*

Fighting
- ☐ tooth raking
- ☐ biting
- ☐ ramming
- ☐ striking

SEXUAL BEHAVIOR
Courtship
- ☐ presenting
- ☐ rubbing
- ☐ display swimming
- ☐ chasing
- ☐ caressing
- ☐ S-shaped posture
- ☐ jaw clapping
- ☐ nuzzling
- ☐ rubbing
- ☐ gentle tooth raking

- ☐ yelping
- ☐ knocking heads
- ☐ *Copulation*

PARENTING BEHAVIOR
- ☐ *Birth*
- ☐ *Nursing*
- *Defending the Calf*
- ☐ echelon formation
- ☐ *Communal Care*

Body surfing

Porpoising

LOCOMOTION   *Sea lions are the original body surfers. When it's time for a breath of fresh air, they leap out of the water like dolphins.*

CALIFORNIA SEA LION

Sea lions are the hambones of circus fame, spinning balls on their nose, twirling in tutus, and patiently performing other rather demeaning acts that were once popular with audiences. Today, though some circuses, zoos, and aquariums still dress "Sparky" up, many more are encouraging sea lions to show off their natural talents in a more dignified manner.

Sea lions don't need coaxing to show off, however. Off the coast of California or Mexico, you can see untrained sea lions diving off rocks, body surfing, leaping, playing chase, and tossing bits of kelp just for the fun of it. One reason sea lions have so much time to play is because they are so proficient in their habitat—able to outdistance predators, maneuver on a dime, and overtake an octopus in a heartbeat.

Their torpedo-shaped body comes complete with a flexible spinal column and strong neck and chest muscles to power their sinuous movements. Like penguins, they use their broad front flippers to row themselves forward, while their hind flippers steer the course. In the interest of streamlining, these limbs have been withdrawn into their bodies over evolutionary time until little more than hands and feet remain. Their ears have been reduced to tiny furls, and even their genitals and mammaries are tucked inside so as not to interrupt the smooth flow of their bodies through the water.

Snorkelers and divers who have cavorted with sea lions in their natural habitat can attest to both their wing-footed prowess and their playful natures. Many a first-time diver has been drafted into a game of chicken, in which the sea

VITAL STATS

ORDER: Pinnipedia

FAMILY: Otariidae

SCIENTIFIC NAME: *Zalophus californianus*

HABITAT: Ocean and rocky or sandy beaches

SIZE: Male length, 6.6–8.4 ft   Female length, 5–6.6 ft

WEIGHT: Male, 440–660 lb   Female, 110–220 lb

MAXIMUM AGE: 20 or more years

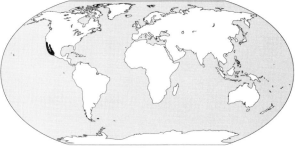

lion dive-bombs the wary diver and then veers away at the last moment, teasingly blowing bubbles or snapping its teeth. The shows you will see in modern zoos incorporate many of these natural moves, serving not only to instruct spectators, but also to keep the sea lion's spirits high.

Still, some of the best shows at sea lion exhibits come between performances. The noisy free-for-alls you'll see on the artificial cliffs are true to natural form. On land, sea lions like to be almost on top of one another, preferably touching. When it comes to defending breeding territories or offspring, however, the same sea lions can become rather antisocial, engaging in elaborate threat ceremonies and occasional fights. By understanding a few of these social behaviors, you'll be better able to appreciate the flurry of barking, chasing, fighting, and comradery you see before you.

BASIC CALIFORNIA SEA LION BEHAVIORS TO WATCH FOR

LOCOMOTION

You can tell true seals from eared seals (sea lions and fur seals) by watching how they propel themselves in the water. True seals fishtail forward with their hind flippers, while sea lions and fur seals row themselves forward with sweeping strokes of their front flippers. They can **swim** at speeds up to 20 miles per hour, and when fleeing, they can increase their speed by **porpoising**, leaping clear of the water in a graceful arc. When traveling together, a string of young sea lions might porpoise one after the other, looking for all the world like the Loch Ness monster roller-coastering through the deep. Sea lions are also the original **body surfers**, using the power of an incoming wave to ferry them to shore or to buoy them up to a rocky ledge.

Once on terra firma, the amphibious sea lion switches gears and starts to **walk**, lifting its body off the ground and moving in a diagonal stride the way other four-legged mammals do. Since only the lower portions of its limbs show (the rest are tucked in for streamlining), a sea lion really walks on its "wrists" and "heels." With each step, it rotates its shoulders and hips and swings its head from side to side.

When they need to scoot, sea lions can also **gallop**, moving first their hind flippers and then their front flippers together, and bobbing their heads up and down. Top speed on a level surface is close to 15 miles per hour, which is almost as fast as they can swim. On a smooth surface or in shallow water, you may see sea lions **stride**; they lift their hind flippers off the ground, rest their weight on their abdomen, and drag themselves forward with long steps of their forelimbs. Territorial bulls (males) stride like this when rushing at intruders.

WALKING   *Using the hands and feet that extend from their body, sea lions walk in a diagonal stride like most other four-legged mammals.*

FEEDING

Catching dinner usually involves a fast, zigzagging **chase** through the deep. The sea lion tracks its quarry—mainly squid, but also anchovy, herring, Pacific whiting, rockfish, hake, salmon, and octopus—by sight and may also use its touch-sensitive whiskers to feel the vibrations created by a swimming form. Some researchers think that sea lions might also use the echoes of their own calls (sonar) to locate prey, but this has not yet been proven.

Also unproven are the claims that sea lions seriously reduce stocks of salmon and other commercial fish. Even with current protective laws, many sea lions are shot by their "competitors" in the fishing industry. While it's true that sea lions in the Rogue River in Oregon, for instance, do eat some salmon as they head upstream, the catch is hardly enough to cripple an industry. For the most part, sea lions feed on noncommercial fish and squid.

GROOMING

The long, rubbery flaps at the ends of the sea lion's hind flippers can be flexed back to expose three toes, complete with claws. The sea lion uses these claws to **scratch** its skin in doglike fashion. By flexing its spine, the sea lion can bend backwards far enough to reach almost everywhere on the front two-thirds of its body. It may also use its foreflipper to **rub** itself, while stretching its head and neck straight up and balancing on the other foreflipper. For even more vigorous rubbing, it may sidle up to a rock, squirm on the sand, or flex its body against another animal. Finally, be sure to watch for **whisker stropping**—rubbing one side of the face and then the other against a rock—when a sea lion exits from the water.

WARMING AND COOLING

The sea lion is a warm-blooded mammal caught between two worlds. In the sea it struggles to keep warm, while on land it often suffers from too much heat. The adaptations that address these dilemmas are worth checking out.

Warm-blooded mammals lose heat 25 times faster in water than they do in air. To cut their losses, sea lions have evolved rather large bodies with less surface area (compared with their body mass) for the heat to leak through. To slow heat leaks even further, their bodies are wrapped in a 4-inch-thick sleeve of blubber. Their flippers are not as generously padded, however, which is why you'll often see flippers sticking up above the water's surface when sea lions are resting. Sea lions conserving heat in this way are said to be **jugging**, since the arc formed by their one front

SCRATCHING *Sea lions use the claws beneath the flexible tips of their flippers to relieve an itch.*

and two hind flippers looks like a jug handle. They can also increase or decrease blood flow to these extremities as a way of conserving heat or keeping cool.

The same blubber that protects like a wetsuit under the water can be a liability on land. To keep cool in the scorching sun, sea lions must periodically **immerse** themselves in water, a need that has shaped much of their behavior. When sea lions come ashore to breed, for instance, they crowd together on coastlines, sticking as close to the sea as possible. Territorial bulls vie for choice oceanfront property so they can go for a cool dip without really leaving their posts. Females that feel the heat must also resort to the sea, bringing their pups down with them when they go. On very hot days, sea lions at a rookery may wait until nighttime to become active.

PLAY

Most of what you'll see at the sea lion exhibit is play, an activity that these aquabats never seem to tire of. The young sea lions are especially active, keeping themselves occupied by shaking and tossing bits of vegetation, gnawing at rocks or banks, surfing in the incoming water jets, and leaping entirely out of the water. The play gets even better when they have a companion.

SLEEPING

When it's time to sleep, sea lions leave the comfort of the water and wriggle up onto rocks. Although it looks terribly uncomfortable, they often sleep propped up on

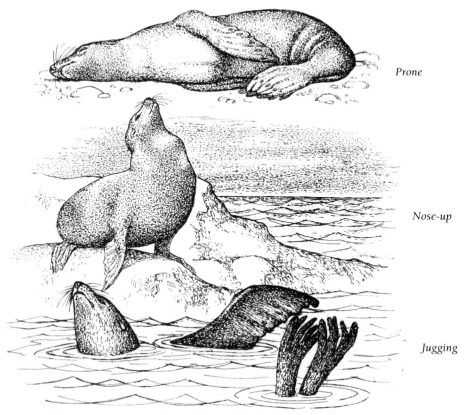

Prone

Nose-up

Jugging

RESTING POSITIONS   *Sea lions take their Z's where they find them.*

their front flippers, with their head thrown back and their nose pointed straight up in the air. They will also sleep on their bellies with their flippers tucked beneath them, or on their sides with their front flippers close to their chest and their hind flippers spread out behind. While resting, the whiskers are folded back, like antennae out of commission. Sea lions can also snooze in the water. If you see a sea lion lying on its side with a jug handle of flippers held above the water, look carefully for signs of dreaming.

FRIENDLY BEHAVIOR

Physical Contact

When they are not defending territories or protecting their young, sea lions can't seem to get enough of one another. They pack themselves so closely together that any incoming or outgoing members have to crawl over a heap of bodies. Almost all sea lion encounters include some form of "face feeling," in which whiskers, eyes, and noses collect information. Watch for this friendly **nuzzling** and **nose rubbing** between pairs or groups of individuals, especially during the nonterritorial seasons of fall and winter.

Social Play

Sea lions also use touch in their social play. The pups ride breakers together, neck wrestle, chase, nip, and playfully mount and lunge at one another. Juveniles will play king-of-the-mountain games, pushing and shoving to gain the summit of a boulder or ledge. These mock jousts look suspiciously like the aggressive and sexual moves that the sea lions will later use to hold territories and pass on their genes.

Communal Defense

One contagious form of behavior to look for is the **alarm reaction**. When frightened by a loud noise, human activity, or the alarm call of gulls, sea lions will make a frantic beeline for the water, even if it means dropping from a cliff onto a rocky beach to get there. The exodus is a group event, and within no time, the entire colony is offshore, barking in alarm. Curiosity eventually prevails, however, and the mob swims back toward shore, heads periscoped out of the water to investigate. Be on the lookout for similar stampedes at the zoo, usually when you least expect it.

NUZZLING *Sea lions exchange greetings with bristling whiskers and moist, sensitive noses.*

CONFLICT BEHAVIOR

Wild colonies off the coast of California are hotbeds of conflict from May to August, when males are setting up breeding territories and defending their turf. Females tend to hang together in small groups during this time, remaining tolerant for the most part, except when warning others to stay away from their pups.

In captivity, males are kept apart during breeding season to avoid violent conflict. You may see some of the following threat behaviors at other times, however, due to the dominant-subordinate relationships that often develop in captivity. The largest male regularly chases and bullies the smaller males, especially when something precious is at stake, such as food, prime resting spots, or access to the pool.

Male Threat

In preparation for breeding in wild colonies, each male tries to secure a territory along the ocean and keep other males from trespassing. They space themselves about every 50 feet, then do everything in their power to stay king of the beach so they can mate with females that wander through their area. Males without real estate are less likely to be able to pass on their genes; to reach a female, they must first confront the short-tempered resident male.

Keeping these suitors at bay is a demanding task, requiring such strict vigilance

Barking

Head shaking

Oblique stare

MALE THREAT   *At the boundaries of their territories, neighbors fill the air with warnings. As the tension mounts, they wag their heads back and forth faster and faster. Finally, they push up chest to chest and turn their heads to either side, confronting each other eye to eye.*

that the resident male barely eats for several weeks. He **barks** loudly and almost constantly to advertise, "I am here and I am willing to take on intruders." He even barks underwater! In addition to his vocal warfare, the male also has a number of visual threats up his sleeve.

Neighboring bulls periodically meet at the edges of their territories to reinforce the invisible fence strung between them. They **rush** toward one another, **barking** excitedly and stretching their whiskers as far ahead as they will go. Before reaching the boundary, they fall flat on their chests and go silent, putting all their energy into vigorous **head shaking** with mouths agape (**open-mouth threat**) and necks slowly snaking from side to side. In a dramatic finale, they rear up on their hind flippers, chest to chest, and turn their heads to stare at each other with one eye (an **oblique stare**). If either bull happens to cross over the boundary, they face in opposite directions and pretend to nip at each other's chest and flippers. The entire scene is as ritualized as a martial arts exercise; though they come very close, neither bull touches the other.

Female Threat

Females reveal their protective natures during pupping season. If another female gets too close, the mother first shifts restlessly around her pup, then rushes toward the intruder. Face-to-face, the two females extend their necks and twist their heads side to side, mouths open in the typical **open-mouth threat** stance. They may weave their heads together like this for some time, squealing and parrying. As tensions mount, listen for a prolonged gagging belch that is a notch more serious than squealing. A harsh, erratic growl signals the most intense form of vocal threat. More sound than fury, these contests rarely go beyond a bite or two, and even then, the wounds are not serious. Most threats end with one of the females simply withdrawing a few feet.

Fighting

Threats between males, however, can escalate into serious fights, especially during breeding season when an intruder ignores the ceremonial cues and crosses into a bull's territory. For the intruding bull, fighting may be the only way to gain access to a mate. In a series of struggles lasting several minutes, the bulls **push** each other chest to chest, delivering quick slashing **bites** to chest, flanks, and flippers. Every now and then the contestants stop and glare, each one appearing to loom larger than the other. These interludes are broken by more **lunging**, sometimes so violent that if the lungee dodges, the lunger lands flat on his face. Brief, explosive fights can also occur in the water, usually turning into a **circle chase** as each bull attempts to seize the other's hind flippers. When one bull has finally had enough, he retreats, whirling frequently to face his opponent and protect his hindquarters.

SEXUAL BEHAVIOR

Once sea lions are gathered together on shore, it pays for them to get both pupping and mating over within the same period. The birth of the pups comes first, and then there's a period of about 27 days before the females will be ready to mate. While the females are tending to their newborns, the males are busy establishing territories so they'll have their real estate ready when the females come into heat.

A Rookery Blueprint

If you could look down on a typical rookery from the air, you'd see territorial bulls spaced evenly in a line along a beach, with dense groups of females in and amongst the bulls. Bachelor bulls gather at the fringes of the colony, occasionally making forays into the rookery. If a male doesn't have a territory himself, he can sometimes borrow a female by sneaking into a territory when the resident bull is sleeping or distracted. If he works quickly, he can mount and even impregnate a cow or two before he is discovered and barked away. If he wants guaranteed access, however, he has to fight, conquer, and displace the territory holder. Even this arrangement is not permanent, though; a bull can rarely hold on to a territory for more than a few weeks before he is displaced.

Courtship

Naturally, before being ousted, males try to mate with as many females as possible. They begin by **sniffing** each female's genital region to see if she is ready to mate. The bull shuffles around her, **barking** and **head shaking** in a style reminiscent of male-male aggression. In a sexual context, however, this behavior has a tentative, searching quality; instead of leading to violent lunges, the bull merely **rubs** his whiskers against the female's body and gently **nips** at her shoulders and sides, presumably picking up scent clues. There are also visual signs of estrus, including a pink flush and slight swelling in the genital area.

HUMANS AND SEA LIONS

Killer whales and sharks have always hunted sea lions, but their harvests did little to diminish the populations. In the same way, thousands of years of subsistence hunting by native peoples barely made a dent in sea lion numbers. It was only when heavy equipment and crews were assembled in the nineteenth century to commercially harvest sea lions for oil and meat that the effects of mass killings could be felt. Today, though the Marine Mammal Protection Act of 1972 protects sea lions from this kind of slaughter, accidental death in fishing nets and deliberate death by shooting continues to take its toll. Far more insidious is the threat that pollutants such as DDT, PCB, and heavy metals will someday find their way into the sea lion's system, where they could accumulate in the blubber layer or liver.

Despite these threats, sea lions seem to be holding their own with an estimated population of 110,000 worldwide. Part of their success lies in their flexibility; if disturbed by motorboats on their rookeries, for instance, sea lions will clear out and go to another site (if a suitable one is available). This is not without cost, however. A new home means that males have to spend their hard-won energy reestablishing territorial rights. If we're not careful, continual disturbance could stress sea lions beyond their limits of tolerance, resulting in fewer births and dwindling populations. Harassment of this nature, especially toward an animal as curious and friendly toward humans as sea lions are, is an unforgivable breach of trust.

Occasionally, a female will present even more compelling evidence that she is in heat by performing the **estrous display**. She lies on her side, belly, or back in front of the bull and solicits his attention by writhing on the ground, looking up at him, and rubbing and pressing her body against his. Sometimes, you may see the female climb over his shoulders and back, as if mounting him. The male may respond by sniffing her genital region, causing her to arch her back and spread her hind flippers. Watch for these preludes to mating in the water or on land.

Copulation

The bull typically mounts the female from the rear, sometimes resting his entire weight on her and grasping her with his front flippers. Mating may last for more than an hour, consisting of several mounts, dismounts, and even a quick nap.

COPULATING *Mismatched in size, sea lion couples copulate on land or in the water, mating intermittently for up to an hour.*

Post-Copulation

The female is the one who finally breaks off the tryst. She suddenly raises her head and forequarters and **bites** the male's neck to pull free. Once she has ended things, she rebuffs both her former partner and any new males that jerk their heads toward her in a desirous way. Once fertilized, the embryo lies dormant in her womb, waiting $3^1/2$ months or so before it resumes its growth. Thanks to this delayed growth, the pups are born in the warm summer season, when food is plentiful and the sea lions are gathered together once again.

PARENTING BEHAVIOR

The pregnant female waits to come ashore until right before she gives birth, which, in California, is in May and June. You can tell when a female is pregnant by watching for her rather awkward, uncomfortable gait and signs of irritability. Since sea lions are normally sleek and frisky, a bulky-looking, slower individual is easy to spot.

Birth

During labor, the female becomes increasingly restless and often turns to nuzzle her genital region. Listen closely and you might hear the female uttering pup contact calls to her yet-to-be-born offspring. The single newborn is a hefty 11- to 14-pound package measuring about $2^1/2$ feet long. As soon as the pup is born, the mother issues her bawling, trumpetlike **pup-attraction call**, and the pup responds with a quavering bleat called the **mother-response call**. They engage in nonstop dialogue for about 20 minutes, long enough for their respective voices to become etched in each other's minds. From then on, each time the mother returns from a feeding trip, she and the pup will

DEFENDING THE CALF   *With an open-mouth threat, females warn other females to steer clear of their pups.*

be able to find each other by exchanging a noisy chorus of these calls.

Defending the Calf

For the first few days, the mother is obsessed with keeping her pup at her side. She frequently grabs it by the loose scruff of its neck and snugs it next to her. If something threatens the pup, she will carry it by the scruff to a safer location. Throughout this time, she is on constant alert, vigorously **barking** and **rushing** at neighbors that get too close or lashing out at them with the **open-mouth threat**.

Calf Grouping

Periodically, the mothers must return to the sea for food, and when they do, the pups on land gravitate toward one another. These playful "pods" have the run of the rookery. They cross territorial boundaries without reproach, usually traveling with one or two pups in the lead and the rest galloping and goofing off behind them. They stop frequently to roll around, spar, and play with any novel object that washes onto shore. They especially love tossing bits of giant kelp in the air and shaking them soundly.

Nursing

When the mother returns from her feeding trip, she begins to call for her pup. If pups other than her own approach her, she rebuffs them with an open-mouth threat or simply grabs them with her mouth and tosses them out of the way. The real pup continues to bleat in response to its mother's calls until, eventually, the two find each other and reaffirm their bonds by nose rubbing and sniffing. The pup nurses noisily, smacking so loud that you can hear it up to 10 yards away. The nursing period lasts at least 3 months, and sometimes up to a year. After pupping season, the males migrate north along the coast, while the breeding females and their young either stay close to the rookeries or migrate south for the winter.

CALIFORNIA SEA LION BEHAVIORS TO LOOK FOR AT THE ZOO OR IN THE WILD

BASIC BEHAVIORS

LOCOMOTION
- [ ] swimming
- [ ] porpoising
- [ ] body surfing
- [ ] walking
- [ ] galloping
- [ ] striding

FEEDING
- [ ] chasing

GROOMING
- [ ] scratching
- [ ] rubbing
- [ ] whisker stropping

WARMING AND COOLING
- [ ] jugging
- [ ] immersing in water

PLAY
SLEEPING

SOCIAL BEHAVIORS

FRIENDLY BEHAVIOR
  Physical Contact
- [ ] nuzzling
- [ ] nose rubbing
- [ ] *Social Play*
  Communal Defense
- [ ] alarm reaction

CONFLICT BEHAVIOR
  Male Threat
- [ ] rushing
- [ ] barking
- [ ] head shaking
- [ ] open-mouth threat
- [ ] oblique stare

Female Threat
- [ ] open-mouth threat

Fighting
- [ ] biting
- [ ] chest pushing
- [ ] lunging
- [ ] circle chase

SEXUAL BEHAVIOR
  Courtship
- [ ] sniffing
- [ ] barking
- [ ] head shaking
- [ ] rubbing
- [ ] nipping
- [ ] estrous display
- [ ] *Copulation*
  Post-Copulation
- [ ] biting

PARENTING BEHAVIOR
  Birth
- [ ] pup-attraction call
- [ ] mother-response call
- [ ] *Defending the Calf*
- [ ] open-mouth threat
- [ ] barking
- [ ] pushing
- [ ] *Calf Grouping*
- [ ] *Nursing*

NORTH AMERICA

HOWLING CHORUS   *Wolves howl to communicate over long distances. They howl in chorus, each one singing in a different pitch.*

GRAY WOLF

When we humans look for parallels to our behavior in the animal world, we invariably turn to the great apes, our evolutionary ancestors. While it's true that early humans probably shared many behaviors with the apes, we can learn just as much from what we *didn't* have in common. Unlike our primate relatives, we weren't strictly vegetarian; in addition to gathering food from the land, we also hunted in packs, stalking and bringing down animals that were larger and usually faster than ourselves. To manage such a feat, early humans had to make an evolutionary leap: we had to cooperate.

So it is with wolves. Like us, they bent their intelligence and ability to work as a team to their own devices and became one of the top predators in their range. Also like us, wolves adapted to a variety of habitats, eventually excelling in just about every environment except deserts and remote mountaintops. Next to *Homo sapiens,* wolves were once the most widely distributed mammal species on earth.

Over thousands of years of living and hunting in the same locales, primitive peoples and their four-legged counterparts developed a rapport. As far as anthropologists can tell, humans first began inviting wolves into their homes some 10,000 to 15,000 years ago. They tamed wolf puppies and bred successive generations for traits such as retrieving, digging, shepherding, or simply carrying themselves elegantly. Thousands of years and generations later, we live with a multitude of dog breeds. Though they differ in looks, it's important to remember that every dog, from the Pekinese to the Irish wolfhound, shares a common ancestry with today's wild wolf. As a result, one of the best ways to learn what makes Fido tick is to visit the wolf pack at your zoo.

Dogs treat other dogs and their human companions in much the same way as wolves treat one another in the wild. In most parts of their range, wolves live and hunt and raise their young in packs averaging from five to eight animals (the larger the prey, the larger the pack). A pack is basically a family group that includes the dominant male and female

VITAL STATS

ORDER: Carnivora

FAMILY: Canidae

SCIENTIFIC NAME: *Canis lupus*

HABITAT: Very adaptable to different habitats, as long as they are remote

SIZE: Male length, 5–6.5 ft   Female length, 4.5–6 ft
Shoulder height, 1.3–3.3 ft

WEIGHT: Male, 45–175 lb   Female, 40–120 lb

MAXIMUM AGE: 10 years in wild, 16 years in captivity

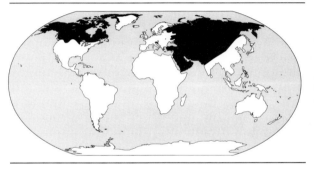

and their descendents of various ages. Members of the pack need one another to successfully run down and kill prey animals, such as moose and deer, that are much larger than they are. They also work together to protect and feed young, sick, or old pack members that aren't able to fend for themselves.

Because teamwork is so central to the pack's survival, much of the behavior you'll see in wolves is designed to promote social harmony. Pack society is organized along a chain of command, or dominance hierarchy, in which two individuals—the alpha male and the alpha female (alpha means first)—are above all others of their gender. Dominance comes complete with perks and responsibilities. As the pack's overall leader, the top male has access to the best food and the opportunity to breed, but he must also initiate pack activities and serve on the front line when it comes to defending the pack from intruders.

Within the group, there are certain classes or castes. The alpha pair comprise the

A LONG-DISTANCE CALL

Howling is the quintessential wolf trademark and, by association, a hallmark of truly wild lands (which makes it all the more wonderful when the air around a zoo is filled with howling). The first time I ever howled to the wolves and got them to howl back, the gooseflesh rose on my arms and the hair on the back of my neck bristled. It was a primitive reaction, perhaps triggered by some ancient memory inscribed on my genes.

Scientists like Fred Harrington have spent many years howling for wolves, gathering data as well as goose bumps. Their research is showing that wolves howl for three reasons: (1) to communicate with separated members of their pack, (2) to announce their presence to neighboring packs, and (3) sometimes just for the heck of it. A typical howling chorus begins with one, then two, then several voices, "like strands braided together." No two wolves howl in the same pitch, and, in fact, they seem to prefer chords! If you enter a chorus with a copycat pitch, the wolf you're copying will automatically switch. The reason for this might be communicative, in that wolf voices are distinct enough (to other wolves) to be like ID tags. Separating their pitches allows them to be heard for

who they are. Neighboring packs may also be able to pick out these differences, count heads, and figure out whether it's worth trying to enter the strange territory.

After a pack howls, it pauses for 15 to 20 minutes, allowing for response time should there be any wolves within hearing distance. Some of the time, wolves decide not to howl back when a neighboring tribe calls across the "fence." The risk of howling is that it locates you, and if a neighboring pack wants to invade, they'll know just where to find you. If a pack is sitting on a kill or protecting pups on a rendezvous site, however, it will be more likely to howl, stating, "We are here and do not want a surprise visit." If it isn't guarding any treasure, the cautious pack may decide to stay silent when the neighbors howl and simply slip away while the getting is good.

Lone wolves hunting for a place to set up a territory may not howl much at all, perhaps because it exposes them too much. A large pack feels more invincible, however, and is likely to call quite a bit. During breeding season, when wolves are apt to throw caution to the wind, they may feel the urge to howl regardless of their pack size.

upper class, the nonbreeding adults are in the middle, and the adult social outcasts are ranked last and must stay at the physical periphery of all group encounters. The up-and-coming wolves (those under 2 years of age) often have their own hierarchies, but will not participate in the adult order until they are sexually mature.

Each wolf expresses its rank through displays: visual displays are designed for face-to-face encounters, while scent marking and howling displays are used to reach packmates and enemies that are out of sight (see A Long-Distance Call and Invisible Beacons, pages 258, 260). If you are patient and observant, you can witness all three forms of communication at your zoo. When the animals in your exhibit are not interacting, however, you can always look for the basic behaviors that all wolves practice, even if they're alone.

BASIC GRAY WOLF BEHAVIORS TO WATCH FOR

LOCOMOTION

The streamlined bodies of wolves are built for trekking. Their long legs hang down from a torso that is as wide at the shoulders as it is at the hips. This means that the front leg and the back leg travel in the same plane, affording them a smooth, swinging stride that seems almost effortless. Traveling at a nonstop, 5-mile-per-hour **trot**, wolves can cover up to 125 miles in a day. When the time comes to finally close in on prey, they can crank their engines to a 45-mile-per-hour **sprint**.

FEEDING

Wolves specialize in tracking and killing large prey such as moose, deer, caribou, elk, bison, musk-oxen, mountain sheep, mountain goats, and in some areas, smaller prey such as beavers. Once they score a kill, they waste no time filling their belly, **wolfing down** as much as 20 pounds at a single feeding. A wolf eating 20 pounds is like a 200-pound man inhaling two large Thanksgiving turkeys at one sitting. For the wolf, however, it's not a matter of gluttony, but the mere fact that its next meal may be some time away.

Though the alpha male is entitled to the choicest parcels of the kill, the rest is up for grabs, and all the wolves get a chance to partake. Even the lowest-rung wolf enjoys a 12-inch sphere of ownership around its mouth; if it can get a tidbit of meat in its jaws, it is granted asylum from hassling. Once the wolf has eaten its fill, it may **bury the leftovers**, a behavior that explains why your dog insists on planting its bone in your flower garden.

ELIMINATION

Urinating and defecating serve double or even triple duty for wolves. Besides ridding the animal of wastes, the act of urinating or diarrhetic elimination may be a signal of fear or submission. And as anyone who walks a dog knows, urine may also be used to scent mark a pack's territory.

ROLLING IN SMELLY STUFF

Why is it that dogs will do just about anything to roll in noxious items such as feces and rotting carcasses? Wolves will make a similar beeline for smelly things, even at the zoo, writhing and rolling in them with eyes closed in a self-satisfied **consummatory face**. There are many theories as to why. Perhaps when the anointed wolf rubs its body against scent posts, the combination of its personal scent and its new perfume will leave an even stronger message. Or perhaps the odors are beneficial in reunions or surprise meetings, when, instead of getting attacked, a smelly wolf will be investigated first. Once the attackers get a close whiff, they may realize that the new wolf is not a stranger after all. A third theory suggests that the odor is in fact used to make the wolves *less* conspicuous by disguising their true scent.

INVISIBLE BEACONS

Even when pack members are spread out in ones and twos over a huge territory, they coordinate their movements as if connected by giant rubber bands. They are able to stay in tune thanks to a long-distance communication network hot-wired through their noses and their ears. Scent marking, like howling, can communicate either friendly messages or threats.

In the wild, a pack's home range can vary from 50 square miles in Minnesota to 5,000 square miles in Alaska. As wolves patrol these ranges for prey, they mark them with personal scents deposited in urine or feces or rubbed directly from their scent glands. Wolves at your zoo are just as interested in marking their range, and it's a behavior you're likely to see.

As a wolf moves at its average speed of 5 miles per hour, it encounters and makes about one scent mark every 2 minutes. If it leaves its mark in the form of feces, it usually scratches the ground with its forepaws, leaving a visual mark along with the chemical one. Sweat glands between the wolf's toes may leave yet another postscript in the disturbed ground. When urinating, both male and female wolves lift a hind leg, which allows them to place the scent on vertical objects well above the still air at ground level, and closer to the recipient's nose.

Wolves leave these odoriferous messages in conspicuous spots throughout their territory—trail crossings, directional changes, bridges—the same places where humans might leave a blaze or a cairn to alert fellow travelers. Specific scent stations include rocks, stumps, logs, snow drifts, or even sticks.

It may be that wolves sniff these posts to help them make directional decisions the next time they pass. Or they may read the scent to find out which wolf made the mark, what its sexual status is, where it was coming from, how long ago it passed, how healthy it is, and whether it is friend or foe. If you are curious about how wolves can pick up such subtleties in the form of smell, consider that in the detection of some substances, canines' noses are a million times better than ours!

Along territorial borders, wolves leave twice as many scent marks, and in addition to identifying the maker, these scents may serve as threats, saying in effect, "This land is occupied, go no further." When traveling wolves encounter a scent mark left by neighboring packs, they usually deposit their own scent on top of the strange one and then head back to their land. In this case, good fences make good neighbors, and both packs are able to avoid dangerous trespasses into another pack's area.

COMFORT BEHAVIORS

Before settling down to **sleep** in the great outdoors or in their naturalistic exhibit, wolves circle round and round, matting down the vegetation into a bed. We see the remnant of this circling behavior in our domestic dogs, even though carpets and bedspreads need no matting down. Wolves rest on their side or sometimes on their abdomen, but when they are in deep sleep, they curl up and tuck their nose beneath their tail for warmth. When they're away from their den in winter, drifting snow forms a protective cover over the rest of their body.

Other ways wolves make themselves comfortable include **panting**, **scratching**, **yawning**, **shaking off water**, **stretching**, and **grooming** (licking and nibbling irritants).

Any animal that can subdue a moose 10 times heavier than itself can obviously do a lot more damage to a packmate its own size. To prevent dangerous blowups, many friendly ceremonies, displays, and gestures have worked their way into the pack's social repertoire. Even when tempers do flare, ritualized displays of dominance help wolves settle most conflicts without coming to blows (or, rather, bites). Even the wolf's anatomy is designed to transmit these messages without misunderstanding. Their tail and face are highly expressive, and dark markings on their tail tip, ears, eyes, and muzzle help dramatize every gesture. Wolves have also evolved a lexicon of noises—whimpers, whines, squeaks, yelps, snarls, howls, and growls—to amplify and clarify the body's language.

Some of the most important peace-keeping displays are acts of submission, in which subordinate wolves "beg" for tolerance from any wolf that is superior to them. You'll see plenty of submissive postures and displays when wolves greet one another on friendly terms, as well as during aggressive conflicts. The flip side of submission is dominance, and the interplay of these two forces keeps the pack stable and secure.

FRIENDLY BEHAVIOR

Individual Greeting

Wolves are constantly checking each other out, using eyes, ears, and nose to determine whether an animal is friend or foe, subordinate or dominant. The most common **social investigation** involves smelling the nose, face, and genitals of other pack members. When two wolves sniff each other, the subordinate holds its tail down while the dominant holds its tail up. To further humble itself, the lesser wolf may lower its body into a crouch, with ears back, tail tucked, and muzzle tilted up, trying to lick the superior's face. This face-licking ceremony, also called **active submission**, may also include vigorous tail

ANAL PRESENTATION *To get a feel for status, wolves read each other's anal scents. The confident dominant raises its tail, while the insecure wolf hides its scent under a lowered tail.*

wagging, dancing in circles, and whining—all intended to elicit friendliness. If this display sounds familiar, it's because your dog performs it every time it greets you, jumping up in an eager effort to reach the face of its "leader."

Group Greeting

Active submission is also the centerpiece of the joyful greeting display that wolves enact when waking from a snooze or reuniting after a separation. All the pack members clamor to be the first to lick and nudge the muzzle of the alpha wolf, who usually stands looking very regal and unconcerned. Wolves also perform a greeting display when they first pick up the scent of a prey animal and after a kill. In zoos, you're most likely to see these ceremonies at feeding time.

Without displays such as these, group encounters could be quite stressful for wolves, given their genetic programming that says, "Attack and drive out strange wolves." The close contact during these free-for-alls gives them a chance to sniff out old friends and remind one another, "We are packmates, not strangers." They also get to pledge loyalty to their leader, who sets the tone for their activities and provides the nucleus that they revolve around.

Snuffling

One of the most tender wolf gestures is that of snuffling—placing the nose through the neck fur and gently touching the skin beneath. Snuffling is a sign of intimate friendship between wolves. Curiously enough, we humans also tend to snuffle (and be snuffled in return) when we hug our dogs!

Paw Raising

Did you ever wonder why it's so easy to teach a dog to raise its paw? It's because wolves instinctively lift a paw when soliciting affection, grooming, or food from another wolf. You'll often hear a high-pitched whining or whimpering along with paw raising.

Snout Bunting

Another way wolves insist on affection is to push upward with their snout against the other wolf's chin. Dogs have modified this slightly to get attention from us. If they're not able to reach our chin, they push upward against our hand, usually, as one author wryly noted, when we're holding a hot cup of coffee.

Scent Marking and Howling

Scent marking performs a friendly function by cuing fellow travelers that a packmate has passed here. This helps pack members stay in contact so they can assemble quickly if need be. Another call for closeness is the infamous wolf howl. This long-

PAW RAISING   *Wolves use this food-begging gesture throughout their lives to solicit affection and friendliness.*

distance communiqué allows individuals to contact their packmates even when they are miles away. The howl helps pack members locate one another and directs them to assemble, especially after members have separated to hunt alone or in small groups.

Social Play

You can learn a lot about wolf society by watching the pups playacting adult behaviors and jockeying for rank, even as tiny balls of fur. Three main elements of play are **chasing**, **ambushing**, and **mock fighting**. These games have an element of aggression, but can be distinguished from actual fighting in several ways. First, playful roles are fluid and ever changing; the pursuer becomes the pursued in the flash of a tail. Second, play sequences are often interrupted by bouts of relatively calm behavior such as eating or gazing. Most telling are the special play-invitation signals that wolves use to coax a companion to play. Most dog owners are familiar with the **play face**, an open-mouthed panting with the lips drawn back horizontally as if grinning. **Bowing** by lowering the front portion of the body on outstretched legs also says, "Let's play," especially when combined with the play face, tail wagging, and barking. Even more irresistible is the **bouncing** of the forelegs while the hind legs swivel. Yet another play invitation gesture is an exaggerated **looking away**: one wolf turns its head sideways and coyly peers over its shoulder at the other, showing the whites of its eyes, flattening its ears, and grinning.

Play is important to pack life because it fuses the affectional ties among pups, especially during the first 5 months of their lives. These bonds will last a lifetime, and so will the urge to play, as owners of frisky geriatric dogs can testify.

Cooperative Hunting

Despite their well-publicized hunting prowess, wolves are no match for many hoofed animals in terms of speed, stamina, or weaponry. Even though wolves can achieve speeds of 45 miles per hour, they have to give up after only 20 minutes or so. If they do happen to close in, they're still not home free; the slashing hoofs of a healthy adult moose, for instance, can easily kill a wolf that dares to get too close.

For this reason, wolf hunts are mostly testing runs designed to measure how fit the prey is. When the wolves first catch wind of prey in their area (even at 1½ miles away), they cock their ears and point in the direction of the scent, whining, wagging their tails, and exchanging excited greetings. After this spontaneous pep rally, the alpha male leads the wolves in single file toward the herd of elk or deer. At a magic distance away from the herd, they break into a group chase, watching carefully for signs of any weak, sick, injured, or old animal that might be easy pickings. The wolves concentrate on this victim, spreading out and circling in much the same way that sheepdogs circle a herd of domestic sheep. Their 2-inch-long canine teeth are the functional equivalent of talons, allowing them to pierce through a flank or close shut a bulbous nose, hanging on until the thrashing prey succumbs.

Most hunts don't go this smoothly, however. In a study of 131 elk hunts, 54 elk were too fast for the wolves to bother chasing. They did pursue and attack 77 elk, but of these, 71 were able to get away or defend themselves. For all their trouble, the so-called "killer" wolves brought only 6 elk to the dinner table. With this rather mediocre 4.6% success rate, you can see why wolves are not likely to wipe out their prey. If anything, they simply act as a pruning mechanism, removing the slower, weaker members and keeping the prey populations from exploding and overgrazing their habitats.

CONFLICT BEHAVIOR

Every wolf in a pack knows its place, having established it early in its career as a pup. At the peak of its reign, a dominant wolf need only signify its status with visual threats that center on the position of its tail, head, ears, and mouth. Once you learn the code, you'll be able to easily pick out the leaders and subordinates in your zoo pack. They play predictable roles in the ritualized threat and submission exchanges that keep the pack sailing on an even keel.

The only time you'll see a serious fight at the zoo is if the pack has decided to reject a member or if they are overthrowing an older alpha male. In the wild, serious fights also occur when strange packs meet.

Low-Intensity Threat

The least intense form of threat is the **direct stare**, in which the dominant wolf simply draws itself up to its full height, raises its tail, and stares intently at the subordinate. This is usually enough to freeze an insecure wolf in its tracks.

The dominant wolf may then approach the subordinate and perform an **anal presentation**, raising its tail and presenting its anal region for inspection by the other wolf. Odors produced by the anal glands tell much about the owner, including its status and individual identification. When the dominant wolf tries to sniff the subordinate's anal region, the lesser wolf covers the area with tucked tail, showing that it is not as confident about having its identity revealed.

RIDING UP *Pups ride up on each other during play, but as they get older, this gesture will express dominance.*

Another pattern of dominance is **standing across**, in which the dominant wolf stands over the upper torso of a reclining individual. The subordinate may even roll over and lick the underside of the dominant wolf. **Riding up** is another common expression of dominance. The top wolf places both forelegs across the shoulders or back of the lesser one, approaching from the side or from behind, as it might do in sexual mounting. It may even symbolically grab the neck of the subordinate in its jaws. These ritualistic bites are "soft," however, and rarely draw blood.

High-Intensity Threat

Since the alpha male is the undisputed "top dog," he rarely needs to resort to high-intensity threats. On the other hand, the lowest-ranked outcasts are far too insecure to perform these threats. This leaves the feisty middle-ranked wolves as the ones most likely to issue high-intensity warnings as a way of beefing up their status.

An example of a high-intensity threat is the **bite threat**, enacted with the head held high, the neck arched, the forehead buckled with muscular tension, and the ears tilted forward. The wolf opens its mouth slightly and retracts its lips vertically to expose the impressive canines. Beneath the wrinkled muzzle, a tongue may dart in and out menacingly. The wolf looks ready to explode into action, with legs stiffened, tail trembling, and fur bristling. Raising the fur around the tail or mane may release some glandular odors, sending an olfactory message that reinforces the visual one. To add even more punch, the wolf may growl or bark as it displays.

Another high-intensity threat is the **ambush**, in which the wolf, sighting the subordinate at some distance, lowers itself to the ground like a lion about to pounce. Both high-intensity threats may escalate into actual attacks if the recipient doesn't retreat or show some form of submission.

Active Submission

A wolf that is feeling submissive or afraid tries to make itself as nonthreatening as possible. It covers its teeth, smooths back its ears and forehead, narrows its eyes into a slant, and pulls the corners of its mouth back into a submissive grin. Its body posture is completely opposite that of a confident wolf; instead of standing erect, it crouches low with its tail lowered or tucked between its legs.

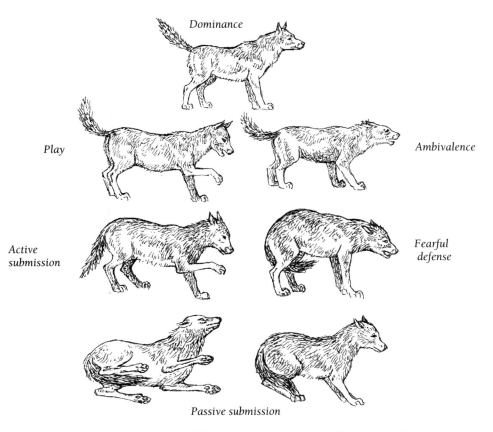

BODY LANGUAGE *Notice how the wolf positions its ears, tail, and body to send a clear message.*

In active submission, the subordinate wolf crouches beneath the dominant one, licking its face and nudging or even tenderly mouthing its muzzle. To spice up the performance, it adds high-pitched whining, squealing, tail wagging, and even some circling dances. This display, also used in the greeting ceremony, is reminiscent of the way pups act when they are trying to induce a parent to regurgitate some food. By putting itself in this infantile posture, the subordinate wolf says, "I recognize that you are superior to me, and I feel nothing but friendliness toward you."

Passive Submission

When a wolf is seriously outranked by a confident superior or by a whole pack of superiors, it resorts to the more extreme form of submission called passive submission. The wolf lies down on its back and splays its legs to expose its genitals to the dominant wolf (much like your dog "asking" to have its belly scratched). It flattens its ears backward and passes its tail between its thighs to cover its anal gland. It may wave its tail slowly, urinate slightly, or lie completely still as if helpless. This passive submission may hark back to the position pups assume so their mother can stimulate them to urinate and defecate.

Defense

A subordinate wolf often resorts to **defensive threats** when its submissive gestures fail to appease the superior wolf or when it finds itself at the mercy of a gang of wolves. It arches its back, slicks back its ears, curves its tail downward, and holds its head backward. Though its eyes are averted, its teeth are bared and snapping, a schizoid pose that reveals both fear and aggression. The two emotions cancel each other out, however, and the snapping never actually leads to biting. The more confident the wolf, the closer its teeth come to the dominant and the more it barks and growls. A deeply subordinate animal won't even come close to the dominant wolf; instead, it simply snaps its jaws from 30 to 40 feet away. When trapped by a pack, an alien wolf may show the ultimate signs of stress such as an inability to run, a tight tucking of the tail, or even spontaneous diarrhea.

Long-Distance Threat

Wolf packs issue warnings by leaving chemical calling cards, or **scent marks**, along the boundaries of their territories that say, "No trespassing." **Howling** between packs serves a similar function by announcing which wolves are present in the area.

SEXUAL BEHAVIOR

In wolf society, the highest-ranking animals are usually the only ones permitted to mate. This behavioral birth control puts a natural and necessary lid on wolf populations, keeping the birthrate down to one litter per pack per season. More pups than this would be difficult to feed and raise in the proper manner.

Inbreeding is also avoided by natural means; subordinate wolves, unable to mate in their original family pack, eventually move away to start their own packs, thus spreading their genes geographically. In the year or two before they disperse, however,

Just as we consider our dogs to be part of our human family, dogs think of us as members of their pack. And, as it turns out, we make fairly good pack members. We share our food and den, and we engage in social interactions, such as grooming, contact sleeping, and play. We even accompany them on long walks around their territory! In return, they shower us with affection and protect us from strangers.

A dog usually treats every member of its household as its superior, but it may consider one member to be the leader of the pack. To keep in the good graces of this leader, the dog shows overt signs of affection, such as greeting, licking, leaning, and jumping up, as well as signs of "remorse" when the leader is not pleased.

Knowing how much people love their dogs, you wonder how the wolf, such a close relative, can be so venomously hated. How could our government have paid hunters to kill hundreds of thousands of these beautiful, intelligent animals? How did it happen that wolves, once so widely distributed, are now extinct over most of their original range and often teetering on the brink where they do occur?

Part of the answer lies in the predatory nature of the wolf and the deep-seated, albeit misplaced, hatred of predators that many people feel. As group predators ourselves, we once believed that wolf packs were in direct competition with us. Wolves preyed on wild species that we coveted, and they occasionally took domestic stock from farms and ranches. Our response was to start exterminating the problem by shooting, trapping, and poisoning every wolf we could find.

We now know that all the programs to exterminate wolves for the benefit of prey species were shortsighted. If left alone, wolves may do deer and moose a service by pruning their populations so they don't explode and eat themselves out of house and home. In fact, we now know that entire ecosystems function best when predator/prey relationships are intact. Of course, the ecosystems that once supported gray wolves were wild and whole, and humans have long ago tamed and fragmented most of them beyond recognition. As a result, wolves occupy only a fraction of their former range: the wilder portions of Alaska, Canada, northwestern Montana, Idaho, northeastern Minnesota, northern Asia, and the mountainous regions of Italy, the Middle East, and South Central Asia.

Legal protection is giving the wolf some breathing room in places where it is barely holding its own. Where packs rub elbows with humans and their livestock, however, wolves are still routinely shot, even though studies have shown that wolves are very *infrequent* predators of livestock. And in parts of Canada and Alaska where the wolf is well distributed, wolves are still being killed in an effort to increase big game populations for sport and sustenance use. In the last decade, $1,000,000 has been spent on "management," with a cost of $1,000 for every wolf destroyed. At the same time, advocates are evaluating plans to reintroduce wolves to such areas as Yellowstone National Park, where howls have not been heard for decades.

Wolves provoke controversy like few animals can. To some the wolf is a gorgeous animal, a longtime friend to humans and kingpin in the food chain, while to others the wolf is no more than a varmint, an ancient nemesis that must be controlled. As these camps debate the wolf's fate, we can only hope that the roadless lands that wolves depend on can somehow escape the domesticating influence of roading, mining, and clearcutting.

these wolves play a significant role by providing food and protection for their younger brothers and sisters born into the pack.

Pre-Courtship

When it comes to mating, wolves definitely play favorites, focusing their courtship rituals only on their lifelong mates. Pre-courtship consists of playful, high-contact sports, including **head rubbing**, **snuffling**, **nipping**, **snout bunting**, and **snout grabbing**, all performed with copious amounts of tail wagging.

Estrus lasts for 5 to 15 days and occurs anytime from January to April, depending on locale. During this time, females produce odors in their menstrual blood, secretions, and urine that greatly arouse the male. After **sniffing** a female's urine, the male will often lick the spot and then urinate on it himself. This ritual helps to form, maintain, and advertise the pair bond and to synchronize the pair's reproductive processes.

SNOUT GRABBING   *Though this part of the pre-courtship ritual looks painful, the bite is controlled and "soft."*

Courtship

Courtship may start months before the actual breeding season. Seeing wolves frolic during this time is a special privilege that zoos afford their visitors. If the wolves have plenty of room to roam, courtship may consist of **mutual chases** as the female solicits and then rejects the male's advances. In more confined quarters, the abbreviated chases look almost like dances. The female wolf's **invitation** may be as subtle as placing her paws or head and neck across the male's shoulders or as dramatic as a full-blown seduction dance. Prominent wolf researcher Rudolph Schenkel described the dance beautifully: "With raised tail, the rutting alpha-bitch moves in a feathery dance step, whimpering or singing tenderly. Meanwhile she moves her genitals in slow, minute, pendulum-like movements in a vertical direction."

Copulation

Few of these dances end in matings, however, because the female usually decides to retreat, tuck her tail, or sit down at the last minute. Only after lengthy courtship, at the height of her heat, does she finally acquiesce. As an invitation to mate, she moves her tail to one side, exposing her genitals and allowing the male to clasp her from the rear. Once he gets into position, you may see him treading his forepaws on her back. This situates his penis deep inside her vagina and allows the pair to accomplish the **tie** that is unique to the canine family. Mating wolves become tied when the enlarged bulb at the base of the penis locks behind the female's strong vaginal muscles. Though the male dismounts and turns the other way, he remains locked to the female, rear to rear, for as long as 30 minutes. This gives him time to safely transport his sperm through a series of ejaculations, while keeping the female away from other males.

PARENTING BEHAVIOR

Den Building

In early spring, the female wolf begins digging a den to prepare for the birth that is still 3 to 5 weeks away. She may dig a new den, enlarge or repair last year's den, or modify a burrow that belonged to a fox, badger, or other animal. In lieu of digging, she may simply set up house in an overturned stump, rock crevice, or hollow log. Pregnant wolves create or look for the same sorts of quarters in captivity. At some zoos, visitors get a chance to see close-ups of birth and pup care thanks to video cameras built into the dens.

As the birth draws near, the mother-to-be and other wolves start carrying meat to the den and burying it for later. Finally, about a day before she's due, the female crawls into the den to wait. The other pack members bring her food, dropping it off in actual chunks or regurgitating it for her.

Caring for the Pups

The female gives birth to an average of five or six blind and helpless pups. If you can get close enough to a den, you'll hear the pups whine when they are hungry, yelp when they are in pain, and grunt when they are in contact with a nipple. At about a month old, they stop **nursing** and begin to eat predigested food that pack members **regurgitate** for them. They quickly learn to mob any adult that enters the den, licking and biting at its muzzle as if to say, "What do you have for me?" (If this seems odd to you, consider the fact that some human tribal societies still feed their babies mouth to mouth by chewing the food and then passing it to their young. In fact, this is how kissing reportedly developed. Gives a whole new meaning to your dog's licks of affection, doesn't it?)

Wolves lavish the pups with attention, **ingesting their wastes** during the first few weeks to keep them clean and carefully **teaching** them what they'll need to know as full-grown wolves. When the pups are 8 to 10 weeks old, they are moved to the first in a series of 1-acre playgrounds called rendezvous sites. Since the pups are too little to go on a hunt, the pack members leave them here to play, returning for a noisy reunion once the hunt is over. Some pups will leave the pack the following season to form their own packs, but others will remain behind to help raise the next litter.

GRAY WOLF BEHAVIORS TO LOOK FOR AT THE ZOO OR IN THE WILD

BASIC BEHAVIORS

LOCOMOTION
- ☐ trotting
- ☐ sprinting

FEEDING
- ☐ wolfing down
- ☐ burying food
- ☐ ELIMINATION

ROLLING IN SMELLY STUFF
- ☐ consummatory face

COMFORT BEHAVIOR
- ☐ sleeping
- ☐ panting
- ☐ scratching
- ☐ yawning
- ☐ shaking water off
- ☐ stretching
- ☐ grooming

SOCIAL BEHAVIORS

FRIENDLY BEHAVIOR
Individual Greeting
- ☐ social investigation
- ☐ active submission
- *Group Greeting*
- ☐ *Snuffling*
- ☐ *Paw Raising*
- ☐ *Snout Bunting*
- ☐ *Scent Marking*
- ☐ *Howling*

Social Play
- ☐ chasing
- ☐ ambushing
- ☐ mock fighting
- ☐ play face
- ☐ bowing
- ☐ bouncing
- ☐ looking away
- *Cooperative Hunting*

CONFLICT BEHAVIOR
Low-Intensity Threat
- ☐ direct stare
- ☐ anal presentation
- ☐ standing across
- ☐ riding up
High-Intensity Threat
- ☐ bite threat
- ☐ ambush
- ☐ *Active Submission*
- ☐ *Passive Submission*

Defense
- ☐ defensive threat
Long-Distance Threat
- ☐ scent marking
- ☐ howling
SEXUAL BEHAVIOR
Pre-Courtship
- ☐ head rubbing
- ☐ snuffling
- ☐ nipping
- ☐ snout bunting
- ☐ snout grabbing
- ☐ urine sniffing

Courtship
- ☐ mutual chases
- ☐ female invitation
Copulation
- ☐ tie
PARENTING BEHAVIOR
- ☐ *Den Building*
Caring for the Pups
- ☐ nursing
- ☐ regurgitating food
- ☐ ingesting wastes
- ☐ teaching

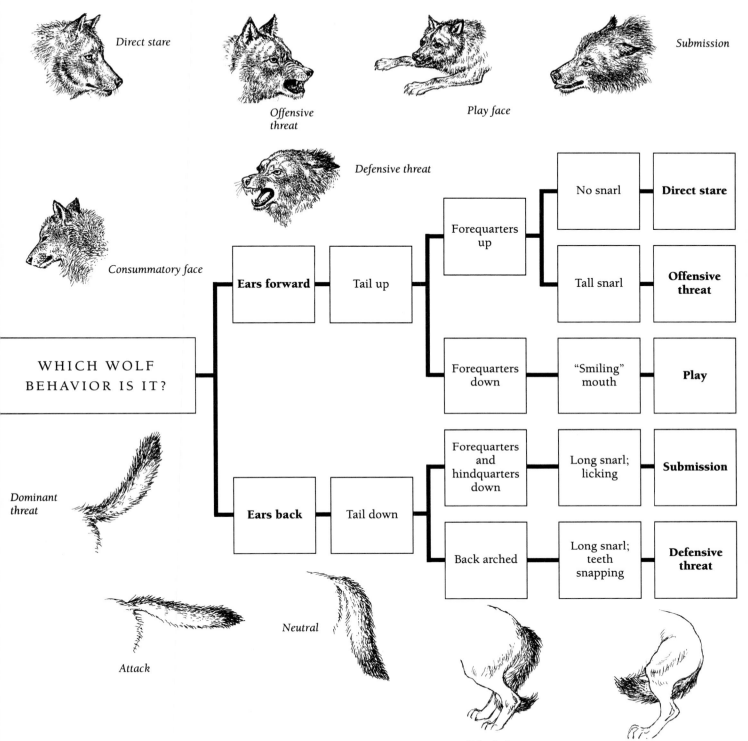

Direct stare

Offensive threat

Play face

Submission

Defensive threat

Consummatory face

Dominant threat

Attack

Neutral

Submission

Fearful defense

WHICH WOLF BEHAVIOR IS IT?

Ears forward → Tail up → Forequarters up → No snarl → **Direct stare**

Tall snarl → **Offensive threat**

Forequarters down → "Smiling" mouth → **Play**

Ears back → Tail down → Forequarters and hindquarters down → Long snarl; licking → **Submission**

Back arched → Long snarl; teeth snapping → **Defensive threat**

THREAT CALL   *An eagle uses a high-pitched call to warn other eagles away from its nest or feeding grounds.*

BALD EAGLE

Eagles may not have always been masters of all they survey. Judging from their present-day behavior, we can speculate that at some point in their evolutionary past, eagles must have pulled through hungry, desperate times. Faced, perhaps, with a shortage of prey or a crush of competition, eagles that were picky about what they ate eventually weakened and died. Those that learned to *diversify* by hunting live prey, scavenging dead prey, and even stealing from other eagles were able to survive and raise eaglets. This resourcefulness was passed down through the generations and became the central theme in the bald eagle's success story. Today, our national emblem is a scavenger, a thief, and still an amazing hunter in its own right, showing a flexibility born of adversity.

The wily eagle further adapted its strategies to find food in the lean winter season, when fish were often sealed under icy wraps. The birds began to congregate where the hunting was good, even if it meant tolerating the presence of hundreds of other eagles. Today these gatherings are among the most spectacular wildlife viewing shows in the world. On the Chilkat River near Haines, Alaska, for instance, more than 3,000 eagles may congregate to feed on salmon that die after spawning. The trees are dark with roosting eagles, the skies full of soaring flocks, and the shorelines noisy with controversy. Wing to wing, dozens of eagles fight over a new catch, while others feed talon to talon on the same carcass. These free-for-alls occur in other places throughout the eagle's range (for example, Glacier National Park in Montana and Skagit River in Washington), but none are as large as the Chilkat aggregation.

Though congregating in groups may once have been foreign, the adaptable eagles developed ways to make it work. They learned to watch for one another, like vultures do, to find food. The mere presence of dozens of eagles circling over a large carcass, for example, signals to other eagles that food is available. The lunch is far from free, however, since eagles that drop in on a carcass have to fight

VITAL STATS

ORDER: Falconiformes

FAMILY: Accipitridae

SCIENTIFIC NAME: *Haliaeetus leucocephalus*

HABITAT: Nests and feeds near water; migrates along mountain ridges; congregates near prolific food sources in fall and winter

SIZE: Length, 34–43 in   Wingspan, 6–7.5 ft

WEIGHT: Male, 8–9 lb   Female, 10–14 lb

MAXIMUM AGE: 50 years in captivity

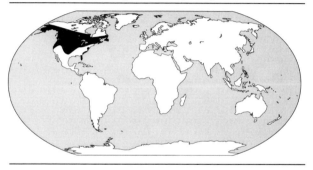

with other hungry pirates for feeding rights. Even so, every persistent eagle will get some share of food, which is more than they would net if they were hunting alone in a less productive area. Because food is not hard to come by in these wintering areas, eagles can spend most of the winter's day (99%) resting on a perch, preserving precious heat. A final fringe benefit of the gatherings may have nothing to do with food; eagles may be evaluating and selecting their mates for the upcoming breeding season.

These winter congregations put more eagles in one place than most people will see in a lifetime. Since zoos can't house that many eagles, you won't be able to see the mass hysteria of an aggregation in captivity. But you will be able to see social interactions on a smaller scale (provided your zoo has more than one eagle) and certainly individual behaviors as well.

BASIC BALD EAGLE BEHAVIORS TO WATCH FOR

LOCOMOTION

The **soaring** flight of the eagle has made many great minds think alike. Almost all the Native American tribes independently came to believe that eagles were messengers to the gods and goddesses, able to soar above the clouds with prayers on their plumes.

If you look closely at their wing design, you'll begin to see why eagles can achieve such heavenly heights. Their flat, 7-foot-wide wingspan allows them to get maximum airlift with minimum effort. When properly adjusted, the fingerlike wing slots (gaps between the large feathers at the tips of the wings) allow the birds to maintain a steady flight pattern, maneuver precisely, and enhance lift so they can transport heavy prey. Their body is also light for its size, thanks to the air-filled skeleton that weighs less than half as much as the 7,000 feathers. All together, the average eagle weighs only 9 pounds — about the size of a well-fed house cat!

Through eons of natural selection, eagles have learned to capitalize on their lightweight design. As they cruise over the landscape, they gravitate toward thermals, columns of warm air that rise rapidly over heated land surfaces. These invisible elevators give eagles a free ride into the heavens where they can see far and wide over the landscape and hopefully spot their next meal. They can also piggyback on updrafts that form as moving air is forced up mountain ranges. Riding thermal to thermal and updraft to updraft, eagles can cover great distances with hardly a wing beat.

When a high-flying eagle spots something edible, it must be able to get to earth quickly. Tucking its wings, the bird turns into a speeding bullet, **diving** at 200 miles per hour. Before hitting the ground, it opens its wings again and braces its tail fan to create drag, much like an airplane adjusting its flaps for landing.

SOARING *Eagles take a free ride on thermals—columns of warm air that rise over heated land surfaces.*

FEEDING

The eagle's trademark flexibility also shines through in its eclectic diet. Though fish are its favorite, mammals, reptiles, amphibians, and birds are all fair game, regardless of species or size. In a single day, for instance, an eagle may snack on something as small as a perch or as massive as a stranded whale carcass.

Their willingness to eat leftovers also opens up a wide range of menu choices—prey that would normally be out of their league. In fact, given a choice, eagles often prefer scavenging or stealing prey to hunting their own. Though we might term it laziness, this hunt-only-if-all-else-fails strategy is really a matter of frugality. By letting others do the work, eagles exert less energy and ultimately need less food to survive. When the need arises or an opportunity presents itself, however, bald eagles can be aggressive predators.

When it's up to them, eagles use five basic hunting methods: (1) hunting in flight, (2) hunting from a perch, (3) hunting on the ground, (4) wading in water, and (5) cooperative hunting. Though you won't be able to see actual hunting at the zoo, you can study related maneuvers by watching eagles in large flight cages and at raptor shows. Once you know what to look for, you may someday be lucky enough to catch their act in the wild.

When **hunting in flight**, eagles soar in circles, combing the landscape with lenses that can see three or four times farther than we can, enabling them to spot small mammals from a mile away. After a breakneck dive, they instantly thrust down their talons, plucking fish from the water, birds from branches, and mammals from the midst of a mad dash. Occasionally, an eagle may even take a bird on the wing. **Perching** is an energy-saving method of hunting; still as a statue, the bird studies the ground or water beneath its perch for hours, waiting for prey to give itself away. When opportunity knocks, the eagle reacts with strobe-light speed, hitting its target with twice the force of a rifle bullet. Eagles will also **walk** to a hunt, sneaking into thickets to flush hiding animals or even **wading** into the shallows to feast on fish. Most interesting of all are the **cooperative hunts**, in which one eagle flushes a small mammal into the waiting clutches of a nearby eagle.

No matter how they hunt, eagles deal the final blow by simply grasping the prey with one superbly adapted foot. When the muscles in their leg contract, the long, sharp talons sink into the prey's body with enough force to stop its breathing and crush its bones. Spiny pads on the undersides of the toes also help keep the captive in place, even if it's wet and wriggling. Smaller prey may be toted in the beak.

HUNTING *Eagles can easily pluck a wriggling fish from the water, killing their prey with one powerful talon.*

COMMUNAL BATHING *At winter gatherings, eagles sometimes bathe together.*

Eagles usually fly to perches to consume their quarry, unless it's too heavy to lift. In the Chilkat Valley, eagles often swim their fat salmon to shore, wings stroking the water like oars. No matter where they eat, eagles must keep a watchful eye out for uninvited guests that might be hungry or bold enough to help themselves to dinner.

Eagles have **stealing** down to a science. A marauding eagle can either commandeer another animal's kill on the ground or harass a fish-carrying bird in flight. Ospreys are especially vulnerable to in-flight muggings, a drama you can actually witness in the wild. The eagle chases and dive-bombs the smaller osprey at close range until the unnerved bird drops its prey. The eagle then plummets after the falling fish, and like a winged Superman, actually catches it before it splashes back into the water. Sea otters make fine targets for looting as well. Their habit of floating on their back and using their chest as a dinner table makes it easy for the eagle to glide in from behind and snatch the otter's newly opened clam or mussel.

The eagle's third dining option is to **scavenge** an unattended carcass. Their high-altitude flybys are ideal for spotting the bodies of animals killed by disease, cold, starvation, or even bullet wounds. Once they touch down, eagles have dominant status at a carcass, scaring away crows, vultures, and other scavengers. Coyotes, domestic dogs, mountain lions, and bears are the only mammalian predators known to challenge the bald eagle's prerogative.

Using their claws to get a firm purchase on the carcass, eagles tear the flesh to pieces with their curved beak. Their large gape allows them to swallow sizable chunks and put away as much as possible before another predator elbows in. When food is abundant, the eagles swallow more than they need, storing it in their crop for later consumption. Look for a bulge in the throat of the eagle at your zoo; up to 2 pounds of food may be stuffed in this outpocketing of the esophagus, enough to keep an eagle satiated for days.

The morning after a meal you may see an eagle cough up a small pellet. This is the trash-compacted remains of the indigestibles in its prey—hair, feathers, and bones.

How Bald Eagles Interact

FRIENDLY BEHAVIOR

Communal Roosting and Bathing

For most of the year, eagles are protective of their territory, their roost, their perches, and their nest area. This territorialism takes a time-out in winter, however, when eagles are actually tolerant of one another. Birds that would normally peck and chase intruders are now willing to roost in the same tree, maintaining only small personal territories around their bodies. The benefits of roosting together seem to outweigh the concessions. It may be, for instance, that roosting birds pick up clues about productive food sites or that they form or renew pair bonds in anticipation of breeding season.

Another social activity seen at large gatherings is **communal bathing**, which, though rare, can be contagious, spreading from one eagle to another until all are splashing together.

Communal Hunting and Feeding

Occasionally, eagle pairs will hunt cooperatively, driving prey toward one another in hopes of being the one to make the final catch. Eagles may also gather to feed on large carcasses, such as whales that have beached themselves. Since no eagle can defend the whole thing, they accept one another's presence, keeping their aggressions down to an occasional squabble. The restraint shown at these gatherings and at winter jamborees is not really an act of overt friendliness, however; the birds are simply tolerating one another so each can have its own piece of the pie.

CONFLICT BEHAVIOR

Conflict arises wherever and whenever the eagle has something precious to defend. Nesting pairs, for instance, defend their territory—including the nest site, favorite perches, and offspring—from intruding eagles or other large predators. In the winter, the defended resource is smaller; very often it is only one corner of a carcass or a single branch on a roost tree. Despite the size of the prize, however, the defensive moves are similar.

Threat

A completely passive form of threat is the **coloration** of an adult eagle. The brilliant white head warns other eagles, even at a distance, that the territory is occupied. To get full use of this advertisement, the energy-thrifty eagle perches in conspicuous places, effectively defending its turf with the minimal amount of fighting.

If an eagle ignores this visual warning and enters a territory anyway, one or both resident eagles may issue a high-pitched **threat call**. If that doesn't work, they'll start a **circling display**, protectively soaring over the interloper and screaming until it leaves the area. The most aggressive and most common way for eagles to discourage an unwanted guest, however, is the direct **territorial chase**, in which both residents fly directly at the intruder and chase it until it leaves or turns to fight. The chasing birds may extend their feet and talons at the intruder to prove they're sincere about the ousting.

Communal roosting trees, common at winter gatherings, are also the site of colorful conflicts, especially early in the evening when the eagles are fighting among themselves for good roosting spots.

PERCH JOSTLING *An eagle looking for a place to roost tries to oust another eagle from its perch.*

If an eagle lands within 2 feet of another, the defending eagle may raise its neck feathers, lower its head, and jab its beak at the intruder in a typical **perch site display**. Once supplanted, the evicted eagle looks for another perch and begins to jostle the resident. One ousting may start a domino effect in the roosting trees, ending only when the last eagle finds an unoccupied spot and settles down.

Food Fights

Most outright attacks are over food, and most of these occur when eagles are massed together. A hungry eagle tries to force another to relinquish its food by using **aerial chases**, **swooping attacks**, **running or walking attacks**, or **leaping attacks**. **Talon displays**, in which the talons are thrust up at the other bird, and **wing displays**, in which the wings are spread wide, are also used to terrify the food owner.

If the owner stands its ground, the pair may begin a skirmish by **locking their talons** and struggling. After breaking away, one eagle may grab the prey with a foot and try to drag it away, prompting the other to recapture it in an avian version of tug-of-war called **taloning**. Eagles may also resort to **pecking** each other with their beaks or trying to **steal** bits of food. Watch how the pilferer keeps its head below that of the food owner each time it sneaks a beakful, perhaps to signal a form of submission that says, "I'm interested in the food, not in fighting with you." If the eagle is a juvenile, its lack of a conspicuous white head also helps to reduce the dominant's aggressive tendencies.

Taloning

Locking Talons

Pecking

Mantling

FOOD FIGHTS   *A marauding eagle drops in at dinnertime, trying to wrest food away from the rightful owner.*

Harassed eagles fight back with many of the same displays their attackers use. They scream, raise the hackles of their head and neck, jab the attacker with their beak, and thrust their wings forward. In **mantling**, the defender spreads its wings over the carcass, as if to shield it from the opponent's sight. If eagles are exhibited together at the zoo, you may see any number of these defiant moves, especially during mealtime.

SEXUAL BEHAVIOR

Courtship

Late winter marks the beginning of pair bonding, when first-time and lifelong mates frolic in the sky, performing displays designed to reduce their aggression and synchronize their desires. The poetry of eagle flight is best showcased in three whirling, heart-stopping shows: (1) the chase display, (2) roller coaster flight, and (3) the cartwheel display.

The **chase display** is like watching two black kites connected by a string, diving, looping, and erratically zigzagging together across the sky. They may change flight positions over and over, as if rolling in a ball down an invisible hill. Occasionally, one eagle will turn upside down and touch the talons of the bird on top.

The **roller coaster flight** is exactly as it sounds. One eagle rises to a great height, peaks, and then plummets to earth, as if on a giant amusement ride. Before hitting the ground, it corrects and begins the steep ascent up the other side. It performs one giant U after another like a Red Baron pilot showing off.

The real barnstormer is the **cartwheel display**, in which two eagles fly to a great altitude, lock their talons together, and then simply free-fall, spiraling down thousands of feet in a great ball of feathers. Just as they are about to crash, they disengage and fly away in a harrowing but beautiful escape. While in the sky or sitting perched, the pair exchange **courtship calls**—high-pitched screams that ring in the air and speak of things wild and untamed.

Copulation

After all that fanfare, copulation lasts only a few seconds, or at most 2 minutes, but may occur several times a day. Usually, the female gets things going with a **solicitation display**—body lowered in a horizontal crouch, head held low, and wings outstretched. The male answers her soft solicitation call, flaps his wings, and moves his tail up and down. As he **mounts** on her back, he curls his talons so as not to injure her, flaps his wings, and calls again. Finally the fe-

CARTWHEEL DISPLAY   *Courting pairs lock talons in midair, then roll over and over in a dizzying free-fall.*

male twists her tail sideways so they can place their cloacas together. After mating, the pair sometimes preens for a while, then flies together to the nest.

HUMANS AND EAGLES

Eagles live a long life, during which they learn about the best places to hunt, nest, and perch, as well as how to find the best updrafts and thermals to take them to their wintering grounds. When they are finally sexually mature, they raise their chicks carefully, only one or two at a time. This long life and relatively low reproductive rate can make them vulnerable, however, especially in a world that we humans are changing so rapidly. The mature forests that eagles prefer are increasingly hard to find, as are undisturbed lakeside nest trees and unpolluted prey sources. We have already seen how poisons such as DDT can build up in the long-lived eagle, leading to nesting failures from thin-shelled eggs. What we don't yet know is how a buildup of heavy metals from acid rain or conta-minated groundwater might affect them. Because eagles are not prolific breeders, they don't have the genetic means to evolve tolerances quickly enough to rebound from these sorts of pressures.

Besides living under the dark cloud of habitat loss and human disturbance, many eagles are still being killed outright despite legal protection. Though the days of killing raptors for bounty money are over, many people still harbor an illog-ical hatred of these birds. Some believe ranchers should have the right to protect their livestock by shooting eagles on their property. As recently as 1970, a systematic effort to exterminate eagles in Wyoming was started on several ranches. Using helicopters and high-powered rifles, the hunters were able to "dispatch" 30 or 40 eagles a day.

Though sanctioned massacres have since been stopped, illegal poaching still feeds a rather lucra-tive black market in eagle feathers. The feathers, used in ceremonial headdresses bought by whites as well as Native Americans, can bring as much as $25 each, while an entire war bonnet of eagle tail feathers may fetch over $5,000. In addition, dozens of eagles are inadvertently killed each year—elec-trocuted on exposed power lines, captured in traps set for other animals, and poisoned by tainted bait or lead shot buried in their prey.

As a result, the breeding population in the con-tinental United States had dwindled to 2,600 pairs by 1990, well below their historic levels. Once found in 45 of the lower 48 states, eagles now breed in only 31 states, and, as of 1982, only 10 of these states accounted for 90% of the eagle nests. Thankfully, many people are working to reverse this downward trend; bald eagle recovery plans and captive breeding successes are the brightest lights of hope right now. As a result of these efforts, bald eagles in the northeastern and northwestern United States may soon be downlisted from endan-gered to threatened. In the Southeast and South-west, bald eagles are still an endangered species.

Zoos give us a rare opportunity to see this aer-ial master at close range. Though we may not see the spiraling flights or the tension-filled hunts, we can stare eye to eye with a bird that has inspired reverence in almost every culture it has soared above. Hopefully, as the eagle blinks down at us these days, it sees that our culture has turned a corner and that our reverence has a single focus: to bring the eagle back.

PARENTING BEHAVIOR

Nest Building

Eagles construct the mansions of North American bird nests. One Iowa nest that was used by various pairs for 40 years was an aching 4,000 pounds before it finally crashed to the ground! With their talons, eagles collect nest-building materials from trees, snatching stout branches and vines as they fly by. The female arranges the branches into place with her beak, weaving a latticework that resists the ravages of wind, rain, and snow. She adds to the structure each breeding season, then adorns it with a fresh, green sprig as if announcing that the nest is open for business. A typical nest in the wild is 5 to 6 feet wide, 2 to 4 feet tall, and placed near the top of the tallest live tree the birds can find. Eagles will also build alternate nests close by, in case they have to jump ship because of harassment or parasites at their main nest.

Caring for the Eggs

Because eagles raise only one or two chicks a season, they take particular care with their eggs. They cover the nest with their wings during wet weather, taking the chill themselves rather than endangering their eggs. Bald eagles can be fierce defenders, as researchers who approached too close to nests in Alaska have found out. An angry eagle is easily capable of swatting a person right out of a tree. In other parts of their range, though, eagles are not as aggressive and will merely circle high overhead and then fly off. It's very important not to test the limits of an eagle's patience, however, no matter where you are. Continual harassment can cause parents to abandon the nest before the eggs even hatch.

Feeding the Eaglets

If all goes well, at least two out of the three eggs will hatch, giving the parents more than enough to handle. If you are lucky enough to have eaglets at your zoo, it's worth visiting often to see them grow up. They will seem to develop before your very eyes, putting on a pound every 4 or 5 days to start, and ballooning from 4 ounces to as much as 11 pounds in only 3 months. To fuel this growth spurt, the parents take turns hunting, bringing a constant stream of nourishment back to the nest. When a new shipment arrives, the chicks issue relentless **begging calls** until one of the parents **shreds the prey** into bite-sized pieces and feeds it to them. If you have a chance to see a film of eaglet feeding, notice how the parents draw down their second eyelid, a protective membrane that covers the eye to guard against an accidental pecking by the eager chicks.

FEEDING THE EAGLETS *The parent shreds the prey into pieces, and the siblings fight over every morsel.*

Sibling Rivalry

Fights between siblings are common, and when one sibling is considerably larger than the other (especially noticeable in the early weeks of life), squabbles can even lead to fratricide. The young eaglets will peck at each other's heads with such vigor that they themselves will topple when they miss their target. The larger one may pincer the smaller sibling's beak, seize its wing, and drag it around the nest until it dies of exhaustion. Interestingly, parents rarely intercede in these fights, perhaps because it's in their best interest to feed only the offspring that is strong enough to survive.

Rehearsing for Real Life

When they are not eating or fighting, you can see the nestlings **preening**, **playing**, **sleeping**, and **exercising**. Just like kids everywhere, eagles jump up and down, use sticks as toy hammers, flap their wings, engage in mock attacks, and play tug-of-war with anything they find in the nest—old feathers, leaves, sticks, leftover prey, even a protesting nestmate. Nestlings will also **steal** a piece of food back and forth from each other until one seems to monopolize it. Usually, the loser is only waiting for another opportunity to race in for a rematch. These early games are valuable practice for adult life, when stealing will be a major source of energy income. When playing alone, they give their new muscles a workout by treading, prancing, trampling, and broad jumping across the nest—moves that they will someday use to dispatch unruly prey.

At 5 or 6 weeks, nestlings will exhibit **defensive behavior** when anything dares to come close to their nest, including a human. They may hiss at the intruder with wings outstretched and ruffled or even strike with talons or beak. As a last resort, a nestling may jump from the nest, usually injuring itself in the process. To prevent a bailout like this, it's best to leave eagle nests in the wild alone and hope that your zoo will provide a discreet one-way window into the nursery.

If the eaglet makes it to the age of 10 weeks, it will be large enough to leave the nest, but not the neighborhood. Juveniles are still dependent on their parents' handouts for up to 3 months until they are savvy enough to hunt on their own.

BALD EAGLE BEHAVIORS TO LOOK FOR AT THE ZOO OR IN THE WILD

BASIC BEHAVIORS

LOCOMOTION
- ☐ soaring
- ☐ diving

FEEDING
- ☐ hunting in flight
- ☐ hunting from a perch
- ☐ hunting on the ground
- ☐ wading in water
- ☐ cooperative hunting
- ☐ stealing
- ☐ scavenging

SOCIAL BEHAVIORS

FRIENDLY BEHAVIOR
- ☐ ***Communal Roosting and Bathing***
- ☐ ***Communal Hunting and Feeding***

CONFLICT BEHAVIOR

Threat
- ☐ coloration
- ☐ threat call
- ☐ circling display
- ☐ territorial chase
- ☐ perch site display

Food Fights
- ☐ aerial chase
- ☐ swooping attack
- ☐ running or walking attack
- ☐ leaping attack
- ☐ talon display
- ☐ wing display
- ☐ talon lock
- ☐ taloning
- ☐ pecking
- ☐ stealing
- ☐ mantling

SEXUAL BEHAVIOR

Courtship
- ☐ chase display
- ☐ roller coaster flight
- ☐ cartwheel display
- ☐ courtship calls

Copulation
- ☐ solicitation display
- ☐ mounting

PARENTING BEHAVIOR
- ☐ ***Nest Building***
- ☐ ***Caring for the Eggs***

Feeding the Eaglets
- ☐ begging calls
- ☐ shredding the prey

Sibling Rivalry
- ☐ fights

Rehearsing for Real Life
- ☐ preening
- ☐ playing
- ☐ sleeping
- ☐ exercising
- ☐ stealing
- ☐ defensive behavior

Head-lowered charge *In this high-intensity threat, an annoyed crane aims its spear and rushes toward an intruder, occasionally nipping at a tail or wing feather.*

SANDHILL CRANE

Two species travel through Nebraska in spectacular numbers each spring. One wears binoculars around its neck, sports two to five legs (with or without tripod), and communicates in loud whispers. Early in the morning, groups of these mammals creep down to the water's edge to see the other migrating species: a handsome, elegant bird honking, dancing, and feasting with 400,000 of its closest companions.

Sandhill cranes use a 70-mile stretch of Nebraska's Platte River as a rest and refueling stop enroute to breeding grounds stretching from Hudson Bay to eastern Siberia. The cranes are drawn to the river because it's wide and shallow (good for stand-up roosting) and because nearby fields are full of corn kernels and juicy insects. The hungry hordes of sandhills start descending in February; by the third week in March, the Platte is "knee-deep in cranes."

The racket made by thousands of cranes has an ancient, untamed quality that humbles a human listener. Their incredibly long windpipes are coiled like French horns to help amplify their sonorous honks and trumpet blasts. The amplification makes sense when you consider that a flock in the Platte may be trying to call in another flock that is flying nearly $2/3$ of a mile above them. They also use their voices to coordinate movements, defend their territories, find mates, and keep lifelong pair bonds alive. When cranes are close enough to see one another, they also rely on sophisticated visual displays that show off their distinctive gray plumage, white cheek patches, and bright red head skin.

VITAL STATS

ORDER: Gruiformes

FAMILY: Gruidae

SCIENTIFIC NAME: *Grus canadensis*

HABITAT: Rests in shallow, open wetlands; feeds in open fields; breeds in sedge meadows, bogs, tundra

SIZE: Length, 34–48 in   Wingspan, 6–7 ft

WEIGHT: 5.7–14.4 lb

MAXIMUM AGE: 24 years in captivity

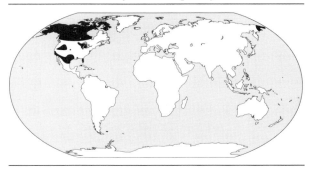

Seeing and hearing so many of these magnificent creatures in one place is like having a box seat to one of nature's grandest events. Even if you can't get to the Platte, you can still experience many of the same sights and sounds at your local zoo. Once you learn to recognize the goings-on in these captive flocks, you'll be ready to book a flight to the big show in Nebraska. What you learn at the zoo will come in handy there.

BASIC SANDHILL CRANE BEHAVIORS TO WATCH FOR

LOCOMOTION

Cranes **walk** in a gait similar to ours, but when they **run**, they bounce a little with each step. When running to build up speed for a take-off, the cranes **flap** their wings to stay balanced. Eventually, the magic moment of lift-off arrives and they head for the heights, which, during migration, can be anywhere from $1/3$ to $2/3$ of a mile straight up. They'll cruise at that altitude for 300 miles or more before coming down for a break.

The energy costs of a flight like that would be prohibitive were it not for the crane's ability to **soar**. Like living gliders, sandhills spread their 7-foot wingspans and soar on thermal after thermal (columns of warm rising air), tapping into the energy of air currents rather than flapping. They usually fly with their neck straight out like a spear and their legs outstretched behind them. If it's very cold, however, they do their goose imitation: they tuck one or both legs forward in their flank feathers.

FEEDING

Wild cranes vanish from their wetland roosts at dawn, traveling in bands of 20 or more to upland fields. They return to the wetland at mid-morning to rest and preen, then head back to the fields for an afternoon meal. In natural grassy openings, they **forage** for grains and berries and **stalk** insects, mice, frogs, lizards, and snakes. They can also use their bill to **probe** for underground delectables such as roots and worms. In agricultural fields, migrating cranes take advantage of free corn kernels that the harvester and the cows have missed.

In the zoo, cranes are fed a mixture of grains and proteins, but they may also hunt for natural foods growing or wriggling in the exhibit. When they **drink**, they look like chickens; they slurp water up and then raise their bill to let it drain down their throat.

PREENING

Like all birds that depend on their feathers for flight and weatherproofing, cranes are meticulous when it comes to **feather cleaning**. They clean one feather at a time, nibbling at its base, then drawing it through their beak to zip together its interlocking barbs. This eliminates ragged gaps that would ruin the airfoil effect of the wings.

PREENING AND FEATHER PAINTING *Sandhills comb their feathers, coat them with oil, then paint them rusty brown with a mixture of grass and mud.*

To help weatherproof themselves, cranes perform **back slicking**: they poke their beak into an oil-producing gland at the base of their tail and then spread the oil everywhere they can reach. The best time to watch for preening is between August and October, when **molting** feathers on the body, neck, and wings are being pushed out by new ones underneath. Once their flight feathers drop, the grounded cranes lie low, hiding in dense cover and preening for hours on end until they can fly again.

For a good overall cleaning, cranes will also take to the water for a bath. **Bathing** cranes crouch in the shallows, flapping their wings and bobbing up and down to get good and wet.

FEATHER PAINTING

Instead of oil, a back-slicking sandhill may occasionally spread a mixture of mud and grass on its feathers. This produces a bright rust-brown stain that puzzles researchers. One camp says the bird uses the stain to camouflage itself, and another says it's a sign of sexual status, but only the cranes know for sure.

STRETCHING

As you watch cranes stretching, you may be able to pick out the four basic moves in their stretching regime. The most common one is the **unilateral stretch**, involving one side of the body (a leg and a wing) at a time. In a **bilateral stretch**, the crane unfurls both wings at once and keeps its neck straight up in the air. In a similar move called **bow stretching**, the bird also raises both wings, but stretches its neck horizontally instead of vertically. A fourth type of stretch is called **jaw stretching**, it looks like a gaping yawn, but is actually just a way to relieve muscle fatigue.

FRIENDLY BEHAVIOR

Flocking

Cranes have long protected themselves from predators and made the most of a patchy food supply by migrating in flocks. Together the birds watch for danger and follow one another's lead when it comes to finding food. Rather than reinvent the wheel, they follow the same migratory paths that their ancestors used, stopping at proven feeding grounds along the route. This way, they waste little energy in wrong turns or aimless food hunts.

One device that helps cranes synchronize their movements is the **location call**, a plaintive cry used to locate and draw in other cranes. If part of the flock lands for feeding, they issue location calls to tell the rest of the flyers, "There's good feeding here." When they are ready to fly again, the cranes stand erect, face into the wind, and signal their plans with the high-frequency **flight intention call**.

Dancing may also have a social function beyond the well-known role it plays in sexual relations. Cranes may dance together to alleviate tension among flock members, for instance, or just to entertain themselves.

Communal Roosting

Turning in at night is another communal activity, and cranes assemble in large flocks for the occasion. With thousands of ears tuned to the slightest noise, predators have

HOW SANDHILL CRANES INTERACT

a hard time raiding these group roosts. Adult cranes sleep standing on one leg or two in shallow water or on the ground. Their bill either faces forward, rests on their breast, or is bent back and tucked beneath their feathers. Rather than standing, resting juveniles sometimes put most of their weight on their ankles, the knobby joints halfway up the leg.

Communal Defense

When dogs, humans, or other suspicious characters approach cranes in the wild or at the zoo, the birds assume a **tall alert posture**, standing rigidly erect with their neck straight up, their bill horizontal, and their body nearly vertical. As soon as one crane assumes the posture, it spreads through the flock like a red alert, rallying the cranes and readying them for flight or fight. **Feather sleeking** goes hand in hand with the alert posture; with feathers sleeked flat against their body, the cranes are ready to fly with a minimum of drag.

When sandhills are curious rather than alarmed, they assume the **alert investigative** posture. This is similar to the tall alert except that the body is either horizontal or vertical, and the head moves back and forth to scan the surroundings. You can also tell the difference in the two postures by watching the responses of other cranes; unlike the tall alert, the alert investigative posture is not contagious.

CONFLICT BEHAVIOR

In the wild, mating pairs separate from the flock during breeding season to establish territories, often delineated by natural barriers such as hills or rows of trees. Zoo managers try to simulate territorial conditions by placing cranes in separate enclosures during breeding season. Without this precaution, the fiercely territorial birds would be apt to injure one another in boundary disputes. You can see glimpses of this fury in the aggressive exchanges that next-door neighbors conduct through the fence.

Outside of breeding time, when cranes are back together in flocks, you may catch them raising their hackles over questions of dominance. In addition to active displays of aggression, be aware that cranes are also communicating with static cues such as coloration. The whiter and brighter the cheek patch, for instance, the more dominant the individual. On the other hand, the lack of a red cap or white cheek patch indicates juvenile status and guarantees immunity from attack.

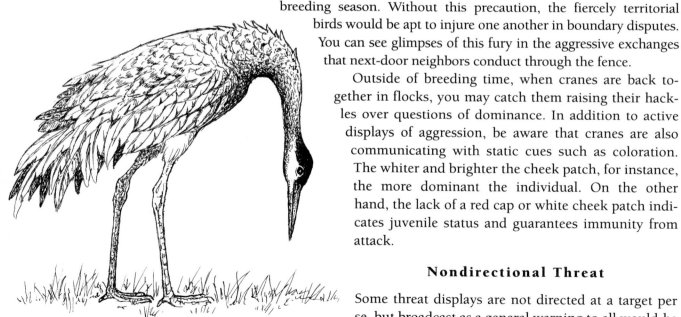

LOW-BOW THREAT   *To issue a general warning, a resident crane lowers its head and manipulates its muscles to expose the bright red of its crown patch. The low bow is also used to cap off a high-intensity threat.*

Nondirectional Threat

Some threat displays are not directed at a target per se, but broadcast as a general warning to all would-be enemies. Cranes sometimes perform the **low bow** when flying into the midst of a flock or when landing next to an intruder in their territory. A challenger also gives the low bow after performing a high-intensity threat (see below). The bird bends its neck steeply downward and tilts its lowered head so its bright red crown is plainly visible. By contracting or relaxing its muscles, the crane can control how much of this red skin area is exposed and how engorged, and thus red, it is. The **ruffle bow** is similar to the low bow except that the crane shakes its feathers as it bows. A third type of nondirectional display used to defend a territory is the bill-up **unison call**, a call that you'll also hear in courtship.

Low-Intensity Threat

The first line of defense directed at a target is the **directed walk threat**. The displaying crane parades around the offending party with its head angled downward, its crown expanded, and its flight feathers raised in a vertical fan shape. It steps stiffly, carefully placing each leg, pumping its head, and showing off its engorged crown. As a finale, the crane may lower its head with a growl and touch its leg or flank in a **mock preening** movement.

High-Intensity Threat

The next step up in aggressiveness is the **head-lowered charge**, in which the crane clamps its wings tight to its sides, extends its neck, and runs toward its opponent, bill aimed like a jousting lance. If the opponent doesn't retreat quickly enough, the charger won't hesitate to nip its wing or pluck some feathers. The charge may also turn into an **aerial pursuit**, in which the aggressor kicks at the fleeing bird.

When a dominant bird wants access to a drinking or feeding spot, it may aim at the other crane with a **bill stab**—a frozen version of the head-lowered charge, without the running. If the subordinate knows what's good for it, it will move aside and act properly submissive.

Submission

When a crane wants to "cry uncle," it lowers its posture and tries to make itself and its bright crown as inconspicuous as possible. In the **neck-retracted submissive posture**, the bird holds its body horizontal, pulls its neck back, and loosely folds its wings. In striking contrast to the dominant's stiff gait and enlarged crown, the submissive bird walks in a loose-jointed fashion and keeps a low crown profile. Rather than risk being misinterpreted, the lesser bird may opt to **retreat**, which is the safest move of all.

UNISON CALL   *In their classic poses, the male and female take turns singing a haunting, staccato song. Usually used as a territorial defense, the duet also helps strengthen pair bonds.*

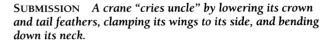

SUBMISSION   *A crane "cries uncle" by lowering its crown and tail feathers, clamping its wings to its side, and bending down its neck.*

Bill Sparring

If two cranes of equal aggressiveness meet and neither submits, you may see some bill sparring. The contests usually occur between males, especially when a resource such as fresh water is limited or when pair formation is in progress. These brief but important duels allow the rivals to show their stuff and win themselves a place in the pecking order. Once rankings are set, the birds can resolve their conflicts with just a subtle show of crown or stiffness of step.

The sparring begins with some face-to-face **bill jabbing**. Next, with their **wings spread** wide and their upraised bill tips almost touching, the cranes vault into the air and **kick** at one another. These kicks rarely log any serious injuries, however, since one bird usually retreats or backs down. As if to get the last word, the dominant may end the sparring episode with a bill stab, a head-lowered charge, or a low bow.

Predator Defense

Cranes use various threats when dealing with mammalian predators (or nosy humans) that invade their flocks. A bird may first perform a bill stab display, and if the predator doesn't retreat, it may aim its spear with a head-lowered charge. If the predator turns the tables and attacks, the crane defends itself with kicks and chases.

Wing-spread display

Bill stab

Kicking

Bill jabbing

BILL SPARRING   *The choreography of a sparring match mixes and matches a few key moves. The stationary bill stab issues a pointed warning. In response, the opponent spreads its wings to appear large. Keeping its wings spread, it thrusts its bill forward in a jabbing motion. Leaping high into the air, both cranes kick on their way down.*

Though you won't see predator conflicts at the zoo, you may see the feisty cranes take on other species of animals displayed in their exhibit.

SEXUAL BEHAVIOR

When cranes leave their parents, they travel with a flock of other juveniles for 2 to 3 years. It is here that they find their lifelong mates. The process of pair formation is a moving one, filled with intricate displays and joyous dancing, all designed to cement the cranes' bond and synchronize their sexual time clocks.

Pre-Courtship

Though it usually functions as a territorial defense, the **unison call**, sung as a duet, is also used to form and strengthen the bond between mates. The birds stand side by side, and as they sing, the female keeps her bill horizontal, while the male tilts his

HUMANS AND SANDHILL CRANES

The dancing of cranes inspires us, perhaps because it seems to express a joy that also resides in the human heart. Throughout the world, native peoples have copied the crane's ballet and used the crane's migration to celebrate the seasons. Today there is something to mourn as well as celebrate when we think of cranes. Seven of the 15 different species are either rare or in danger of becoming extinct.

For the moment, sandhills seem to be holding their own, though we can't say the same for their habitats. Consider the Platte River, where 75% of North America's sandhill crane population stops over before migrating north. This river used to be perfect for cranes: in 1866, the channel was about a mile wide, with a shallow, meandering flow and plenty of room for the cranes to roost mid-river without worrying about land predators. Fields on both sides provided ample food during their daytime excursions.

Today it's simply not the same river. In a little over a century, 70% of the flow has been diverted, mostly for irrigation. The main channel west of Overton, Nebraska, now averages less than 300 unobstructed feet from bank to bank, and it's get-

ting smaller with each passing year. When the flow diminishes during the drier months, seedlings of cottonwoods and willows are able to establish themselves on the exposed sandbars. In the past, intruding seedlings were never a problem because springtime flows were strong enough to scour them away and keep the channel open. Today, even at full capacity, the flow is too weak to dislodge the seedlings, and stands of woody vegetation are clogging the crane's roosting channel. The fear is that the cranes will eventually have to abandon the Platte for less productive areas.

Another threat looms simply because of the crane's mass lifestyle. With so much of the population in one place at one time, a storm, disease outbreak, toxic chemical spill, or other environmental accident would spell major tragedy. Adding to their vulnerability is the fact that cranes are slow breeders, raising an average of less than one chick per pair per year. For all these reasons, critical habitat protection is essential for crane survival. Should our cautions fail, however, captive breeding programs in zoos may offer the only hope of rekindling populations before they flicker out.

straight up. Typically, the female utters the first note, touching off a series of long, pulsed, staccato calls that alternate between male and female. Though both sexes keep their outer wing feathers tight to their body through the ceremony, the male may raise his inner wing feathers, causing a ruffling over the tail.

Courtship

Three attraction-getting displays are part of the springing, whirling dance that cranes are known for. In the **upright wing-stretch display**, the male raises his wings while walking in full alert posture in front of the female. In the **horizontal head pump**, both male and female spread their wings and bring their heads down to the ground and back up again, repeating the motion several times in a row. Another attention-getting display is the **vertical toss**, in which the birds lower their heads, grab a stick or piece of vegetation from the ground, then fling it into the air on the way back up.

In the **bow display**, the birds look as if they are warming up for the leaping portion of the dance. Beginning in a crouch, with neck retracted and legs bent, each bird tilts its body forward and extends or lifts its wings. Suddenly, like a jack-in-the-box popping, it stands up straight with neck extended, holds still a moment, then bows back down into the crouch position. Between bows, each crane may take a surprise jump.

Bowing

Vertical leap

Tossing

COURTSHIP DANCE *Like ballet dancers warming up at the barre, bowing birds crouch, pop up, and crouch again. When the dance begins, one partner springs into the air while the other symbolically tosses a piece of vegetation.*

Once they are fully inspired, the cranes explode into the **vertical leap display**, the heart of the crane dancing. Each bird coils into a semicrouch and then springs as high as 15 feet into the air, with its wings spread wide and its legs dangling in front for balance. Holding its head high like a ballet dancer, it flaps its wings up to five times before coming down to earth. The leap of one crane in a flock may set others to leaping, until the entire flock is bouncing up and down like birds on a trampoline. The enthusiasm that fills the dancing flock is contagious, and as you will see, it can easily sweep away human hearts as well.

Luckily, breeding season is not the only chance you'll have to see dancing at the zoo. Young cranes dance to develop motor skills, submissive cranes dance to thwart a dominant's aggression, threatened cranes dance when they're nervous, and some cranes just dance because they feel like it.

Copulation

Either sex may start the preliminaries of mating, but very often they both get the same idea. The male walks toward the female in a **parade march**, with his flight feathers erect, his bill jutting straight up to the sky, and his crown bright red as in "ready." The female postures in much the same way, and when the male approaches her closely, she holds out her wings and droops them at the wrist. Listen for the series of purring notes known as the **pre-copulatory call**. As the male inches nearer, he lowers his bill, extends his wings, and leaps onto the female's back, flapping to maintain his balance. After a brief mating, he dismounts by stepping backward or by hopping over the female's head.

Post-Copulation

Contrary to the mating practices of most animals, the crane's ritual doesn't end abruptly after mating. For about 20 seconds afterward, the birds stand side by side with their crowns expanded and their bills pointed skyward. Then, in perfect unison, they may perform a **low bow** display or a **ruffle bow**. To wind down, they may **preen** or engage in a little celebratory **dancing**.

PARENTING BEHAVIOR

Nesting

Cranes build their 3- to 5-foot-diameter nests in wide, lonely wetlands filled with sedges, cattails, or bog plants. The dishlike nests are usually perched safely above floodline on a mound of sticks, mosses, dead reeds, and rushes. If you see the cranes at your zoo **building a nest**, be sure to listen for the low moaning **nesting call** uttered during construction or while the female is **incubating** her one or two eggs.

Caring for the Chicks

Crane chicks are precocial, which means that they are fairly well developed when they are born. A mere 2 or 3 days after hatching, they are strong enough to follow their parents to feeding grounds where they try their hand at stalking and probing. Until their accuracy improves, the apprentices must rely on their parents for the bulk of

DISTRACTION DISPLAY *One parent acts as a decoy, calling attention to itself to lure the predator away from the nest.*

their nutrition. You'll no doubt hear their peeping **food-begging call** at the zoo or see the parents and chicks **bill touching**, a sign that food is being begged or delivered. Thanks to the steady stream of handouts, the gangly juveniles are able to grow large enough in 3 months to join the flock for the long migration back to their winter haunts.

Besides dishing out dinner, crane parents are also kept busy on **predator-mobbing** patrol. Working as a team, the pair surrounds intruders, such as dogs, foxes, or humans, and barks out loud single-syllable **guard calls** until they retreat. If eggs or chicks are in the nest, either parent may perform a spread-wing **distraction display** to refocus the enemy's attention.

Because the chicks are able to wander at quite a young age, **contact calls** are an extremely important part of their protection package. The chicks utter these purring calls constantly while foraging or being brooded to assure their parents that they are close-by and well. Should a chick become separated, chilled, or hungry, it summons its parents with a loud, unbroken **stress call**.

Throughout their 10-month rearing period, the chicks are learning the ins and outs of being a sandhill crane: the best feeding sites, roosting sites, food items, migration routes, and even which sex to be attracted to. They are also rehearsing the moves they will need in later life, so be sure to look for dancing, bowing, and stick tossing in their play.

SANDHILL CRANE BEHAVIORS TO LOOK FOR AT THE ZOO OR IN THE WILD

BASIC BEHAVIORS

LOCOMOTION
- ☐ walking
- ☐ running
- ☐ wing flapping
- ☐ soaring

FEEDING
- ☐ foraging
- ☐ stalking
- ☐ probing
- ☐ drinking

PREENING
- ☐ feather cleaning
- ☐ back slicking
- ☐ molting
- ☐ bathing

FEATHER PAINTING

STRETCHING
- ☐ unilateral

- ☐ bilateral
- ☐ bow stretching
- ☐ jaw stretching

SOCIAL BEHAVIORS

FRIENDLY BEHAVIOR
Flocking
- ☐ location call
- ☐ flight intention call
- ☐ *Communal Roosting*
  Communal Defense
- ☐ tall alert posture
- ☐ feather sleeking
- ☐ alert investigative posture

CONFLICT BEHAVIOR
Nondirectional Threat
- ☐ low bow
- ☐ ruffle bow
- ☐ unison call
Low-Intensity Threat
- ☐ directed walk threat
- ☐ mock preening
High-Intensity Threat
- ☐ head-lowered charge
- ☐ aerial pursuit
- ☐ bill stabbing

Submission
- ☐ neck-retracted submissive posture
- ☐ retreating
Bill Sparring
- ☐ bill jabbing
- ☐ wing-spread display
- ☐ kicking
- ☐ *Predator Defense*
SEXUAL BEHAVIOR
Pre-Courtship
- ☐ unison call
Courtship
- ☐ upright wing-stretch display

- ☐ horizontal head pump
- ☐ vertical toss
- ☐ bow display
- ☐ vertical leap display
Copulation
- ☐ parade march
- ☐ pre-copulatory call
Post-Copulation
- ☐ low bow
- ☐ ruffle bow
- ☐ preening
- ☐ dancing

PARENTING BEHAVIOR
Nesting
- ☐ nest building
- ☐ nesting call
- ☐ incubating eggs
Caring for the Chicks
- ☐ food-begging call
- ☐ bill touching
- ☐ predator mobbing
- ☐ guard call
- ☐ distraction display
- ☐ contact call
- ☐ stress call

THE POLES

COURTSHIP SWIMMING   *With characteristic grace, courting belugas "dance" together.*

BELUGA WHALE

T he fossil folks tell us that the beluga's ancestors were warm-blooded, air-breathing mammals that prowled the earth on four legs. Some 200 million years ago, they made their way into the sea, and natural selection has been perfecting their aquatic adaptations ever since. Today they live in Arctic and sub-Arctic seas surrounding the North Pole and are perfect examples of how living with salt, water, and ice can ultimately dictate how a species looks and behaves.

After years of fighting water resistance, beluga bodies have become more streamlined than their ancestors' were. But streamlining alone can't explain why belugas travel faster than mathematical models predict they should. Physicists eager to learn the beluga's secret had to put away their formulas and look through a microscope. They found that the cells in the beluga's smooth skin are constantly forming and falling off, creating a lubricant layer that squirts them through water without the usual amount of friction. This slipperiness, combined with the submarine-like sleekness of their body, makes for nearly effortless travel.

Lungs—holdovers from their land-dwelling days—require belugas to stay within commuting distance of the water's surface. To make breathing easier, their nostrils have migrated to the top of their head and fused into one blowhole. A muscular pad opens and closes this hole as the whale breaks the surface, letting out used air and taking on new air without taking on water. The used air comes out in a big "blow" that looks more like a fountain of water spray

VITAL STATS	
ORDER:	Cetacea
FAMILY:	Monodontidae
SCIENTIFIC NAME:	*Delphinapterus leucas*
HABITAT:	Arctic and sub-Arctic waters; typically spend winter in pack ice, moving to coastal waters and river estuaries in summer
SIZE:	Male, 12–16 ft Female, 11–14 ft Maximum size, 22 ft
WEIGHT:	Average male, 3,300 lb Average female, 3,000 lb Maximum weight, 4,400 lb
MAXIMUM AGE:	30 years

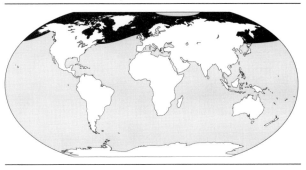

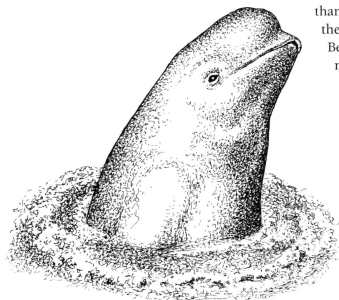

SPY-HOPPING *Curious belugas rise out of the water to take a good look around.*

than air. This mist forms when warm, pressurized air from the lungs expands and condenses in the cool outside air. Belugas can also exhale underwater when they need to be more discreet. To refill, they slide their blowhole barely above the surface, a movement that may be difficult for predators to detect.

It's amazing that a 15-foot, 2,200-pound animal could do anything in the ocean without someone noticing, but they do quite often. Researchers tell tales of belugas slipping unseen up to their boats, raising their heads over the side to peer at them, and then disappearing, all without disrupting the boat's position. Much of this noteworthy grace is due to the beluga's flexibly jointed bone structure. You'll notice as you watch belugas (and as they watch you) that they can turn their head from side to side without twisting their entire body. They can also maneuver at extremely slow speeds, paddling their rounded flippers forward or in reverse to get in and out of tight spots that would strand less adept whales. This agility boosts not only their popularity in zoos, but also their survival odds in the pack ice and in the shallow, rocky waters where they fish.

Offshore, belugas are able to negotiate depths, thanks to specially adapted controls that slow their heart, ration blood to vital organs, and make the most of every cubic inch of air. To extend their stay underwater, belugas have evolved a tolerance to high levels of carbon dioxide, allowing them to hold their breath for up to 20 minutes. They can also tolerate quantities of salt in their food and water that would stymie most mammals. To get rid of excess salt, they have evolved huge kidneys composed of 400 lobes, each with its own excretory duct. These lobes increase the filtering surface area and help the belugas strain salt from their system.

Besides adapting to salt and to water, belugas have evolved ways to beat the frigid temperatures of polar seas. They have a 6- to 8-inch-thick layer of insulating blubber that accounts for 40% of their weight, covered by a thick skin that not only insulates, but also keeps them from lacerating themselves on the ice. In fact, the skin has grown so thick near the crown of their head that they can break through a solid ceiling of ice to breathe. The brilliant snow-white color of their skin is a beautiful complement to this iceberg slush. The camouflage helps fool not only their traditional enemies, killer whales and polar bears, but their modern enemy, humans, as well.

Belugas also have keen senses that allow them to zero in on feeding and breathing opportunities in the constantly shifting blocks of ice. Sensory areas of the beluga's mouth are thought to be able to detect a chemical change that means open water (and easy breathing) ahead. Even more accurate is their special sonar system that enables them to "see with their ears" the way bats and dolphins do. How these senses contribute to long-distance navigation we are not yet sure.

We do know that certain populations of belugas leave the pack ice like clockwork each spring, finding their way to coastal estuaries and rivers where it's safe and warm enough to raise their young. Some belugas have been known to travel as far as 1,260 miles up rivers in Siberia to give birth. For the people fortunate enough to live near a beluga summering place, the return of the whales is a magnificent way to mark the passage of time.

BASIC BELUGA WHALE BEHAVIORS TO WATCH FOR

LOCOMOTION

Belugas, whether they are **swimming** backward, forward, or upside down, are a picture of athletic grace. Next time you're at a zoo with belugas, plant yourself at an underwater observation window and prepare to be mesmerized by the everyday movements of these small whales. They often come by the window, turning their bodies in all directions and even rotating their heads beside the crowd of onlookers. Using their tail flukes for power, they swim at a leisurely 2 to 5 miles per hour, but are capable of bursting up to 14 miles per hour when frightened or trying to escape. Females with calves in tow must slow down to a top speed of 6 to 7 miles per hour.

BREATHING

Although belugas are capable of staying underwater for up to 20 minutes, they usually surface for a breath at least every 3 to 5 minutes when feeding. When migrating, they surface every 10 to 40 seconds, pausing a scant 3 seconds before resuming their long journey. Since whales are adapted to stay suspended in the salinity levels of ocean water, they have to work harder to bring themselves up to the surface in the less salty estuaries and inland rivers that they frequent during calving season. In winter, a ceiling of ice that's too thick to break through presents another kind of obstacle. Belugas are adept at finding natural breaks in the ice, however, and if need be, they can swim up to 2 miles to find the next breathing hole.

FEEDING

For most of the year, belugas feed in the shallow, iceberg-studded waters of the continental shelf. In their wanderings, however, belugas also forage in both deep (up to 2,090 feet) and shallow ocean waters and in the fresh waters of inland rivers. This extensive feeding range offers a varied menu, which the beluga, the most adventurous eater of all the whales, takes full advantage of. It feeds on bottom-dwelling as well as open-sea organisms, from sandworms to octopuses, and is limited only by the size of its mouth and esophagus.

Its highly flexible neck allows the hungry beluga a wide sweep; when cruising the ocean floor, it can either **vacuum** up prey animals with its puckered lips or squirt a **jet spray** of water to dislodge them. Groups of belugas may cooperate to drive schools of fish up sloping shores or into the shallows, where the fish soon panic and grow exhausted.

In a typical aquarium or zoo, belugas enjoy a smorgasbord of fish, including herring, capelin, smelt, and mackerel, along with squid and a sampling of vitamin supplements. A well-exercised beluga will finish off as many as 60 pounds of these delectables in a day.

RUBBING

One day you may notice that the ice-cream white belugas at your zoo are starting to look a little yellow. It's nothing to worry about. Belugas shed a layer of skin each year, sloughing off the old yellowed garb to reveal a fresh layer of white underneath. To help speed the process of shedding, they will often rub against objects in the tank, as if scratching a body-sized itch.

If you're ever beachcombing in the beluga's wild haunts, watch for signs of a "rubbing party" of belugas in the shallows right offshore or in the shallow inlet of an estuary. You may not see whole bodies, but you'll see plenty of tail flukes thrashing as the whales rub their molting sides, backs, and bellies along the pebbly grit of the shore bottom.

RUBBING   *Each year, belugas shed their yellowed skin, rubbing it on the rocks to expose a new, snow-white layer.*

RESTING

If your beluga looks a little less than "with it," it may simply be taking a whale-sized catnap. Dozing belugas remain in the same spot for a long time, gently waving their flukes to stay close to the surface. Every now and then (not as often as when awake) they dreamily rise to the surface to breathe, submerging slowly to begin treading water again. One side of their brain is alert (complete with one open eye) while the other side rests. Alternating sides allows the belugas to fully rest, but still be conscious enough to take intermittent breaths.

How Beluga Whales Interact

In the wild, belugas gather in large herds that may range from 100 to up to 10,000 individuals, depending on the season. The summer birthing season, for instance, traditionally brings whole populations of belugas together in estuaries and rivers in Siberia and northern Canada.

Within these herds, there are family groups and male groups. Family "pods" are composed of females, their newborns, and one or more older calves. Early family life is a kind of social and navigational training camp for these young apprentices. During migration, calves stick close by their mothers to learn the routing secrets of generations of whales. Adult males form their own separate pods of 8 to 16 individuals.

In deep water, the groups swim in a formation that may help them fish cooperatively for gregarious or schooling fish. When they are in shallows or near banks, the belugas disperse, presumably to feed on scattered bottom-dwelling organisms. Herd members communicate with one another via facial expression, flipper touch, rubbing, mouthing, and even more important, through sounds. Whistles and squawks can travel farther than visual or tactile cues, and other whales can pick them up even

VOCALIZING   *When whistling or squawking, the beluga contracts the muscles above its jaw and moves its bulging melon back and forth.*

in murky waters. Sound works so well, in fact, that belugas have become downright loquacious. Their large repertoire of calls (many of which humans can hear) led early sailors to dub them the "sea canaries."

FRIENDLY BEHAVIOR

You can tell when belugas are relaxed or in a friendly mood by watching the melon on top of their head. A **flaccid melon** set far back signals that they are mellow. This laid-back comradery is typical in group interactions. Content belugas are notoriously playful, and in the protected realm of the zoo, you're likely to see **play chases** and other favorite whale games. **Cooperative hunting** is another form of friendly, tolerant behavior made possible by the beluga's rich and communicative social life.

CONFLICT BEHAVIOR

Threat

We don't know much about the beluga's language, but we can interpret some of their visual signals. You can spot an annoyed beluga, for instance, by looking for a **turgid melon** set far forward on its head. (If the melon starts to move back and forth, however, the whale may not be annoyed; it may simply be vocalizing.) To threaten an opponent, a beluga will **gape** with open mouth to show off its 40 peglike teeth. If its

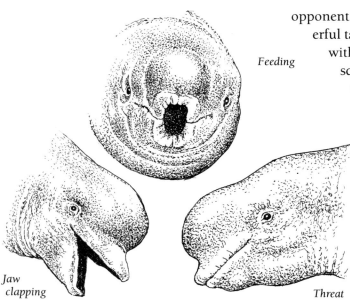

Feeding

Jaw clapping

Threat

FACIAL EXPRESSIONS *The beluga's lips are flexible enough to vacuum up prey from the ocean bottom. When angered, the beluga makes a statement by sharply clapping its jaws together. A turgid, thrust-forward melon is a sign of displeasure.*

opponent shows no signs of submitting, it may pump its powerful tail and launch into an open-mouth **chase**, punctuated with a series of **jaw claps**. As the jaws slam together, the square, worn-out teeth strike one another with a sound that says, "I am not amused." Belugas may also announce their displeasure by **slapping the water** surface with their flippers, flukes, or their entire body. Many a keeper who's been a few minutes late with the food bucket has been drenched by these tantrums of impatience. In the wild, nonvocal slaps and claps, in addition to a host of barks, groans, yelps, squeaks, and squawks, allow group members to settle most disputes quickly.

Biting

When all else fails, annoyed belugas resort to a more indelicate approach: they bite. Since their teeth are rather blunt, biting gets the message across without any real bloodshed. Because their skin is soft, however, the bites leave souvenir rake marks—two or three parallel grooves that you can plainly see.

SEXUAL BEHAVIOR

In the wild, breeding takes place offshore or in the pack ice sometime between February and June, just before the whales travel to their coastal summering grounds. Dominant male belugas appear to mate with several females. Sexual behavior can also be seen in zoo environments. You can tell males because they're about 20 inches longer than females, have larger melons, and are somewhat lighter in color. To identify females, look for the gray coloring around their blowhole and eyes.

Courtship and Mating

Look for lots of **chasing**, **rubbing**, **mouthing**, **nuzzling**, **posturing**, and **jaw clapping** in beluga courtship. The male pesters the female, performing what look like threat gestures as he chases her. She may **solicit** and rebuff him many times before mating. Rolling on her side, one flipper out of the water, she flashes her underside toward him or rubs against him until his large erection becomes obvious. They **mate** underside to underside, then gracefully fall away from each other.

BITING *In a fight, belugas may rake each other's soft skin with their blunt teeth.*

Belugas tend to be more curious than cautious when it comes to human beings. Since we are not a traditional part of their watery habitat, they have never bothered to evolve a fear of us. Instead, we are a novelty, and once they get used to the look and sound of us, they usually accept the hand that reaches out to stroke or feed them. This trusting nature enables these whales to form exceptionally close bonds with their favorite zookeeper. All it takes is one harassing incident, however, and belugas become suspicious and self-protective around humans. Individuals that have lost their kin to whaling boats, for instance, learn to make themselves scarce in the waters that whalers frequent. If they do run into nets, they are often smart enough to find their way out, leading the rest of the herd out with them.

Unfortunately, even these Houdinis of the whale world can't always escape human persecution. In the eighteenth and nineteenth centuries, whalers would force mass strandings of these whales, killing hundreds to top off their supply of whale oil from their primary target, the bowhead whale. In the early part of this century, members of the fishing industry claimed that belugas were directly competing with them for fish. In one government-sponsored "control" program in the 1930s, bombs and sharpshooters annihilated belugas seeking safe harbor in the Gulf of Saint Lawrence.

Today, with only 62,000 to 88,000 belugas left in the world, biologists want to keep a close eye on the remaining populations. An international treaty called CITES (Convention of International Trade in Endangered Species) has established a ban on commercial hunting of belugas in hopes that the whales can rebuild their populations. Despite the ban, many belugas are still being killed commercially by nontreaty whalers from Norway and the former USSR.

Even if all whaling countries would agree to the treaty, it still wouldn't protect the whales from the invisible dangers that threaten sea creatures. PCB, DDT, and mirex—toxic compounds so deadly they have been banned in the United States and Canada—have shown up in high concentrations in whales that frequent the St. Lawrence Seaway. It turns out that these substances, which are not soluble in water, dissolve wondrously in blubber. Because belugas are long-lived and near the top of their food chain, they accumulate dangerous quantities of these chemicals and then may pass them on to their newborns through fat-rich milk. As writer Jon Luoma points out, "For the St. Lawrence and indeed for the entire Great Lakes system, the lipid-rich and hence toxic-assaulted beluga, once called the 'canary of the sea,' may also be the canary in the mine."

Oil rigs and hydroelectric dams in beluga habitat may also pose invisible but deadly threats. Heating up the water, for instance, could turn the warm havens of estuaries into dangerous hot tubs. For that matter, our mere presence in their traditional breeding estuaries may be stressing belugas in ways that are hard to measure. Though the oft-photographed whales seem to be acclimating to the freighters and pleasure boats in their midst, they may not be as comfortable as they look. As we've learned from the decline of other species, tolerating a bad situation is not the same as thriving in it.

PARENTING BEHAVIOR

So far, attempts to breed belugas in captivity have produced some calves, but none have survived more than a few weeks or months. Each birth teaches us something about how to create an environment conducive to healthy calving. Males, for instance, are now excluded from the tank during and after birth due to their tendency to attack calves and worry females. In this and many other ways, researchers are striving to perfect captive breeding techniques. Until they're successful, the best place to see parenting behavior among belugas may be on nature programs or from the decks of whale-watching boats.

Wild belugas leave the pack ice in early spring and migrate into shallow bays, estuaries, and rivers, where they will stay from mid-June to mid-September. The shallow water protects them from predators and provides a warm place to raise their calves. Here, the calves have time to put on a thick layer of fat before they have to face the sapping cold of the Arctic seas. Adult females benefit from the warmer surroundings as well. Since they don't need as much fat to keep heat in, they can transform their blubber stores to rich milk for nursing. Migrating into the coast for warmth also saves them the trouble of migrating thousands of miles farther south to reach temperate oceans.

Birth

After a 14- to 15-month pregnancy, a Beluga cow will usually break away from the main herd to give birth, allowing only one nonbreeding or immature female to accompany her. Whether these animals are an actual help to the mother or whether they are simply curious is open to question. Researchers believe some of them may be tag-along older calves trying to milk their maternal ties a little longer.

ECHELON FORMATION   *Newborn belugas stick close to mom, drafting in the pressure wave caused by her movement.*

Caring for the Calf

As maternal ties go, the beluga's is indisputably strong. If a newborn calf is having trouble breathing, the female will turn on her side and **lift the newborn** to the surface with the flat of her tail. If the calf dies, the mother will repeatedly try to revive it or simply push it ahead of her wherever she goes. Later, she may continue to shower maternal affection on a surrogate object such as a log.

When the calf is healthy, it will "shadow" the mother for the first few weeks, swimming close beneath, on top of, or to the side of her. By hitching a ride in the slipstream created by the mother's movement, the calf may save precious energy that it needs to grow. This type of **echelon swimming** also buys the calf protection. A beluga mother will position herself between her calf and any strange object such as a boat. If the intruder gets too close, females with calves are also willing to **charge** and **bite**.

The calf **nurses** underwater, riding at right angles to the mother. Milk provides sustenance for up to 2 years, supplemented in the second year by easily captured prey such as mollusks, annelids, and crustaceans. This long nursing period may give the young apprentice ample time to learn the ropes of belugahood—the feeding and migration strategies, the dangers, and the escape routes. You'll notice that young belugas, besides being lots smaller than adults, are also darker. Born with sepia-brown or slate-black skins, they will lighten up as they mature, becoming pure white by the age of 9 (males) or 11 (females).

NURSING   *The dark brown or gray calf may suckle for up to 2 years.*

BELUGA WHALE BEHAVIORS TO LOOK FOR AT THE ZOO OR IN THE WILD

BASIC BEHAVIORS

LOCOMOTION	FEEDING	☐ RUBBING
☐ swimming	☐ vacuuming	☐ RESTING
☐ BREATHING	☐ jet spraying	

SOCIAL BEHAVIORS

FRIENDLY BEHAVIOR	CONFLICT BEHAVIOR	SEXUAL BEHAVIOR	☐ posturing	*Caring for the Calf*
☐ flaccid melon	*Threat*	*Courtship and Mating*	☐ jaw clapping	☐ lifting the newborn
☐ play chases	☐ turgid melon	☐ chasing	☐ soliciting	☐ echelon swimming
☐ cooperative hunting	☐ gaping	☐ rubbing	☐ mating	☐ charging and biting
	☐ chase	☐ mouthing	PARENTING BEHAVIOR	☐ nursing
	☐ jaw claps	☐ nuzzling	*Birth*	
	☐ slapping the water			
	☐ *Biting*			

THREAT *The open-mouth stare says, "This bear means business."*

POLAR BEAR

T he Arctic is a land of extremes. The sun never sets in summer nor rises in winter, the ice cap never melts, and the shadows cast by caribou reach clear to the horizon. The Arctic is also home to one of the world's largest carnivores, a bear that is more powerful, potentially more dangerous, and in many ways gentler than any other.

Even so, you wouldn't want to test the limits of the polar bear's patience. Standing 12 feet tall on its hind legs, a polar bear could look an elephant in the eye. With its foot-wide, 50-pound paws, it can yank a seal through a small hole in the ice, crushing all its bones in the process. Our human constructions don't offer much more resistance, as you can see in this description of a bear-trashed cabin:

> Bedsprings are ripped apart, bedding lies in tatters, the stove tipped over and the pipes crushed, bottles broken, tinned goods torn or perforated with tooth marks, and any plastic materials shredded. There seems to be mischief in their make-up. In a final burst of fin, they smash through the wall to exit, even though the door hangs wide open.*

Obviously, when polar bears interact with one another, it pays for them to be prudent. A blow from the sledgehammer paw of an opponent could knock them out, and a slash from sharp teeth or claws could easily leave a fatal wound. Instead, polar bears treat one another gingerly, exerting great control despite their strength. Other animals in the Arctic have also learned to respect the big bears, as well as to capitalize on their hunting prowess. Arctic foxes and ravens follow in the bears' footsteps, lucky to find leftovers in such a hostile land. Curiously, these treasure-seeking animals are rarely the object of the polar bear's hunt.

Native Arctic peoples regard the bear with similar gratitude and reverence. Traditionally, they killed polar bears

*Richard C. Davids. *Lords of the Arctic: A Journey Among the Polar Bears.* New York: Macmillan Publishing Co., Inc. 1982.

VITAL STATS

ORDER: Carnivora

FAMILY: Ursidae

SCIENTIFIC NAME: *Ursus maritimus*

HABITAT: Edge of pack ice; some populations summer on land

SIZE: Male length, 6.5–8.2 ft   Male shoulder height, 5.3 ft   Female length, 6–6.5 ft

WEIGHT: Male, 661–1,764 lb   Female, 330–660 lb

MAXIMUM AGE: 30 years in wild, 41 years in captivity

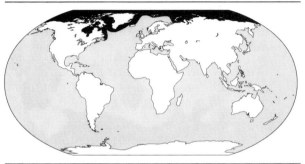

for meat and other products, but they also accepted the rare occasions when polar bears ate one of their clan. It was actually considered fortuitous if a bear found an elderly person who had been left in an igloo to die. If the bear ate the body, clan members believed they could rekindle the old person's spirit by killing and eating the bear. They also believed that bears were shamans—liaisons with the spirit world and therefore wise with the knowledge of the ages. On a physical plane, this masterful predator provided a role model for native peoples, teaching them by example how to hunt, navigate, and stay warm in the Arctic. You too can learn a great deal about keeping warm and well fed at the top of the world, even if the polar bears you are watching happen to be in San Diego.

Basic Polar Bear Behaviors to Watch For

LOCOMOTION

For some reason, polar bears, like many native Arctic peoples, are **bowlegged**. Their massive paws point inward, giving them a photogenic, pigeon-toed innocence that melts human hearts. When they're up and **trotting**, they move the legs on each side of their body in unison, like the famous pacing horse Dan Patch. They can cover 12 to 18 miles in an hour when scrambling over uneven pack ice, but on level ground they can really make tracks. Bears **running** across the tundra have been clocked at a racy 35 miles per hour.

SNIFFING THE AIR *When tracking prey by scent, bears often reach up as far as they can and move their head from side to side to locate airborne odors.*

Also like native Arctic hunters, bears take long, solitary treks in search of food. They first pick up a scent in the driving air, then follow it like a signal beam back to its source. The beam is probably rather narrow, and to keep it in focus, the great bears occasionally stand upright and sniff, moving their head from side to side. They can easily cover 43 miles or more in a day, and one bear that was tracked off the coast of Alaska logged 694 miles in a year.

These denizens of sea and ice are also competent **swimmers**, bear-paddling with their front paws at a steady 4 miles per hour. In the wild, they can paddle nonstop for up to 100 miles, as long as they resurface every 2 minutes or so for a gasp of air. Many zoos have installed underwater portals so you can watch polar bears in their watery realm. With a great splash, they **dive** headfirst into the water, digging in with their front paws and steering with their hind paws. You'll notice that they keep their eyes open, their nostrils shut, and their ears flattened to their head.

FEEDING

Polar bears snack on fish, mammals, and even grass, but nothing gets them going like a good slab of seal blubber. They travel north and south, dutifully following the advancing and retreating pack ice to be where the seals are. The best place to find their favorite prey—the 5-foot-long, 200-pound ringed seal—is in the zone where the ice is constantly cracking and moving to the tune of wind and sea currents. This kind of action is most common across the mouth of bays and at the seaward edge of the land-fast ice. As ice blocks separate, they create intermittent lanes of open water, which then refreeze. Since the newly iced cracks are more stable than most, they're perfect places for seals to create holes through which they can breathe or haul themselves out on the ice to rest.

After eons of hunting, polar bears have become savvy to the seal's ways. They can detect a haul-out or breathing hole just by the lingering stench of the seal's foul breath. Bears have learned to **still hunt** at these holes, sitting statue-still for hours, then swatting with trigger-quick precision when a sleek head appears. They have also become expert **stalkers**, flattening themselves like a rug and snaking forward on their belly toward basking seals. They take their time, advancing when the seal naps, freezing when it wakes, and ultimately closing in with a lethal rush. If the stalk is **underwater**, the bear lowers itself feet first into the drink without a ripple, then cuts through the water like a slow, silent submarine.

PREPARING TO DIVE   *The lumbering polar bear is a graceful swimmer, beating the chill of Arctic waters with an insulating coat of fur and blubber.*

During the seal's breeding season, the polar bear also looks for pressure ridges, cliffs of ice piled up by currents and wind. The seals dig their pupping dens on the leeward side of these ridges, working as quickly as they can and keeping one worried eye out for polar bears. Even after the lair is dug, the seals still aren't safe. A hungry bear can smell a seal pup through as much as 6 feet of snow. They break in paws first, **pouncing** like foxes on molehills.

Both seal and bear, trapped in the Arctic together, are engaged in a classic war of adaptations, each keeping the other on its toes. Observers report that some polar bears go so far as to hold a paw over their snout when stalking, perhaps to cover the conspicuous black nose. In turn, seals have evolved equally wary natures, using their wide-angle eyes to scan for moving shadows and their phenomenal hearing to detect the slightest shuffle.

The bear's reward for outwitting the wily seal is blubber, an incredibly rich energy source that fattens polar bears quickly and gives them a personal pantry to draw from during summer food shortages. Though bears are quite capable of "inhaling" an entire seal (their stomach holds up to 200 pounds), they usually consume only the blubber, demurely peeling back the skin and leaving the remains to the foxes and crows that follow in their footsteps.

DRINKING

Though polar bears need fresh drinking water just as we do, they've found a way to exist on the sea ice with nothing but salt water for as far as the eye can see. Early Arctic explorers knew the trick too; since salt has a way of eating down through ice over time, the bears look for old ice and drink from the puddles of sweet water on top.

GROOMING

After feasting on blubber, a polar bear will **lick** its giant paws and smooth the fur on its face like a cat doing its washing. If dinner dribbles on its coat, the meticulous bear will often rub and roll itself clean in a bed of dry snow. You can watch for the same cleaning moves after feeding time at the zoo. Don't blame poor grooming, however, if your bears seem to have a yellowish cast to their fur. It's quite natural for bears to have seasonal and individual variations in their coat color.

GROOMING *Like a big cat, the polar bear is careful to wash its paws after eating.*

WARMING UP AND COOLING DOWN

You'd need a scanning electron microscope to see why polar bears don't freeze in the winter. If you could magnify one of the hairs, you'd see a hollow, transparent shaft without a hint of white pigment. The hairs appear white to us for the same reason a snowflake looks white: the rough inner surfaces of the hollow tubes reflect visible light. Ultraviolet light from the sun travels down the core of each hair, where it is greedily soaked up and stored by the bear's coal-black skin. Finally, the hairs trap this heat like a million panes of greenhouse glass. Under the fur is a tough hide and a blubber layer that may be up to 4½ inches thick. On days when vicious winds manage to cut through even this armor, polar bears **plow** into snowdrifts, covering their upper body with an igloo of insulation and leaving their rump in the air for a windbreak.

Because their body is built to hold heat in, polar bears often suffer when temperatures rise. The struggle is most intense in the summer, when the pack ice melts and many bears are forced inland. A hot bear is an uncomfortable one, and as you will see in the zoo, they beat the heat by **panting**, **sprawling** on their belly or sides in wet sand, or **waving** their paw pads in the air. These pads have very little fur, but are heavily supplied with blood vessels. Hot blood is ferried both here and to a sheet of muscle in the shoulders, so the heat can radiate out, cooling the bear in the same way a car radiator cools an engine.

When temperatures become unbearable, polar bears in the Manitoba–Ontario coastal region dig shallow pits in cool sand dunes or peat hummocks. They use the same burrow year after year, **digging** a little farther each time to find fresh permafrost (which melts when exposed) and eventually creating elaborate tunnels going back 15 feet or more.

SLEEPING

Living at the top of the world presents polar bears with at least two challenges: regulating their body temperature and storing enough energy to last between meals. Sleeping helps on both accounts. When polar bears sleep, they are not as exposed to cold wind, and they don't burn as much precious energy. By the same token, when outside temperatures are warm, moving around is the last thing a massive, bulky bear wants to do. A good sleep puts the polar bear on ice, so to speak, allowing it keep cool and save its energy.

SLEEPING   *Lots of downtime helps polar bears stretch their energy budget.*

Sleepy bears adopt a variety of postures, depending on whether they want to conserve heat or lose it. When they are cold, they'll plow face first into the snow or wrap a furry paw around their heat-radiating nose. When they are warm, they'll press their belly to the ice or sprawl on their back with their paws up.

Polar bears share with other bears the distinction of being able to crank down their metabolism for deeper-than-ordinary sleeps. When they move inland during summer, they dig burrows and sleep in a kind of summer hibernation called estivation, during which their metabolism and heartbeat slow down and their appetite stalls, allowing them to live nearly rent-free during a period of scarce food. Even when they are active, they don't eat or drink for months on end, living solely off their own fat stores. Conveniently, this internal digestion produces only carbon dioxide and water as a byproduct, leaving nothing that needs to be excreted. Nursing females turn an even greater profit; their entire milk supply comes from fat combustion.

Polar bears use this ability on and off during the winter season (September to March) when they need to conserve energy. When asleep in temporary daybeds, their pulse rate slows from its normal 53 to 85 beats a minute to a lazy 8 beats. In an emergency, the bears can waken from this lethargy in a moment, which is why it's referred to as **walking hibernation**.

Scientists are currently searching for the hormone or other substance that induces this hibernation state in bears. Imagine if we humans could use such a substance to give our ailing bodies a rest or to allow ourselves to sleep through long space flights! Hibernation study is not the only reason to preserve bears, but it certainly is a compelling one.

STEREOTYPED MOVEMENTS

Members of the bear family are particularly likely to exhibit repetitive, stereotypic behavior in captivity. In the small zoo near my former home, the polar bear has a large outside exhibit complete with a deep pool for swimming. Even with plenty of room, the bear swims the same oval route over and over, pushing off with one paw in exactly the same spot each time he turns. Other common stereotyped movements are **figure eights**, **backward walking**, **pacing**, and **head swinging**. Although movements like these provide some exercise for the bears, they can become detrimental, especially if the bear begins to hurt itself. Devising ways to keep bears occupied in their exhibits so they can break these habits is challenging, but it can and must be done. For all the hours of sheer pleasure they give us, the least we can do for these magnificent creatures is to give them stimulating environments.

HUMANS AND POLAR BEARS

No one really knows when female polar bears first began using their traditional denning grounds, but they have returned to these safe sites generation after generation. Unfortunately, the guarantee of safety may no longer hold. Outposts of human civilization are cropping up in the Arctic, bringing garbage, dogs, boom boxes, and wood smoke with them. It's an old story; in our search for oil and other consumables, we're punching paths of "progress" into the fragile permafrost. Just as we don't know how long the bears have been denning here, we also don't know how much it will take to scare them away.

Farther out in the coastal waters, pollution or commercial fishing could someday threaten the ringed seal populations. Because polar bears depend so heavily on blubber, they will certainly feel any pinch the seals feel.

Polar bear experts tell us that this kind of habitat destruction is the single greatest threat to the bears themselves. There are, optimistically, 30,000 polar bears now alive on earth, and these bears are not reproducing at what you could call speedy rates. Given the long rearing period for each cub, most females have only enough time to produce two litters in their 20- to 30-year lifetime, a characteristic that would make it hard for them to bounce back from any major reduction in their numbers.

The *Red Data Book* of the International Union of Conservation of Nature and Natural Resources (IUCN) lists the bear as vulnerable, which is only one notch away from endangered. Threats that could push the bear a notch too far include overexploitation, habitat deterioration, and human disturbance. Thankfully, five nations (Canada, Norway, Denmark, the former USSR, and the United States) have signed a treaty that (1) bans all wanton killing, including shooting from aircraft and large motorized boats (traditional subsistence hunting by native peoples is still allowed in some nations), (2) commits each nation to study the polar bears' habits, and (3) protects the areas that are important to polar bear survival. If it works, we may be able to keep the polar bear from slipping and make sure the mistakes made on other continents with other species will not be repeated here.

FRIENDLY BEHAVIOR

Greeting Ceremony

Watch two polar bears as they approach each other. The smaller one will try to position itself downwind of the larger one so it can get a feel, or actually a smell, for the situation. If the scent is familiar, it may attempt a greeting. At first there is an uneasiness as one bear pads softly around the other bear, ears back, moving with the grace of a dancer. After one or two turns, both bears close in until they are touching noses. Next, they sit down, heads askew and mouths gaping, and begin clasping each other's jaws, yawning deeply, or gently chewing each other's necks. Their movements are as slow and as silent as mime, perhaps because neither wants to incite the other to violence.

Social Play

Outside of breeding season, bears will go out of their way to avoid a fight. After greeting each other, they usually work off tensions by playing. One bear will **invite** the other to play by wagging its head quickly from side to side. They may then stand on their hind legs, **pushing**, **biting**, and **swatting** each other with their paws. Though these exchanges may look like confrontations, they are in fact friendship rituals

Circling

Sniffing

Neck chewing

GREETING CEREMONY   *Each bear tries to position itself downwind of the other, in the path of informational odors. After taking stock of each other, they gape and grab at each other's jaws and necks with "soft bites," showing that they mean no harm.*

Play invitation

SOCIAL PLAY *Bears invite each other to play by wagging their heads from side to side. Up on their hind legs, they lock forearms and push, bite, and swat, all in silent, playful slow motion.*

Play-fighting

among male bears. Polar bears tend to seek out a certain play-fighting partner (usually a sibling) and may even travel with this companion during the summer and fall. One bear that won't play-fight, however, is an adult female with cubs in custody; she's more intent on keeping bears away from her young than on building good relations.

CONFLICT BEHAVIOR

Fighting

Not all polar bear fights are in jest; when males vie for the right to mate with a female, they can be deadly serious. Though their moves resemble play-fighting, they take on a frightening power, speed, and efficiency when the conflict is real. You can tell the difference even at a distance, the same way you can tell when teenage boys switch from horseplay to a real brawl.

At the start of a typical showdown, both bears keep their heads lowered and their upper lip thrust straight out. Bristling with tension and staring defiantly, the normally quiet bears suddenly become quite vocal. With froth dripping from their muzzles, they **chuff** (by contracting their chest and abdomen), **blow**, **chomp** their jaws, **growl**, **roar**, and **hiss** like cornered cats. A sudden charge breaks the spell, and the bears rise up on their hind legs and begin **boxing**, **biting**, and **clawing** at each other's neck and shoulders with enormous claws. Even though the skin in these areas is thickened to withstand punishment, it's amazing that more bears are not killed outright during these battles.

SEXUAL BEHAVIOR

When a female comes into heat, she begins **marking** her trail with frequent urinations. The scented hormones announce that she is ready to mate, and in no time, she has a male or two following her. This is when males are most apt to wage war on one another for the right to mate.

The winner engages in a rather leisurely liaison with the female, traveling with her for days or sometimes weeks. Despite the fact that the female is so much smaller than the male, she doesn't hesitate to spurn him if she is not quite ready to mate. By the same token, once she is receptive she may continue to be cagey as part of her courtship strategy. When the male finally mounts, he grabs the scruff of her neck in his jaws. They mate repeatedly and sometimes so violently that the male's penis bone can snap under the pressure. Breeding may occur anytime in April, May, or June, and the cubs are born sometime in December or January.

PARENTING BEHAVIOR

Birth

Pregnant females leave the seal-hunting grounds and come ashore in October and November to **dig maternity dens** into snowdrifts or slopes. They build a heat-trapping entrance sill, then dig a 5- to 7-foot tunnel into the snow, sloping it upward so warm air will be trapped in the birth chamber at the tunnel's end. This floor plan seems to work well; scientists plunging a thermometer down into bear dens found them to be as much as 40 degrees Fahrenheit warmer than outside. Even in captivity, polar bear moms give birth to cubs only if they have solitude, security, and a den that is snug enough to be heated with the female's own body heat.

In midwinter, the sleeping bear wakes up to give birth to one to three blind, naked cubs that are no bigger than rats. The cubs gain up to 30 pounds apiece in the next 3 to 4 months, **humming** as they **nurse** on the female's 33%-fat milk. Because she doesn't eat or drink during this time, the milk must come solely from the nutrients stored in her fat.

Play

By spring, the maternity den has several alcoves carved by the playful young. They emerge from the den ready for mischief, and like cubs and pups of any species, they're a delight to watch. The snow provides a natural medium for **sliding play** and a soft cushion for ambushes and rolling bouts of **play-fighting**. Polar bears are unusually noisy at this age, given to **whimpering**, **purring**, and even **growling** occasionally. They may also smack their lips together with a loud **popping sound** that will startle you. If handled, they usually protest with catlike **shrieks** and babylike wails and squalls.

Defending the Cubs

Cubs spend the next 2 to 5 years learning how to be polar bears but relying on their mother to feed and protect them from most anything that moves, including adult male polar bears. Though males kill cubs only occasionally, the female doesn't take chances. She avoids males completely, and if the cubs happen to stray from her sight, she **chuffs** and **roars** until they are safely underfoot again.

CARING FOR CUBS   *The mother lowers her head and stares directly at a would-be enemy. The cubs are tucked safely behind her.*

POLAR BEAR BEHAVIORS TO LOOK FOR AT THE ZOO OR IN THE WILD

BASIC BEHAVIORS

LOCOMOTION
- ☐ bowlegged walk
- ☐ trotting
- ☐ running
- ☐ swimming
- ☐ diving

FEEDING
- ☐ pouncing
- ☐ still hunting
- ☐ stalking
- ☐ underwater stalking

- ☐ **DRINKING**
GROOMING
- ☐ licking
WARMING UP AND COOLING DOWN
- ☐ plowing into snowdrifts

- ☐ panting
- ☐ sprawling
- ☐ paw waving
- ☐ digging into dens
SLEEPING
- ☐ walking hibernation

STEREOTYPED MOVEMENTS
- ☐ figure eights
- ☐ backward walking
- ☐ pacing
- ☐ head swinging

SOCIAL BEHAVIORS

FRIENDLY BEHAVIOR
- ☐ *Greeting Ceremony*
Social Play
- ☐ play invitation
- ☐ pushing
- ☐ biting
- ☐ swatting

CONFLICT BEHAVIOR
Fighting
- ☐ boxing
- ☐ clawing
- ☐ biting
- ☐ chuffing
- ☐ blowing
- ☐ jaw chomping
- ☐ growling
- ☐ roaring
- ☐ hissing

SEXUAL BEHAVIOR
- ☐ scent marking
PARENTING BEHAVIOR
Birth
- ☐ den building
- ☐ nursing
- ☐ nursing hum

Play
- ☐ sliding play
- ☐ play-fighting
- ☐ whimpering
- ☐ purring
- ☐ growling
- ☐ shrieking
- ☐ lip popping

Defending the Cubs
- ☐ chuffing
- ☐ roaring

LANDWARD HO!   *Penguins "squirt" out of the water when scaling icy cliffs.*

ADÉLIE PENGUIN

P enguins have no idea how funny they are. Waddling around their icy colonies, so formal and full of themselves, they look as if they're imitating us! Put a hundred or so penguins together, said one observer, and you've got a lobby full of businessmen at a ballet intermission.

Of course Adélie penguins, like other animals, can't survive on comic charm alone. Their endearing qualities are actually shrewd adaptations to an icy, aquatic way of life in the Antarctic. Take the upright walk, for instance. These birds must stand up to walk because their legs, unlike those of ducks and geese, are set far back on their bodies. Though it makes them a bit more awkward on land, this rear orientation allows them to be graceful and efficient underwater. The new penguinariums in many zoos allow you to peer into tanks and watch penguins zigzagging through the water, using their legs like steering rudders. They zigzag like this in the wild to track down quick-moving krill, the shrimplike creatures that make up most of their diet.

In another swimming adaptation, penguins' wings have narrowed into rigid paddles that stick straight out from their body and propel their underwater flights. The body itself— tapered at the head and widest about a third of the way down—is cunningly shaped to allow water to slip by with little drag. (When asked to design the perfect body for a diving bird in cold water, engineers revved up their computers and arrived at the very same shape.) The black and white tuxedo that wraps the penguin's body may also be a design coup,

VITAL STATS

ORDER: Sphenisciformes

FAMILY: Spheniscidae

SCIENTIFIC NAME: *Pygoscelis adeliae*

HABITAT: Seawater, ice, rock, islands, coast

SIZE: Height, 2 ft

WEIGHT: 30 lb

MAXIMUM AGE: 18 years in captivity

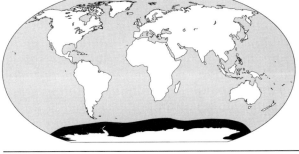

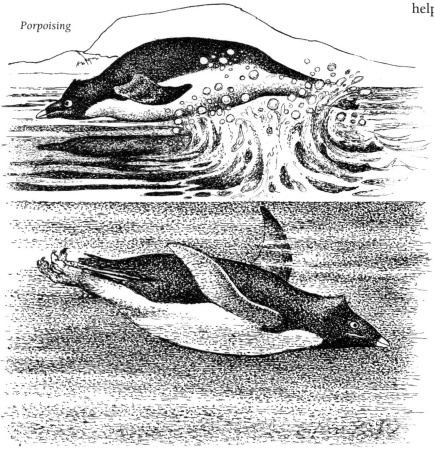

Porpoising

Swimming

LOCOMOTION *Without breaking its stride, a penguin leaps above the surface to take a breath, then resumes its rapid underwater flight.*

helping to warm the bird while camouflaging it. When a penguin swims near the surface, its black back absorbs the weak Antarctic sun while its white belly looks like sky to predators and prey that peer up from below.

Even the gregarious nature of penguins at a rookery is a brilliant adaptation. The Antarctic offers a short season and limited coastline for breeding. Since all the Adélies must hit the same beaches at the same time to take advantage of the mild weather, it makes sense for them to behave tolerantly toward one another. This tolerance is what enables your zoo to display penguins in a group.

In the wild, being part of a group comes in handy at breeding time. Any predator that tries to attack from the air faces a forest of upraised, angry bills—much more of a deterrent than one or two bills would be. Also, since a thousand eyes are better than two, the group can keep a sharp watch on nearby waters where leopard seals and killer whales lurk. Even when predators do get lucky, they have thousands of prey to choose from, giving each individual a better chance of surviving. When the breeding season is over, security-minded penguins group up again to feed and to migrate to their wintering grounds farther north.

Obviously, penguins stand to profit if they can overcome their aggression and keep the peace. One way they've managed to do this is by evolving a complex language of communication gestures designed to attract, appease, and settle disputes quickly. The good news is that you can witness most of these interactions at your zoo, and you don't even need mukluks to enjoy the show.

BASIC ADÉLIE PENGUIN BEHAVIORS TO WATCH FOR

LOCOMOTION

Although penguins make do on land, **waddling** with a Chaplinesque flair or **tobogganing** on their belly, they really come to life in water. When torpedoing after prey, they press their feet close to their body so swirls and eddies won't slow them down. To build up speed, they flap their flippers up and down like wings, beating against water rather than air. The ballast for their **dives** is in their bones, which are solid instead of honeycombed with holes the way most birds' are.

Cruising at 11 miles per hour, the penguin doesn't even slow down to take a breath. It simply leaps out of the water like a porpoise, arcing through the air and sneaking a breath before plunging back in. This "disappearing act" might also help confuse its major underwater predator, the leopard seal. Even though the penguin doesn't have to worry about leopard seals at the zoo, it still uses its **porpoising** trick for a quick breath.

FEEDING

It's a good thing that the Adélie's favorite food is so very abundant. Each Adélie must capture at least one shrimp every 6 seconds to fuel its movements, keep itself warm, and feed its chicks. When 5 million Adélies get together for nesting, they way they do on Laurie Island in the South Orkney Islands, they can polish off 8,800 tons of krill and small fish a day! That's like having 70 modern fishing trawlers plying the waters at once. But Adélies are not the only kind of penguin in town, nor probably in the exhibit at your zoo. Various species of penguins make up almost 90% of the bird biomass in the Antarctic and may consume as much as 70% of the region's underwater food.

In the interest of coexisting, these different species of penguins don't compete in the same places for the same foods. Some species haunt the feeding grounds in-shore, while others swim out to exploit an entirely different niche. Diving capacity can also determine how they divvy up the habitat. For instance, chinstraps, which can dive down to 250 feet, mine the upper levels with Adélies while gentoos, which can plunge to over 400 feet, head for the depths. The fact that breeding cycles of the various penguins are somewhat staggered also helps alleviate competition during peak demand periods. Adélies breed 2 weeks before gentoos, which breed 2 weeks before chinstaps. By the time chinstaps start requiring large amounts of krill for their broods, the Adélie chicks have already fledged.

MOLTING

Once a year, you can catch penguins with their wetsuits down. Since feathers provide 80% of a penguin's insulation, it's essential that the old, ragged set be replaced each year. During the 3 weeks

MOLTING *The penguin's tuxedo looks a little ragged during the molt, when new feathers are pushing out the old.*

that it takes for the new feathers to push out the old, the penguin is at the mercy of the elements. While its insulation is being repaired, it dares not enter the water to fish, even though its body is burning twice as much energy as usual to make the new feathers. With no way to take the edge off its appetite, a molting penguin can lose up to half its body weight during this time.

It's easy to tell when the penguins at your zoo are about to molt; their plumage loses its luster, becomes brown-tipped in places, and looks puffier than normal. Their flippers get so engorged with blood that keepers may have to loosen the birds' identification bands for better circulation.

FEATHER CARE

Once the new feathers are in, **cleaning**, **oiling**, and **combing** them out takes top priority. Watch for preening each time a penguin exits the water. Since diving pushes air out of the feathers, the penguin must reset its insulators by fluffing them up and realigning them. Other body care moves to watch for are **scratching** skin irritations and **shaking** to remove excess water.

COOLING DOWN

Because penguins are equipped to keep heat in, their major grief begins when temperatures soar to 32 degrees Fahrenheit and they suddenly have to release heat. Since flippers and feet have less blubber insulation, they make perfect radiators for excess heat. Penguins also **pant** with open beaks, cashing in on the cooling powers of evaporation. On warm days, panting chicks lie on the ground, belly down, with legs outstretched and the undersides of their feet pointing up.

SLEEPING

A wide-awake penguin is ever watchful. In contrast, you can tell the sleeping ones by their unusual stillness. Even though they are standing, their feathers are relaxed, flippers limp at their sides, neck withdrawn, and eyes shut. They may sleep face forward, with bill horizontal, or they may turn their head sideways over one shoulder and tuck the bill under one flipper. At the zoo, you may also see penguins stretched out on their stomach to catch some shuteye. When they wake, watch for **stretching** and **yawning**, which helps them stir sluggish circulation and speed oxygen to the brain and muscles.

HOW ADÉLIE PENGUINS INTERACT

In an environment where social is safer, penguins have learned to fish together, migrate en masse, and breed side by side in huge onshore rookeries. These oases of rock and ice give the penguins some protection from their main underwater predator, the leopard seal. Here they build nests made of stones, fight for their rights, mate with the same partner year after year, and raise footballs of fluff called chicks. Zoo designers have worked hard to emulate the rookery environment (right down to the lapping waves), and penguins have responded by settling in and raising families. When you take your ringside seat for the festivities, use the following guide to interpret all the action.

FRIENDLY BEHAVIOR

After breeding season, penguins begin to head north to the belt of drifting pack ice that lies between the stable ice of the continent and the vast open sea. Here they can **fish** in the open for krill, yet climb up on floating ice blocks to escape the voracious leopard seal. To reach this belt, penguins **migrate together** over the ice cap in single-file trains. If one penguin gets down on its belly and plows a trough in the snow, the others will save energy by following in its track rather than breaking trail themselves. When they reach a lead—a lane of open water—they all line up at the edge, jostling one another like school kids at the edge of an icy swimming hole. None of the penguins wants to be the first to take the plunge, not because it's cold, but because they know that leopard seals might be lurking below. Once one penguin lands in the drink without attracting a hungry seal, the rest feel safe enough to follow.

HUMANS AND PENGUINS

For some reason, we tend to believe that because we don't live there, we aren't despoiling the Antarctic. In the same way, we once believed that the world's oceans were immune to our assaults. They were a vast frontier, and according to early accounts, they were "broiling with whales, and the air [was] heavy with the oily tang of their breath." Today we've learned just how far-reaching our impacts can be. Where whales once "broiled," they are now rarely seen from year to year.

Ironically, the destruction of krill-eating baleen whales actually boosted penguin populations in some parts of the world by removing the penguins' competition. Today penguins have a new, two-legged competitor whose burgeoning population is hungry and looking for new sources of protein. In some areas, large-scale commercial krill fishing has already put a strain on penguin populations. So has human visitation of Antarctic habitats, which brings disturbance and leaves exotic predators such as dogs and new diseases. The most insidious threat comes from air and water pollution that makes its way to the bottom of the world. Researchers have found that Adélies already contain levels of chlorinated hydrocarbon pesticides that resemble those found in animals at northern latitudes. The threat of oil spills and ozone depletion further jeopardizes this unique species.

In addition to these indirect assaults, there is still the possibility that someone may attempt to turn penguins into profit. As recently as 1982, plans to harvest penguins commercially have cropped up again in South America, where penguins were once routinely slaughtered for their skin, meat, and oil.

This time, with luck, penguins will have plenty of people willing to champion their cause. Throughout the world, millions of visitors have been touched, in the heart and the funny bone, by penguins that they've seen at the zoo. Hopefully, instead of cashing in on these birds, more of us will fight to protect them, both for their beauty and for the many things they can teach us about overcoming adversity.

Even in the thick of serious migrating, penguins rarely miss an opportunity to **play**. Jumping aboard a passing ice floe is a favorite game. When the floe lands, the penguins hurry back upshore to wait for another free ride. Researchers, searching for survival value in this behavior, have had to admit that penguins might just floe-surf for fun. In the zoo, where predators don't prowl and the living is easy, you're even more likely to see penguins at play.

CONFLICT BEHAVIOR

Anytime five million animals mate, give birth, and raise their young on the same patch of snow and ice, tensions are bound to erupt. Even so, with predators all around, it pays to nest as closely as your temperament will allow. Penguins space their stone nests so that if next-door neighbors were to stretch toward each other as far as they could, their bill tips would barely touch. This cuts down on spontaneous pecking, but it doesn't completely prevent flare-ups. When intruders try to steal nest stones, or disoriented penguins return to the wrong nest, the residents rabidly defend their small piece of turf. They work out their differences in a noisy détente of threat and appeasement displays, all of which can be seen at the zoo. Once you learn what the displays mean, penguin watching can be addictive, so be sure to save plenty of time for this exhibit.

Appeasement

Unless their nest is right at the water's edge, most penguins have to walk through the colony when they go to the water to feed. In the **slender walk**—an appeasement display that says, "I mean no harm"—a penguin assumes a nonthreatening pose and shimmies through the gauntlet as quickly as it can. It keeps its feathers sleeked, its flippers held to its sides or behind, and its bill pointing up, symbolically taking its weapon out of commission. Depending on how far into the colony the penguin's nest is located, the slender walker may have to pass as many as 90 nests.

The residents at these nests are only too aware of intruders and passersby. When a stranger approaches, they display a sequence of threat behaviors that systematically gets hotter depending on the intruder's response.

SLENDER WALK   *Hustling through a colony of suspicious nesters, penguins must adopt this pose as a way of waving the peace flag.*

Low-Intensity Threat

Starting at the low-intensity end of the spectrum, an Adélie performs the following displays when defending its territory from strangers or neighbors:

- **Bill-to-Axilla Display**: The penguin stoops forward at 45 degrees, turning its head to one side and touching its bill to the axilla, or armpit, where the flipper meets the body. It may roll its head up and down or switch sides during the display, all the while keeping its crest feathers erect and eyes rolled (The iris slides to the corner so that a portion of white shows). At full intensity, the penguin may even growl and beat its flippers in rhythm with the rolling.
- **Sideways Stare**: At the next phase of agitation, the bird simply turns its head to the side and keeps it there, staring with one eye at the intruder. The flippers are held fast to the body and the bill may be pointed downward or obliquely at the intruder.
- **Alternate Stare**: If the sideways stare doesn't prompt the intruder to leave, the penguin may resort to turning its head from one side to the other, pointing one eye at a time at the intruder. This display can take place from a standing or crouched position. As the penguin turns from side to side, as many as 18 times, it waves its flippers slowly up and down, lifts its feathers, and rolls its eyes.

Bill to axilla *Sideways stare* *Alternate stare*

Point *Gape* *Charge*

THREAT *Threat displays, in order of increasing intensity. Notice how the bill (the weapon) becomes more prominent in the later displays.*

Penguins use these first three displays to deter intruders. If the offending party doesn't take the hint and move on, the territory holder pulls out all the stops and begins the high-intensity repellent threats.

High-Intensity Threat

In this more active phase, the resident penguin signals that it's ready to wage warfare. The following threats are in order of intensity from lowest to highest:

- **Point**: The penguin stops, stretches its neck forward (from a standing or prone position), and points its closed bill directly at the offender. It rolls its eyes, erects its crest feathers, and keeps silent except for an occasional growl.
- **Gape**: The gape is a taste of things to come. By opening its bill, the penguin signals, "I am ready to bite if need be." It calls harshly and may lean forward and inch toward the intruder with small jumps. In a gaping duel, two penguins open their bills at right angles to each other.
- **Charge**: At this point, the penguins have issued their warnings and are on a fast track to a fight. To indicate readiness, the challenger takes a few steps toward the offender with its neck craning forward and its bill (open or closed) tilted down or straight forward.

A WARM-BLOODED BIRD IN COLD WATERS

When penguin ancestors left the skies and traded their wings for flippers, they had two new dangers to face. The first was the frigidness of the Antarctic water. Because birds are warm-blooded, they must heat their body with an internal furnace, just as we do. To conserve their hard-won warmth in achingly cold temperatures, penguins evolved two forms of insulation: a layer of blubber and a tight, overlapping coat of feathers and down. This skull-cap-smooth covering traps a layer of air next to the penguin's body, allowing it to brave the iceberg-chilled waters in relative comfort.

The second danger occurs because the penguin, despite its dependence on the sea, has never evolved gills. It has to carry a supply of oxygen when it dives and therefore faces the same problems human scuba divers do. As water pressure changes on the way down, the air inside the penguin compresses by 50% for every 33 feet of descent. To make the most of this limited air, the penguin expands its airpipe like a concertina to shuttle air between the lungs and special air sacs at the base of the neck. Each time the air is fanned against the richly veined lung surface, the blood absorbs more oxygen. Penguins also carry an extra serving of oxygen in their muscles, bonded to a special respiratory pigment called myoglobin. Finally, penguins actually burn less oxygen in the water than they do on land. When they dive, their heart slows from 100 beats a minute to 20, allowing them to sip rather than gulp oxygen, so they can stay underwater longer.

Fighting

Sometimes threat gestures are not enough to prevent an actual fight. Fighting penguins spar bill to bill, **pecking** and twisting, pulling, or grabbing each other's bodies with their beaks. Once they get a grip, they may **strike** with their flippers or use their chests to knock their opponent off balance. The trounced penguin "cries uncle" by lowering or hiding its bill or by escaping completely. The victor may then **chase** the opponent for a while. Notice how the chaser erects its crest feathers and pushes out its chest; the chasee's posture is just the opposite.

Predator Defense

To see a penguin get really uptight, wait until a predator tries to drop in on the colony. In the wild, seabirds called brown skuas feed on eggs and chicks, if they can get a beak in edgewise. The **alarm call** goes out as soon as one is spotted. If it attacks from the air, the skua faces a militia of upturned penguin beaks issuing raucous screams. If it tries to walk into the colony, it will be chased on foot by enraged penguins—feathers erect, eyes rolled, and bills gaping. Because of this tight defense, the only fair game for the skua are eggs and chicks that are not guarded, or chicks that are sick or wandering. Because skuas are successful only a fraction of the time, a colony of 100,000 penguins usually provides good hunting for no more than 10 pairs of these predators.

Leopard seals, which do their hunting from the sea, are masters of surprise. They lurk under thin, newly frozen lanes of ice, poking their head up when a tasty penguin crosses overhead. Fully aware of the danger, penguins normally waddle across these thin-ice areas as fast as their webbed feet will take them. Open water can be even more treacherous. Leopard seals wait under ice ledges, emerging like lethal submarines as soon as a penguin plops into the water. To avoid these attacks, penguins will often wait at an open channel for hours until the ice deepens enough for a safer crossing. Or if an ice floe floats by, they may hop on and take it to shore, even if it adds hours to their trip. When penguins must go in the water to migrate or feed, they remain together in a group. At the first sign of a leopard seal, they scatter quickly, reuniting when the danger is past.

SEXUAL BEHAVIOR

Pre-Courtship

In mid-October (the Antarctic spring), penguins travel south to the solid edge of the continent, where they have formed breeding rookeries for ages. They return to the areas where they were hatched and usually choose the same nest site year after year. Penguins in captivity are also faithful to their nest sites.

The males arrive at the rookery first, claiming their nest sites and defending them from all would-be takers. The expressive **ecstatic display**, performed again and again, is a male's way of announcing ownership and attracting females to his nest. The ecstatic performer stands upright with his head reared back and his bill pointing straight to the sky. He fluffs out his crest feathers and rolls his eyes back until the whites show. His chest vibrates with a soft pumping that grows into a harsh

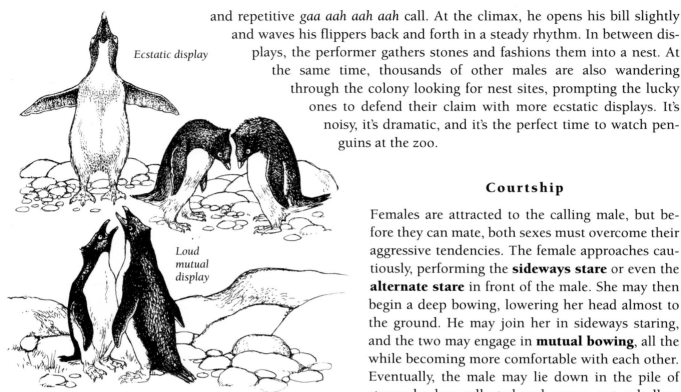

Ecstatic display

Loud mutual display

COURTSHIP *Males advertise and attract females to their nest by waving their flippers and giving a loud* gaa aah aah aah *call. After bending close together, the birds raise their bills toward the sky, calling and waving.*

and repetitive *gaa aah aah aah* call. At the climax, he opens his bill slightly and waves his flippers back and forth in a steady rhythm. In between displays, the performer gathers stones and fashions them into a nest. At the same time, thousands of other males are also wandering through the colony looking for nest sites, prompting the lucky ones to defend their claim with more ecstatic displays. It's noisy, it's dramatic, and it's the perfect time to watch penguins at the zoo.

Courtship

Females are attracted to the calling male, but before they can mate, both sexes must overcome their aggressive tendencies. The female approaches cautiously, performing the **sideways stare** or even the **alternate stare** in front of the male. She may then begin a deep bowing, lowering her head almost to the ground. He may join her in sideways staring, and the two may engage in **mutual bowing**, all the while becoming more comfortable with each other. Eventually, the male may lie down in the pile of stones he has collected and scrape out a hollow with his feet.

Among the most beautiful bonding rituals are the **loud mutual display** and the **quiet mutual display**. Though they look exactly alike, one has audio and the other does not. Facing each other, the male and female raise their bills skyward and weave their heads back and forth, slightly out of phase with each other. If the birds are strangers, they will perform the quiet mutual, calling very softly through closed bills. If they are partners from a previous season, however, they will reaffirm their lifelong connection by performing the well-named loud mutual display. You can hear the sounds of this rowdy reunion up to a mile away.

Copulation

Once the penguins have overcome their reticence, they are free to copulate. The male begins the **arms act** by standing before the reclining female and holding his vibrating flippers down and in front of himself. He then lowers his head as he sidles up and **mounts**. At this, the female reaches up and places the tip of her vibrating bill at the base of his bill. The male **vibrates** both his bill and flippers, **wagging his tail** as he **treads backward** on the female's back. (Muddy tread marks are a good way to identify the females at your zoo.) At the climax of the act, the female raises her tail and the male lowers his so that their cloacas line up. It's all over in less than a minute, and the male jumps off and stands beside the female. They may stand quietly side by side for some time before copulating again.

Post-Copulation

Even after actual mating, the pair continues to perform bonding displays that will boost their confidence in each other and prepare them for the journey into parenthood. As you will see, this journey will require sacrifice from both parties, putting the bonds they form here to the test.

PARENTING BEHAVIOR

Egg Laying and Incubating

In the wild, the female mates and then remains with the male at the nest site for a couple of weeks. When she finally lays her 2 eggs, 2 to 3 days apart, she heads for the water in search of food. The male, who has been at the nest site for up to 3 weeks by now, stays on, incubating and protecting the eggs from predators. To keep them off the cold ground, he puts an egg on each foot and then lies down so that the brood patches (two special warm spots below his chest) cover them. If intruders threaten at this time, the incubating bird performs an appeasement gesture known as the **withdrawn crouch**, in which he lays very still, sleeks

INCUBATING   *Penguin parents nestle their eggs against a brood patch, an area of thin feathers that provides direct access to warm skin.*

his feathers, and hides his bill against the side of his body. He says, in effect, "I am not interested in fighting with you." At this point a fight would be too risky because it would take the penguin away from the nest and expose the eggs.

After her 2-week feeding trip, the female returns to relieve the male and let him break his 5-week fast. She brings a stone as an offering, and they greet each other with **loud mutual displays**. Over the next few weeks, they will take turns incubating, gradually shortening their tours of duty and making more frequent fishing trips. Fueling up frequently allows both birds to be strong enough for the hard work of raising a hungry family.

In the zoo, penguins adapt a slightly different parenting strategy. Since they have abundant food close at hand, they have no need to go on long feeding trips that would separate them from their mates. Instead, they have shorter incubation bouts, more frequent relief exchanges, and more time to interact with one another than they would in a wild environment.

Caring for the Chicks

By the age of 3 weeks, the chicks are so large and eat so much that both parents must leave the nest to search for food. In the wild, where winds whip and skuas swoop, chicks from nearby nests form crèches—groups of 20 to 30 unrelated chicks that huddle together for protection. Though you may see these crèches on nature pro-

FEEDING CHASE *Full of goodies from its fishing trip, the adult penguin runs from a group of hungry chicks, some of which are not its own. After a while, the unrelated chicks will give up the chase, and the parent will feed only its own offspring.*

grams, you probably won't see them in the protected exhibit at the zoo.

One parental display you will see occurs at feeding time. When the parent returns to the nest after eating, it lets loose a **loud mutual call**. The chick recognizes the call and comes running, usually accompanied by unrelated chicks. To avoid wasting food on the strange chicks, the parent won't regurgitate until it performs the **feeding chase**. During this display, the parent runs away from the hungry horde of chicks. Only the true offspring, assured of a meal, will bother to follow the retreating parent.

Play

Penguin chicks are the VIP members of their colony. Their distinctive gray coloration grants them immunity from territorial rebuffs and gives them a free pass to **play** anywhere they like. You too will recognize the chicks; they're the fluffy ones racing around the exhibit pecking, flapping their flippers, and bumping into one another. They love to play a kind of king-of-the-mountain game in which they vie for the highest spot—a pile of thrust-up ice or even a sleeping Weddell seal. This game prepares them for later life when they will have to climb real ice cliffs to get away from real enemies.

In the wild, chicks are the last to leave the rookery, following the adults already headed for their molting grounds farther north. The journey to the pack ice is treacherous, and only 10% to 15% of the chicks will make it. The survivors will spend 2 to 5 years maturing on the pack ice before returning to the rookery to begin the cycle anew.

ADÉLIE PENGUIN BEHAVIORS TO LOOK FOR AT THE ZOO OR IN THE WILD

BASIC BEHAVIORS

LOCOMOTION
- ☐ waddling
- ☐ tobogganing
- ☐ diving
- ☐ porpoising
- ☐ **FEEDING**
- ☐ **MOLTING**

FEATHER CARE
- ☐ cleaning feathers
- ☐ combing feathers
- ☐ oiling feathers
- ☐ scratching
- ☐ shaking

COOLING DOWN
- ☐ panting

SLEEPING
- ☐ stretching
- ☐ yawning

SOCIAL BEHAVIORS

FRIENDLY BEHAVIOR
- ☐ group feeding
- ☐ group migrating
- ☐ play

CONFLICT BEHAVIOR

Appeasement
- ☐ slender walk

Low-Intensity Threat
- ☐ bill-to-axilla display
- ☐ sideways stare
- ☐ alternate stare

High-Intensity Threat
- ☐ point
- ☐ gape
- ☐ charge

Fighting
- ☐ pecking
- ☐ striking with flippers
- ☐ chasing

Predator Defense
- ☐ alarm call

SEXUAL BEHAVIOR

Pre-Courtship
- ☐ ecstatic display

Courtship
- ☐ sideways stare
- ☐ alternate stare
- ☐ mutual bow
- ☐ quiet mutual
- ☐ loud mutual

Copulation
- ☐ arms act
- ☐ mounting
- ☐ bill vibrating
- ☐ tail wagging
- ☐ treading backward
- ☐ *Post-Copulation*

PARENTING BEHAVIOR

Egg Laying and Incubating
- ☐ withdrawn crouch
- ☐ loud mutual display

Caring for the Chicks
- ☐ loud mutual call
- ☐ feeding chase
- ☐ *Play*

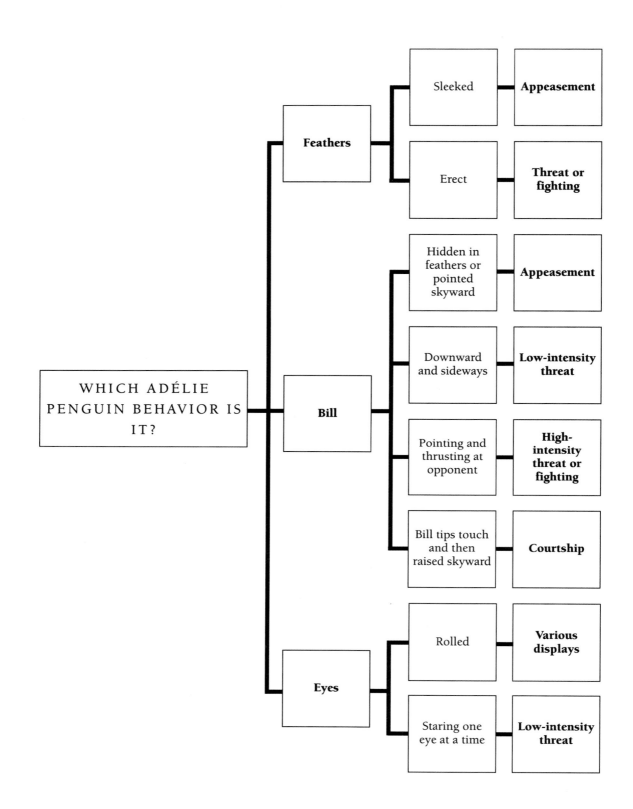

WHICH ADÉLIE PENGUIN BEHAVIOR IS IT?

Feathers
- Sleeked → **Appeasement**
- Erect → **Threat or fighting**

Bill
- Hidden in feathers or pointed skyward → **Appeasement**
- Downward and sideways → **Low-intensity threat**
- Pointing and thrusting at opponent → **High-intensity threat or fighting**
- Bill tips touch and then raised skyward → **Courtship**

Eyes
- Rolled → **Various displays**
- Staring one eye at a time → **Low-intensity threat**

Low-intensity threat

High-intensity threat

Courtship

In innovative zoo exhibits, chimpanzees "fish" for food with sticks the way they do in wild termite mounds.

ZOO CRITIQUE

Any discussion of zoos is incomplete without a look at the bad along with the good. Scarcely 10% of all animal exhibitors in this country pass the test that gives them accreditation by the American Association of Zoological Parks and Aquariums (AAZPA). Only these top-flight exhibitors are making the financial investments and policy shifts needed to properly care for the physical and mental health of their animals. That leaves 90% of animal exhibitors in the "twilight" category that includes petting zoos and roadside attractions. These operations often use animals as a way to lure people and their dollars in the door for some other purpose, like the grand opening of a hardware store. More likely than not, the keepers have neither special training, nor any particular regard for the well-being of their animals. The horror stories of substandard facilities and cruel treatment drag down the good efforts being made on the part of legitimate zoos. The AAZPA is dedicated to rewarding good zoos, while the Humane Society of the United States works to rout out those that are still treating animals like inmates rather than honored guests. You can find out which zoos receive the failing rating by contacting the Humane Society, 2100 L Street NW, Washington, DC, 20037.

You can also do your own evaluation. Remember that you need not be an expert to spot shady dealings at zoos; you can usually tell when an animal is underfed, overfed, diseased, or worse. Mentally stressed animals act strange; they perform repetitive, stereotypic behaviors such as pacing, weaving, self-mutilation, and trance, as well as excessive grooming, hypersexuality, and mistreatment of young. Some of these behaviors may be due to past abuses, but others may be linked to design flaws in the current exhibit. As you size up the exhibits at your zoo, consider the following.

WHAT DOES IT MEAN TO BE FREE?

Most people automatically assume that any animal in captivity is longing for freedom on the open plains or in the jungle. The truth is that animals are not really free to go wherever they want in the wild. They operate within strict limits imposed by the seasons, the paucity or richness of their habitats, and by their territorial status. How much energy they can afford to spend also constrains them. Notoriously thrifty, animals are constantly balancing their energy checkbooks, and if they don't have to travel far to eat or to find a mate, they won't. A squirrel, for instance, may spend its whole life in a few hundred square feet of forest, as long as it can harvest enough nuts, find a mate, and hide from predators there. Wolves, on the other hand, travel long distances, not because it's fun, but because moose are a lot harder to locate than acorns are.

Since an animal's basic needs are met at the zoo, a well-designed exhibit need not be vast to be a home. In fact, the boundaries that we humans associate with a lack of freedom are sources of safety to many species, symbolizing the line over which intruders (human visitors) will not pass. Once an animal marks a zoo territory as its own, it is usually reluctant to leave it for any reason.

WHAT SHOULD A GOOD EXHIBIT INCLUDE?

What really counts is the *quality* of space, not the amount of it. The exhibit can be as large as a roller rink, but if it doesn't allow the animal to express its normal behavior patterns, it may as well be a prison. To judge the quality of an exhibit, consider the life of the animal in the wild. Does it forage on the ground, nest in the trees, burrow under the soil, or swim in the sea? Is it a day-tripper, a night owl, or a twilight-active species? Is it social or solitary, sedentary or exploratory? (Use the chapters in this book to help you answer these questions.) As you evaluate an exhibit, try to match the realities of the animal's wild lifestyle with the realities of its zoo environment.

First look at the physical features of the exhibit, and see if they address the animal's species-specific needs. Zebras, for instance, need posts to rub their hide against; hippos need water to bathe in; beavers need water to defecate in; lions and tigers need posts on which to rake their claws; reptiles need a heat and light source to help them warm up after a chilly night; and rhinos and giraffes need a hard, abrasive floor that will wear down their hoofs before they grow to unnatural heights. Being on the ground terrifies some animals, while others feel insecure without wet places in which to wallow. The more responsive the exhibits are to these needs, the more you'll see an animal "doing its thing."

The physical layout of an exhibit may also have social or psychological ramifications. Each species has a certain flight distance, for instance, that it needs to escape from intruders. In a cramped exhibit, humans and other animals are constantly invading this flight distance, and the resident, having nowhere to go, can easily go mad. When building escape routes, exhibit designers must keep in mind the way that an animal naturally flees. Monkeys need vertically generous enclosures since they usually head up and into the branches to feel safe. Exhibits for plains animals, however, must be horizontally deep enough to allow them to sprint off at great speeds without crashing. Atmosphere can also affect stress levels. The human concept of cleanliness is not always the best policy for wild animals. Overzealous disinfecting can be maddening to a territory holder that must scent the area again and again to claim it.

WHAT ABOUT THE SOCIAL SETTING?

Social needs are just as important as physical ones. For gregarious animals like female elephants or zebras, being alone is a highly artificial setup. Arranging a social grouping is like choosing compatible table-mates at a dinner party; it takes a thorough knowledge of an animal's behavior. Still, the appropriate company in one season may be deadly in another. During the breeding season, for instance, tempers can flare between two dominant males or among females if there is *not* a male present.

Whenever a new individual joins the group, it usually has to vie for its place in the hierarchy, giving us some interesting behavior watching before it's all over. Keepers must monitor these interactions carefully, however, since undue stress is not good for any of the animals involved. One antidote is to provide inferior animals with visual screens or places where they can hide from their dominant exhibit-mates. Without these screens, the constant threat of attack might sicken or even kill a sub-dominant animal.

ARE THE ANIMALS SPIRITED?

Part of the beauty of a wild animal is its indomitable ability to adapt to feasts and famines, dangers and opportunities in wild existence. In the zoo, where we attend to the animal's every need, the spikes and troughs of its life are not nearly as deep or challenging. The least we can do when we remove an animal from the wild is to find fair substitutes for these stimuli. Zoo animals that confront a sterile, unchanging world day in and day out are the most likely to develop abnormal behaviors.

Many animals, especially the more intelligent ones, need novelty to keep their minds sharp and their spirits healthy. One of the main benefits of elephant, dolphin, and whale shows, for instance, is the mental exercise that it gives these brainy mammals. Birds also like stimuli and will spend hours playing with wooden balls and other objects. Primates have perhaps the greatest craving for novelty and stimulation. To make an old exhibit new, some zoos change the rope patterns regularly, giving swinging apes a chance to devise new routes.

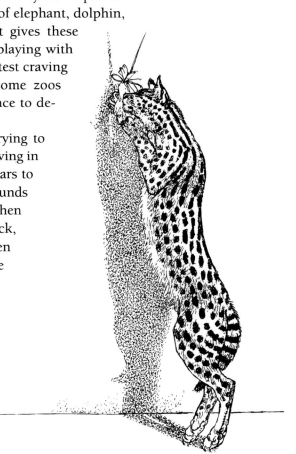

Other zoos are experimenting with new ways of feeding, trying to capitalize on the hundreds of different ways animals make their living in the wild. Putting honey inside a log, for instance, encourages bears to dig out a "hive" as they would in the wild. Stuffing termite mounds full of food encourages chimps to fish the food out with sticks. When sea otters are given clams and rocks, they will float on their back, place one rock on their chest, and use a second rock to smash open a clam. Meat suspended from a string will remind large cats of the effort it takes to haul down live prey, while primates will spend hours searching for food that is scattered in the grass. Research at these self-feeding exhibits has shown that animals would rather *work* for their food, even when a pile of free food is standing by. Like us, they may feel more in control if they can see things change because of their efforts. The challenge facing zookeepers is to keep these opportunities coming, promoting a wellness that goes beyond simple nutrition and sanitation. As zoo visitors, we gain by getting to watch what we used to have to read about: tigers pouncing, otters diving for fish, beavers building a lodge, and penguins causing a ruckus at the nest.

When the serval's food is served on a swinging rope, mealtimes are more active affairs.

The following Do-It-Yourself Exhibit Critique serves as a checklist of exhibit components that may or may not be vital to the animals you are watching. The list will at least give you questions you can bounce off the nearest zookeeper. Your concern may inspire the keeper to do a better job of providing what the animal needs.

DO-IT-YOURSELF EXHIBIT CRITIQUE

ENCLOSURE DESIGN

Do the exhibits have the following elements, if appropriate to the animal? To gauge appropriateness, you need to understand the biology of the animal. When in doubt, ask a zookeeper.

Space	enough for comfort, with a generous flight distance from humans and other animals
Sanitation	clean, but not scent-free
Lighting	including sun or heat lamps to sunbathe in
Temperature and humidity	appropriate to the species; not too hot, too dry, too cool, or too wet
Nature of ground	soft for animals that dig, hard and abrasive for those that must wear down their hoofs
Dust or sand	for animals that like to take dust baths
Water	to feed in, play in, bathe in, or build a dam or lodge in
A place to haul out of the water	even aquatic animals need to rest sometimes
Plants or substitutes that can provide	1. *food:* such as leaves, seeds, bark, roots, stems 2. *support:* for climbing, shelter, roosting, nesting 3. *abrasive material:* for sharpening incisors, tusks, beaks, claws 4. *comfort:* for rubbing the skin 5. *signal posts:* for leaving scent marks 6. *stimulus substitutes:* for redirecting aggression 7. *playthings:* for mental and physical stimulation 8. *tools:* for acquiring food, fanning away insects 9. *building materials:* for nesting 10. *cleansers:* for removing velvet from antlers 11. *cover:* for hiding, reducing tension 12. *camouflage:* for concealing the body
Rubbing posts	for keeping horns, claws, and hides in shape
Climbing apparatus	for exercising or escaping
Social groupings	appropriate to the species; exhibit-mates of different species should also be compatible
Privacy	visual shields, hiding places
Challenging enrichments	playthings, food items presented in a new way, interaction with trainer

EDUCATIONAL VALUE	
Authenticity	Is the exhibit a valid representation of the habitat from the animal's perspective (as closely as we can determine)? Human perceptions, though not as important as the animal's, also come into play. Does the exhibit make the hair on the back of your neck stand up? Do you feel like you are *in* the habitat, unsure of whether the animals could reach you?
Display signs	Do they entice you to read them, and do they answer some questions while prompting others?
Shows	Do they showcase the natural behaviors of the species without degrading or humanizing the animal?
Zoo personnel	Are they willing to answer your questions? Are they knowledgeable, and do they display obvious enthusiasm and respect for the animals?
EFFECTS ON THE ANIMALS	
Here again, it's important to know the biology of the animal. Your zookeeper can help.	
Natural behaviors	Are the animals exhibiting natural behaviors?
Stereotypic behaviors	Are they free of stereotypic behaviors?
Health	Do they seem healthy?
Longevity	Are they living long lives?
Captive breeding	Are they reproducing?

If you can answer yes to the questions in the Exhibit Critique, it sounds as if your zoo is trying hard. It may still need improvements in areas, but on the whole it looks like a reputable institution. If you answered no to any of the questions, your zoo may need an overhaul. Consult the following list to see what you can do to make it happen.

HOW YOU CAN IMPROVE YOUR ZOO

IF YOUR ZOO IS DOING A POOR JOB

1. **Ask hard questions**. There is no reason animals have to live in dangerous, boring, or frightening environments. Animals can be emotionally and physically comfortable in captivity. If the animals at your zoo are not, ask the director why this is so, and don't give up until you get answers.

2. **Complain to the funding agency**. In the case of a municipal or state zoo, contact the city or state officials, members of the legislature, the mayor's office, the parks department, the local animal control agencies, or the local chamber of commerce. If your zoo depends on private endowments, present yourself to the foundation's board of directors or to the individual donors. People who contribute generously to what they believe is a good cause just might be interested to hear what you have to say.

3. **Contact USDA APHIS**. The Animal and Plant Health Inspection Service (APHIS) has set minimum standards for anyone who exhibits animals. These

standards, established by the Animal Welfare Act, cover the basics of food, water, light, and so on. If your zoo is lacking in even the bare essentials, APHIS has the power to put them on probation or even to close them down. Keep in mind, however, that inspectors can interpret the law liberally and may not demand that the zoo attend to the finer points of social comfort we have discussed. If you've taken the federal law as far as it will go and you're still uneasy, contact an animal rights organization such as the Humane Society of the United States.

4. **Inform the Humane Society**. The Humane Society is dedicated to preventing cruelty to animals, including zoo animals. They have an investigations branch that can follow up on your complaints. Besides conducting inspections, they have the means to drum up negative publicity and put pressure on the powers that be.

5. **Influence public opinion**. You can influence public opinion yourself by speaking at town meetings, writing letters to the editor, circulating petitions, or inviting reporters to see what you've seen. Since the public accounts for most of the ticket sales, zoos can't afford much negative press.

6. **Don't give up**. Remember the moral of the tuna and the dolphin. The tuna industry in the United States recently did an about-face and started fishing in ways that wouldn't harm dolphins. Evidently, it wasn't just the economics of the tuna boycott that caused them to comply; it was the constant, irrepressible bombardment of sincere letters, legislative appeals, and negative publicity that finally got to them. Complaining until you are blue in the face just may work. Don't give up!

HELPING YOUR ZOO KEEP UP THE GOOD WORK

1. **Become a friend of the zoo**. Friends-of-the-zoo organizations allow you to contribute money and/or time to improving your zoo. Some offer a program through which you can adopt a particular animal and pay to feed it for a month or a year. You can also help by volunteering as a docent, a person who leads tours of the zoo or stands by exhibits to answer questions. Volunteering allows you to indulge that secret wish to get behind the scenes at the zoo. You'll see the animals off-hours (when some of them are quite busy) and be able to watch zoo professionals care for them. The more friends a zoo has, the better equipped it is to educate visitors.

2. **Help raise money for the good zoos**. P. T. Barnum maintained that if you want to tell people your story, you first have to get them in under the tent. Zoos have an important story to tell, but they are competing nowadays with the likes of Disney World and Great Adventure amusement parks for the public's attention. Unlike museums and dance companies that can be federally funded through the National Endowment for the Arts, most zoos rely solely on grants from local governments, private donations, and ticket sales. Naturally, it costs a bit more to house and feed a black rhinoceros than it does to shelter and dust a painting for a year. One would think that the world's last golden-lion tamarins were at least as precious as a Rembrandt, but the allocation priorities do not always come out in

favor of the animals. Your efforts to raise funds, therefore, and to get more people under the tent would be wildly appreciated.

3. **Make sure some of the money is going toward habitat conservation**. Many organizations are jumping on the conservation bandwagon, but you should make sure that your zoo is putting its money where its publicity is. Some of the dollars you raise should be going toward the ultimate solution: stemming the flood of habitat destruction. Conservation breeding is admirable work, but if we have no intact habitats to reintroduce species to, it will become a tragically moot point.

FURTHER READING

Studying animal behavior is delightfully addictive. If you find yourself craving more information about the animals in this book, check out the following sources. If you get the urge to delve deeper, your librarian can lead you to dozens of research journals and natural history magazines that describe ongoing studies. Browse through a few of these each month (they're not too difficult to read), and you can keep up with the continuing quest to learn what makes animals—including ourselves—tick. Happy exploring!

WHAT'S NEW WITH ZOOS?

Arrandale, Tom. 1987. *Zoo Renaissance.* Washington, DC: Congressional Quarterly.

Benginer, Robert. 1981. *The Fall of the Wild, The Rise of the Zoo.* New York: E.P. Dutton.

Bruns, Bill. 1983. *A World of Animals: The San Diego Zoo and Wild Animal Park.* New York: Henry N. Abrams, Inc.

Campbell, Sheldon. 1978. *Lifeboats to Ararat.* New York: McGraw-Hill.

Cherfas, Jeremy. 1984. *Zoo 2000—A Look Behind the Bars.* London: Parkwest Publications.

Diolé, Philippe. 1974. *The Errant Ark: Man's Relationship with Animals.* New York: Putnam.

Elspeth, Joscelin Grant. 1981. *Whipsnade: Captive Breeding for Survival.* London: Collins.

Gibbons, Gail. 1987. *Zoo.* New York: Crowell.

Gold, Don. 1988. *Zoo: A Behind-the-Scenes Look at the Animals and the People Who Care for Them.* Chicago: Contemporary Books.

Hancocks, David. 1971. *Animals and Architecture.* New York: Praeger.

Hediger, Heini. 1955. *Studies of the Psychology and Behaviour of Captive Animals in Zoos and Circuses.* London: Butterworths Scientific Publications.

————. 1964. *Wild Animals in Captivity.* New York: Dover Publications.

————. 1969. *Man and Animal in the Zoo: Zoo Biology.* New York: Delacorte Press.

Johnson, James Ralph. 1971. *Zoos of Today.* New York: D. McKay Co.

Livingston, Bernard. 1974. *Zoo: Animals, People, Places.* New York: Arbor House.

Luoma, Jon. 1987. *A Crowded Ark: The Role of Zoos in Wildlife Conservation.* Boston: Houghton Mifflin.

Markowitz, Hal, and Victor Stevens, eds. 1978. *Behavior of Captive Wild Animals.* Chicago: Nelson-Hal.

Markowitz, Hal. 1982. *Behavioral Enrichment in the Zoo.* New York: Van Nostrand Reinhold Co.

McKenna, Virginia, Will Travers, and Jonathan Wray, eds. 1987. *Beyond the Bars: The Zoo Dilemma.* Rochester, Vt.: Thorsons Publishing Group.

Meyer, Alfred P., ed. 1979. *A Zoo for All Seasons.* Washington, DC: Smithsonian Institution.

Meyer-Holzapfel, Monica. 1968. *Abnormal behavior in zoo animals.* In *Abnormal Behavior in Animals,* ed. Michael W. Fox. Philadelphia: Saunders.

Morris, Desmond. 1964. The response of animals to a restricted environment. *Symposium of the Zoological Society of London* 13: 99–118.

Mullan, Bob. 1987. *Zoo Culture*. London: Weidenfeld & Nicolson.

Page, Jake, and Franz Maier. 1990. *Zoo—The Modern Ark*. New York: Facts on File.

Paige, David. 1978. *Behind the Scenes at the Zoo*. Chicago: A. Whitman.

Polakowski, Kenneth J. 1987. *Zoo Design: The Reality of Wild Illusions*. Ann Arbor: University of Michigan, School of Natural Resources.

Rensberger, Boyce. 1977. *Cult of the Wild*. New York: Anchor Press, Doubleday.

Scott, Jack Denton. 1978. *City of Birds and Beasts: Behind the Scenes at the Bronx Zoo*. New York: Putnam.

Stewart, Alva W. 1988. *The Renaissance of American Zoos: A Brief Checklist*. Monticello, Ill.: Vance Bibliographies.

Stott, R. Jeffrey. 1981. *The American Idea of a Zoological Park: An Intellectual History*. Ph.D dissertation. University of California, Santa Barbara.

Tongren, Sally. 1985. *To Keep Them Alive: Wild Animal Breeding*. New York: Dembner Books.

Tudge, Colin. 1992. *Last Animals at the Zoo: How Mass Extinction Can Be Stopped*. Washington, DC: Island Press.

White, Anthony G. 1983. *Zoos and Zoological Facilities: A Selected Bibliography*. Monticello, Ill.: Vance Bibliographies.

Wilson, Edward O. 1984. *Biophilia: The Human Bond with Other Species*. Cambridge: Harvard University Press.

Woodroffe, Gordon. 1981. *Wildlife Conservation and the Modern Zoo*. Hindhead, Surrey, England: Saiga Publishing Co.

Zuckerman, Lord, ed. 1980. *Great Zoos of the World: Their Origins and Significance*. Boulder, Colo.: Westview Press.

HOW ANIMALS BEHAVE: A PRIMER

Alcock, John. 1989. *Animal Behavior: An Evolutionary Approach*. 4th ed. Sunderland, Mass: Sinauer Associates.

Attenborough, David. 1990. *The Trials of Life: A Natural History of Animal Behavior*. Boston: Little, Brown.

Barash, David P. 1982. *Sociobiology and Behavior*. 2nd ed. New York: Elsevier.

Barnett, Samuel A. 1981. *Modern Ethology: The Science of Animal Behavior*. New York: Oxford University Press.

Beck, Benjamin B. 1980. *Animal Tool Behavior: The Use and Manufacture of Tools by Animals*. New York: Garland STPM Publishers.

Bekoff, Marc, and Dale Jamieson, eds. 1990. *Interpretation and Explanation in the Study of Animal Behavior*. Boulder, Colo.: Westview Press.

Bell, William J. 1991. *Searching Behaviour: The Behavioural Ecology of Finding Resources*. London and New York: Chapman and Hall.

Birkhead, Mike. 1990. *The Survival Factor*. New York: Facts on File.

Bonner, John Tyler. 1980. *The Evolution of Culture in Animals*. Princeton, N.J.: Princeton University Press.

Bright, Michael. 1984. *Animal Language*. London: British Broadcasting Corporation.

Brown, Vinson. 1987. *The Secret Languages of Animals*. Englewood Cliffs, N.J.: Prentice Hall.

Burton, Maurice, and Robert Burton. 1977. *Inside the Animal World: An Encyclopedia of Animal Behavior*. New York: Quadrangle/New York Times Book Co.

Burton, Maurice. 1978. *Just Like an Animal*. New York: Scribner's.

Catchpole, Clive. 1979. *The Animal Family*. New York: Norton.

Chauvin, Rémy. 1968. *Animal Societies from the Bee to the Gorilla*. New York: Hill and Wang.

Cohen, Daniel. 1971. *Watchers in the Wild: The New Science of Ethology*. 1st ed. Boston: Little, Brown.

Cousteau, Jacques Yves. 1973. *Invisible Messages*. New York: Danbury Press.

David, McFarland, ed. 1982. *The Oxford Companion to Animal Behavior*. Oxford and New York: Oxford University Press.

Dawkins, Marion S., Tim R. Halliday, and Richard Dawkins. 1991. *The Tinbergen Legacy*. London and New York: Chapman and Hall.

Dawkins, Marion Stamp. 1986. *Unravelling Animal Behavior*. Harlow, Essex, England: Longman.

Dawkins, Richard. 1976. *The Selfish Gene*. New York: Oxford University Press.

Dewsbury, Donald A. 1978. *Comparative Animal Behavior*. New York: McGraw-Hill.

Drickamer, Lee C., and Stephen H. Vessey. 1986. *Animal Behavior: Concepts, Processes, and Methods*. Boston: Prindle, Weber, & Schmidt.

Ehrlich, Paul R. *The Machinery of Nature*. 1986. New York: Simon and Schuster.

Elia, Irene. 1988. *The Female Animal*. New York: Henry Holt.

Ellis, Derek V. 1985. *Animal Behavior and Its Applications.* Chelsea, Mich.: Lewis Publishers, Inc.

Evans, Peter. 1987. *Ourselves and Other Animals.* New York: Pantheon Books.

Fagen, Robert. 1981. *Animal Play Behavior.* New York: Oxford University Press.

Ferry, Georgina, ed. 1984. *The Understanding of Animals.* Oxford: Basil Blackwell Ltd.

Forsyth, Adrian. 1986. *A Natural History of Sex: The Ecology and Evolution of Sexual Behavior.* New York: Scribner's.

Gould, James L. 1982. *Ethology: The Mechanisms and Evolution of Behavior.* New York: Norton.

Grier, James W. 1991. *The Biology of Animal Behavior.* 2nd ed. St. Louis: Mosby.

Griffin, Donald R. 1984. *Animal Thinking.* Cambridge: Harvard University Press.

Grzimek, Bernhard, ed. 1977. *Grzimek's Encyclopedia of Ethology.* New York: Van Nostrand Reinhold.

Hailman, Jack. 1977. *Optical Signals: Animal Communication and Light.* Bloomington: Indiana University Press.

Halliday, Tim R., and Peter J. B. Slater, eds. 1983. Communication. Vol. 2, *Animal Behaviour.* London: Blackwell Scientific Publications.

Harre, Rom, and Roger Lamb, eds. 1986. *The Dictionary of Ethology and Animal Learning.* Cambridge: MIT Press.

Hess, Eckhard H., and B. Slobodan, eds. 1977. *Imprinting.* Stroudsburg, Pa.: Dowden, Hutchinson, and Ross.

Hinde, Robert A., ed. 1972. *Non-Verbal Communication.* Cambridge and London: Cambridge University Press.

———. 1982. *Ethology: Its Nature and Relations with Other Sciences.* Oxford and London: Oxford University Press.

Honore, Erika K., and Peter H. Knopler. 1990. *A Concise Survey of Animal Behavior.* San Diego: Academic Press.

Johnsgard, Paul A. 1967. *Animal Behavior.* Dubuque, Iowa: W.C. Brown Co.

Johnson, Cecil E., ed. 1970. *Contemporary Readings in Behavior.* New York: McGraw-Hill.

Kerker, Ann E., ed. 1975. *A Selected List of Source Material on Behavior with Emphasis on Animal Behavior.* West Lafayette, Ind.: Veterinary Medical Library, Purdue University.

Kevles, Bettyann. 1986. *Females of the Species: Sex and Survival in the Animal Kingdom.* Cambridge, Mass.: Harvard University Press.

Krames, Lester, Patricia Pilner, and Thomas Alloway, eds. 1975. *Nonverbal Communication of Aggression. Vol 1: Advances in the Study of Communication and Affect.* New York: Plenum Press.

Krebs, John R., and Nicholas B. Davies, eds. 1991. *Behavioral Ecology: An Evolutionary Approach.* 3rd ed. Oxford: Blackwell Scientific Publications.

Lorenz, Konrad. 1982. *The Foundations of Ethology: The Principal Ideas and Discoveries in Animal Behavior.* New York: Simon and Schuster.

Monaghan, Pat, and David Wood-Gush, eds. 1990. *Managing the Behaviour of Animals.* London and New York: Chapman and Hall.

Morris, Desmond. 1990. *Animalwatching.* New York: Crown Publishers.

National Geographic Society. 1972. *The Marvels of Animal Behavior.* Washington, DC: National Geographic Society.

Robinson, David. 1980. *Living Wild: The Secrets of Animal Survival.* Washington, DC: National Wildlife Federation.

Robinson, Michael H., and Lionel Tiger, eds. 1991. *Man & Beast Revisited.* Washington, DC: Smithsonian Institution Press.

Schein, Martin W., ed. 1975. *Social Hierarchy and Dominance.* Stroudsburg, Pa.: Dowden, Hutchinson, and Ross.

Sebeok, Thomas, ed. 1977. *How Animals Communicate.* Bloomington: Indiana University Press.

Shaw, Evelyn, and Joan Darling. 1985. *Female Strategies.* London: Walker and Co.

Slater, Peter J., ed. 1987. *Encyclopedia of Animal Behavior.* New York: Facts on File.

Smith, W. John. 1977. *The Behavior of Communicating: An Ethological Approach.* Cambridge: Harvard University Press.

Sparks, John. 1982. *The Discovery of Animal Behaviour.* Boston: Little Brown.

Stephens, Gretchen. 1982. *Animal Ethology Bibliography.* West Lafayette, Ind.: School of Veterinary Medicine, Purdue University.

Walters, Mark Jerome. 1988. *The Dance of Life: Courtship in the Animal Kingdom.* New York: William Morrow.

GENERAL MAMMAL BOOKS

Brown, Richard E., and David W. MacDonald. 1985. *Social Odours in Mammals.* Oxford: Clarendon Press.

Burton, Maurice. 1975. *How Mammals Live.* London: Elsevier-Phaidon.

Dewsbury, Donald A., ed. 1981. *Mammalian Sexual Behavior: Foundations for Contemporary Research.* Stroudsburg, Pa.: Hutchinson Ross Pub. Co.

Estes, Richard Despard. 1991. *The Behavior Guide to African Mammals: Including Hoofed Mammals, Carnivores, Primates.* Berkeley: University of California Press.

Ewer, R. F. 1968. *Ethology of Mammals.* London: Logos Press.

Grzimek, Bernhard, ed. 1990. *Grzimek's Encyclopedia of Mammals.* 5 vols. New York: McGraw-Hill.

Gubernick, David J., and Peter H. Klopher, eds. 1981. *Parental Care in Mammals.* New York: Plenum Press.

Kingdon, Jonathan. 1971. *East African Mammals: An Atlas of Evolution in Africa.* London and New York: Academic Press.

Leuthold, Walter. 1977. *African Ungulates: A Comparative Review of Their Ethology and Behavioral Ecology.* Berlin: Springer-Verlag.

MacDonald, David W., ed. 1984. *The Encyclopedia of Mammals.* New York: Facts on File.

Moss, Cynthia. 1982. *Portraits in the Wild: Behavior Studies of East African Mammals.* 2nd ed. Chicago: University of Chicago Press.

Nowak, Ronald M., and Ernest P. Walker. 1991. *Walker's Mammals of the World.* 5th ed. Baltimore: Johns Hopkins University Press.

Peters, Roger. 1980. *Mammalian Communication: A Behavioral Analysis of Meaning.* Monterey, Calif.: Brooks/Cole Publishing Co.

Poole, Trevor B. 1985. *Social Behaviour in Mammals.* Glasgow: Blackie.

Various authors. *Mammalian Species.* New York: The American Society of Mammalogists. (A series of short, definitive monographs about hundreds of species; use index to find articles.)

GENERAL BIRD BOOKS

Alcorn, Gordon Dee. 1991. *Birds and Their Young: Courtship, Nesting, Hatching, Fledging, the Reproductive Cycle.* Harrisburg, Pa.: Stackpole Books.

Brown, Leslie, Emil K. Urban, and Kenneth Newman, eds. 1982. *The Birds of Africa.* London and New York: Academic Press.

Catchpole, Clive. 1979. *Vocal Communication in Birds.* 2nd ed. Baltimore: University Park Press.

Cody, Martin L., ed. 1985. *Habitat Selection in Birds.* Orlando, Fla.: Academic Press.

Forsyth, Adrian. 1988. *The Nature of Birds.* Camden East, Ontario: Camden House.

Jellis, Rosemary. 1977. *Bird Sounds and Their Meaning.* London: British Broadcasting Corporation.

Perrins, Christopher and Alex L. A. Middleton, eds. 1985. *Encyclopedia of Birds.* New York: Facts on File.

Perrins, Christopher M. 1976. *Birds: Their Life, Their Ways, Their World.* New York: Abrams.

Ruppell, George. 1977. *Bird Flight.* New York: Van Nostrand Reinhold.

Silver, Rae, ed. 1977. *Parental Behavior in Birds.* Stroudsburg, Pa.: Dowden, Hutchinson, and Ross.

Skutch, Alexander Frank. 1989. *Birds Asleep.* Austin: University of Texas Press.

Welty, Joel C. 1975. *The Life of Birds.* 2nd ed. Philadelphia: Saunders.

GENERAL REPTILE BOOKS

Bellairs, A. d'A and C. B. Cox, eds. 1976. *Morphology and Biology of Reptiles.* London: Academic Press.

Carr, Archie Fairly. 1972. *The Reptiles.* New York: Time-Life Books.

Gans, Carl, and others. 1969-1982. *Biology of the Reptilia.* Vols. 1–13. London: Academic Press. A continuing series now published by Wiley, New York.

Goin, C. J., O. B. Goin, and G. R. Zug. 1978. *Introduction to Herpetology.* 3rd ed. San Francisco: W. H. Freeman and Co.

Halliday, Tim R., and Kraig Adler, eds. 1986. *The Encyclopedia of Reptiles and Amphibians.* New York: Facts on File.

Heatwole, Harold. 1976. *Reptile Ecology.* St. Lucia, Q. Queensland, Australia: University of Queensland Press.

Honders, John, ed. 1975. *The World of Reptiles and Amphibians.* New York: Peebles Press.

Huey, Raymond B., Eric R. Pianka, and Thomas W. Schoener, eds. *Lizard Ecology: Studies of a Model Organism.* Cambridge: Harvard University Press.

Mattison, Christopher. 1989. *Lizards of the World.* New York: Facts on File.

Richardson, Maurice Lane. 1972. *The Fascination of Reptiles.* New York: Hill and Wang.

Roth, Gerhard, Dr. 1987. *Visual Behavior in Salamanders.* Berlin and New York: Springer-Verlag.

Spellerberg, Ian F. 1982. *Biology of Reptiles: An Ecological Approach.* Glasgow: Blackie.

Tyning, Thomas F. 1990. *A Guide to Amphibians and Reptiles.* Boston: Little, Brown.

GORILLA

Altmann, Stuart A., ed. 1967. *Social Communication Among Primates.* Chicago: University of Chicago Press.

Box, Hilary O. 1984. *Primate Behaviour and Social Ecology.* London and New York: Chapman and Hall.

Cousins, Don. 1990. *The Magnificent Gorilla: The Life History of a Great Ape.* Sussex, England: Book Guild.

Dixson, Alan F. 1981. *The Natural History of the Gorilla.* New York: Columbia University Press.

Erwin, Joseph T., Terry L. Maple, and Gary D. Mitchell, eds. 1979. *Captivity and Behavior: Primates in Breeding Colonies, Laboratories, and Zoos.* New York: Van Nostrand Reinhold.

Fossey, Dian. 1983. *Gorillas in the Mist.* Boston: Houghton Mifflin.

Godwin, Sara, and Gerry Ellis. 1990. *Gorillas.* New York: Mallard Press.

Goodall, Alan. 1979. *The Wandering Gorillas.* London: Collins.

Groves, Colin P. 1970. *Gorillas.* New York: Arco.

Harre, Rom and Vernon Reynolds, eds. 1984. *The Meaning of Primate Signals.* Cambridge and New York: Cambridge University Press.

Hooff, Johan A. R. A. M. van. 1962. Facial expressions of the higher primates. *Symposium of the Zoological Society of London* 8: 97–125

Kavanagh, Michael. 1984. *A Complete Guide to Monkeys, Apes and Other Primates.* New York: Viking Press.

Maple, Terry L., and Michael P. Hoff. 1982. *Gorilla Behavior.* New York: Van Nostrand Reinhold.

Meyers, Susan. 1980. *The Truth About Gorillas.* New York: Dutton.

Nichols, Michael, George B. Schaller, and Nan Richardson. 1989. *Gorilla: Struggle for Survival in the Virungas.* New York: Aperture Foundation.

Patterson, Francine, and Eugene Linden. 1981. *The Education of Koko.* New York: Holt, Rinehart & Winston.

Schaller, George B. 1976. *The Mountain Gorilla: Ecology and Behavior.* Chicago: University of Chicago Press.

———. 1988. *The Year of the Gorilla.* Chicago: University of Chicago Press.

Tuttle, Russell. 1986. *Apes of the World: Their Social Behavior, Communication, Mentality, and Ecology.* Park Ridge, N.J.: Noyes Publications.

Williams, Jean Balch. 1985. *Behavioral Observations of Feral Gorillas: A Bibliography.* Seattle: Primate Information Center, Regional Primate Research Center, University of Washington.

Willoughby, David P. 1978. *All About Gorillas.* South Brunswick, N.J.: A. S. Barnes.

Zuckerman, Solly. 1981. *The Social Life of Monkeys and Apes,* 2nd ed. London and Boston: Routledge & K. Paul.

LION

Rundai, Judith A. 1973. *The Social Life of the Lion.* Wallingford, Penn: Washington Square East.

Schaller, George B. 1972. *Serengeti: A Kingdom of Predators.* New York: Alfred A. Knopf.

———. 1972. *The Serengeti Lion: A Study of Predator–Prey Relationships.* Chicago: University of Chicago Press.

Jackman, Brian. 1983. *The Marsh Lions.* Boston: David Godine Publishers.

AFRICAN ELEPHANT

Barbier, Edward B. 1990. *Elephants, Economics and Ivory.* London: Earthscan.

Buss, Irven O. 1990. *Elephant Life: Fifteen Years of High Population Density.* Ames: Iowa State University Press.

Cumming, D. H. M., R. F. Du Toit, and S. N. Stuart, eds. 1990. *African Elephants and Rhinos: Status Survey and Conservation Action Plan.* Gland, Switzerland: International Union for Conservation of Nature and Natural Resources.

DiSilvestro, Roger L. 1991. *The African Elephant: Twilight in Eden.* New York: Wiley.

Douglas-Hamilton, Iain. 1975. *Among the Elephants.* New York: Viking.

Eltringham, Stewart K. 1982. *Elephants.* The Blandford Mammal Series. Dorset: Blandford Press.

Freeman, Dan. 1981. *Elephants: The Vanishing Giants.* New York: Putnam.

Hanks, John. 1979. *A Struggle for Survival: The Elephant Problem.* London: Country Life Books.

Kunkel, Reinhard. 1982. *Elephants.* New York: Abrams.

Laws, Richard M, and I. S. C. Parker, R. C. B. Johnstone. 1975. *Elephants and Their Habitats: The Ecology of Elephants in North Bunyoro, Uganda.* Oxford: Clarendon Press.

Moss, Cynthia J., and Joyce H. Poole. 1983. Relationships and social structure in African elephants. In *Primate Social Relationships: An Integrated Approach,* ed. Hinde, Robert A., 315-25. Oxford: Blackwell.

Moss, Cynthia. 1988. *Elephant Memories: Thirteen Years in the Life of an Elephant Family.* New York: William Morrow.

Nasimovich, A. A. 1975. *The African Elephant.* Moskva: Nauka.

Orenstein, Ronald, ed. 1991. *Elephants: The Deciding Decade.* San Francisco: Sierra Club Books.

Sikes, Sylvia K. 1971. *The Natural History of the African Elephant.* New York: American Elsevier.

Thornton, Allan, and Dave Currey. 1991. *To Save An Elephant: The Undercover Investigation into the Illegal Ivory Trade.* London and New York: Doubleday.

PLAINS ZEBRA

MacClintock, Dorcas. 1976. *A Natural History of Zebras.* New York: Scribner's.

Waring, George. 1983. *Horse Behavior. Sexual Behavior of Stallions.* Park Ridge, N.J.: Noyes Publications.

BLACK RHINOCEROS

Balfour, Daryl. 1991. *Rhino.* London: New Holland.

Martin, Esmond, and Chryssee Bradley. 1982. *Run Rhino Run.* London: Chatto & Windus.

Martin, Esmond Bradley. 1983. *Rhino Exploitation: The Trade in Rhino Products in India, Indonesia, Malaysia, Burma, Japan and South Korea.* Hong Kong: World Wildlife Fund.

Penny, Malcolm. 1988. *Rhinos: Endangered Species.* New York: Facts on File.

Pink, Pattie, ed. 1988. *Zimbabwe: Last Stronghold of the World's Black Rhinoceros (Diceros bicornis).* Causeway: Zimbabwe National Conservation Trust, Rhino Survival Campaign.

Rookmaaker, L. C. 1983. *Bibliography of the Rhinoceros: An Analysis of the Literature on the Recent Rhinoceroses in Culture, History, and Biology.* Rotterdam: A.A. Balkema.

Schenkel, Rudolf. 1969. *Ecology and Behavior of the Black Rhinoceros.* Hamburg: Paul Parey.

U. S. Congress, H. Committee on Science, Space and Technology. Subcommittee on Natural Resources, Agriculture Research, and Environment. 1988. *Technologies for Conserving Species: Saving the Endangered Rhinoceros.* Washington, DC: GPO.

GIRAFFE

Dagg, Anne Innis, and J. Bristol Foster. 1976. *The Giraffe: Its Biology, Behavior, and Ecology.* New York: Van Nostrand Reinhold.

MacClintock, Dorcas. 1973. *The Natural History of Giraffes.* New York: Scribner's.

Spinage, Clive A. 1967. *The Book of the Giraffe.* Boston: Houghton Mifflin.

OSTRICH

Sauer, E. G. Franz and Eleonore M. Sauer. 1964. Social behaviour of the South African ostrich, *Struthio camelus australis. Proceedings of the Second Pan-African Ornithological Congress,* 1964, Pietermaritzburg, Natal. Ostrich. Supplement 6: 183-91.

GREATER FLAMINGO

Allen, Robert Porter. 1956. *The Flamingos: Their Life History and Survival,* Research Report no. 5. New York: National Audubon Society.

Brown, Leslie. 1973. *The Mystery of the Flamingos.* Nairobi: East African Pub. House.

Kear, Janet, and Nicole Duplaix-Hall, eds. 1975. *Flamingos.* Berkhamsted, England: T. & A.D. Poyser Ltd.

Ogilvie, Malcolm Alexander, and Carol Ogilvie. 1986. *Flamingos.* Gloucester: A. Sutton.

Olson, Storrs L., and Alan Feduccia. 1980. *Relationships and Evolution of Flamingos (Aves, Phoenicopteridae).* Washington, DC: Smithsonian Institution.

Sprunt, Alexander. 1988. The greater flamingo. In *Audubon Wildlife Report,* 1988–89, 553–64. New York: National Audubon Society.

NILE CROCODILE

Cott, Hugh B., and A. C. Pooley. 1972. *Crocodiles: The Status of Crocodiles in Africa.* IUCN Publication, New Series, Supplementary Paper no. 33. Morges, Switzerland: IUCN.

Guggisberg, Charles A. W. 1972. *Crocodiles: Their Natural History, Folklore and Conservation.* Harrisburg, Pa.: Stackpole Books.

International Union for Conservation of Nature and Natural Resources. 1990. *Crocodiles: Proceedings of the Tenth Working Meeting of the Crocodile Specialist Group of the Species Survival Commisson of the IUCN, The World Conservation Union, convened at Gainesville, Florida, 23 to 27 April 1990.* Gland, Switzerland: IUCN.

Neill, Wilfred. 1971. *The Last of the Ruling Reptiles: Alligators, Crocodiles, and Their Kin.* New York: Columbia Unversity Press.

Ross, Charles A., ed. 1989. *Crocodiles and Alligators.* New York: Facts on File.

Webb, Graham, S., Charlie Manolis, and Peter Whitehead, eds. 1987. *Wildlife Management: Crocodiles and Alligators.* Sydney: Surrey Beatty.

GIANT PANDA

British Museum (Natural History). 1975. *The Giant Panda: Ailuropoda melanoleuca.* British Museum (Natural History) Zoology leaflet, no. 3 Dorchester, England: The Museum.

Catton, Chris. 1990. *Pandas.* New York: Facts on File.

Jin, Xuqi. 1986. *The Giant Panda.* New York: Putnam.

Jing, Zhu, and Li Yangwen, eds. 1981. *The Giant Panda.* Beijing, China: Science Press.

McDearmon, Kay. 1986. *Giant Pandas.* New York: Dodd, Mead & Co.

Morris, Ramona, and Desmond Morris. 1981. *The Giant Panda.* New York: McGraw-Hill.

Roots, Clive. 1989. *The Bamboo Bears: The Life and Troubled Times of the Giant Panda.* Westport, Conn: Hyperion Press.

Schaller, George B., Hu Jinchu, Pan Wenshi, and Zhu Jing. 1985. *The Giant Pandas of Wolong.* Chicago: University of Chicago Press.

Taylor, David. 1990. *The Giant Panda.* London: Boxtree.

PEACOCK

Bergmann, Josef. 1980. *The Peafowl of the World.* Hindhead, England: Saiga Pub. Co.

Johnsgard, Paul, A. 1986. *The Pheasants of the World.* Oxford and New York: Oxford University Press.

Lavine, Sigmund A. 1982. *Wonders of Peacocks.* New York: Dodd, Mead & Co.

Martin, Lynne. 1975. *Peacocks.* New York: William Morrow & Company.

Mukherjee, Ajit Kumar. 1979. *Peacock, our National Bird.* New Delhi: Publications Division, Ministry of Information and Broadcasting, Govt. of India.

Thankappan, Nair, P. 1977. *The Peacock: The National Bird of India.* Calcutta: Firma KLM.

KOMODO MONITOR

Auffenberg, Walter. 1981. *The Behavioral Ecology of the Komodo Monitor.* Gainesville: University Presses of Florida.

Hopf, Alice Lightner. 1981. *Biography of a Komodo Dragon.* New York: Putnam.

Lutz, Richard L., and Judy Marie Lutz. 1991. *Komodo: The Living Dragon.* Salem, Oreg.: Dimi Press.

BOTTLENOSE DOLPHIN (*see also* Beluga Whale)

Andersen, Harald T., ed. 1969. *The Biology of Marine Mammals.* New York: Academic Press.

Baker, Mary. 1987. *Whales, Dolphins, and Porpoises of the World.* New York: Garden City.

Bonner, William Nigel. 1989. *Whales of the World.* New York: Facts on File.

Bryden, M.M., and Richard J. Harrison, eds. 1986. *Research on Dolphins.* Oxford: Clarendon Press.

Coffey, David J. 1977. *Dolphins, Whales, and Porpoises: An Encyclopedia of Sea Mammals.* New York: Macmillan.

Cousteau, Jacques Y., and Philippe Diolé. 1975. *Dolphins.* New York: Doubleday.

Doak, Wade. 1981. *Dolphin, Dolphin.* Dobbs Ferry, N.Y.: Sheridan House.

Donoghue, Michael, and Annie Wheeler. 1990. *Save the Dolphins.* Dobbs Ferry, N.Y.: Sheridan House.

Heintzelman, Donald S. 1981. *A World Guide to Whales, Dolphins, and Porpoises.* Tulsa, Okla.: Winchester Press.

Jouventin, Pierre, and A. Cornet. 1980. *Advances in the Study of Behavior,* Vol. 2, *The Sociobiology of Pinnipeds,* 121–141. New York: Academic Press.

Leatherwood, Stephen, and Randall R. Reeves, eds. 1990. *The Bottlenose Dolphin.* San Diego: Academic Press.

Lowell, Robert. 1973. *The Dolphin.* New York: Farrar, Straus & Giroux.

McIntyre, Joan, ed. 1974. *Mind in the Waters: A Book to Celebrate the Consciousness of Whales and Dolphins.* New York: Scribner's.

Norris, Kenneth. 1966. *Whales, Dolphins, and Porpoises.* Berkeley: University of California Press.

Perrin, W.F., ed. 1989. *Dolphins, Porpoises and Whales: An Action Plan for the Conservation of Biological Diversity, 1988–1992.* Gland, Switzerland: IUCN.

Pryor, Karen, and Kenneth S. Norris. 1991. *Dolphin Societies: Discoveries and Puzzles.* Berkeley: University of California Press.

Purves, P. E. and G.E. Purves. 1983. *Echolocation in Whales and Dolphins.* London and New York: Academic Press.

Robson, Frank D. 1988. *Pictures in the Dolphin Mind.* Dobbs Ferry, N.Y.: Sheridan House.

Sammons, V.O., ed. 1972. *Dolphins.* LC science tracer bullet; TB 72-15. Washington, DC: Library of Congress, Science and Technology Division, Reference Section.

Schusterman, Ronald J., Jeanette A. Schusterman, Forrest G. Wood, eds. 1986. *Dolphin Cognition and Behavior: A Comparative Approach.* Hillsdale, N. J.: Lawrence Erlbaum Associates.

Senate Select Committee on Animal Welfare. 1985. *Dolphins and Whales in Captivity.* The Parliament of the Commonwealth of Australia, Parliamentary Paper, no. 498/1985. Canberra: Australian Government Publication Service.

Shane, Susan H. 1988. *The Bottlenose Dolphin in the Wild.* Fenton, Calif.: Dolphin Books.

Truitt, Deborah. 1974. *Dolphins and Porpoises: A Comprehensive, Annotated Bibliography of the Smaller Cetacea.* Detroit: Gale Research Co.

Winn, Howard E., and Bori L. Olla , eds. 1979. *Behavior of Marine Animals: Current Perspectives in Research.* Volume 3: Cetaceans. New York: Plenum Press.

CALIFORNIA SEA LION

Aurioles-Gamboa, David. 1988. *Behavioral Ecology of California Sea Lions in the Gulf of California.* Ph.D. dissertation. University of California, Santa Cruz.

Bonner, William Nigel. 1990. *The Natural History of Seals.* New York: Facts on File.

Haley, Delphine, ed. 1986. *Marine Mammals of Eastern North Pacific and Arctic Waters.* 2nd ed. Seattle: Pacific Search Press.

Hanggi, Evelyn Betty. 1988. *Social Behavior and Kin Recognition in Captive California Sea Lions* (Zalophus californianus). MS. thesis. University of California, Santa Cruz.

Harrison, Richard J. and others, eds. 1968. *The Behavior and Physiology of Pinnipeds.* New York: Appleton-Century-Crofts.

Heath, Carolyn B. 1989. *The Behavioral Ecology of the California Sea Lion,* Zalophus californianus. Ph.D. dissertation. University of California, Santa Cruz.

Peterson, Richard S., and Bartholomew, George A. 1967. *The Natural History and Behavior of the California Sea Lion.* Stillwater, Okla.: American Society of Mammalogists. Special Publication no.1.

Ridgway, Sam H., and Sir Richard Harrison, eds. 1981. *Handbook of Marine Mammals. Vol. 1. The Walrus, Sea Lions, Fur Seals, and Sea Otter.* New York: Academic Press.

Riedman, Marianne. 1990. *The Pinnipeds: Seals, Sea Lions, and Walruses.* Berkeley: University of California Press.

GRAY WOLF

Bogan, Michael A., and Patricia Mehlhop. 1983. *Systematic Relationships of Gray Wolves* (Canis lupus) *in Southwestern North America.* Albuquerque: Museum of Southwestern Biology, the University of New Mexico.

Fox, Michael W. 1971. *Behavior of Wolves, Dogs and Related Canids.* New York: Harper & Row.

————. 1980. *The Soul of the Wolf.* Boston: Little, Brown.

Frank, Harry, ed. 1987. *Man and Wolf: Advances, Issues, and Problems in Captive Wolf Research.* Boston: Dr. W. Junk Publishers.

Gasaway, William C. 1983. *Interrelationships of Wolves, Prey, and Man in Interior Alaska.* Washington, DC: Wildlife Society.

Ginsberg, J. R., and David W. MacDonald, eds. 1990. *Foxes, Wolves, Jackals, and Dogs: An Action Plan for the Conservation of Canids.* Gland, Switzerland: IUCN.

Hall, Roberta, and Henry Sharp, eds. 1978. *Wolf and Man: Evolution in Parallel.* New York: Academic Press.

Harrington, Fred H., and Paul C. Paquet, eds. 1982. *Wolves of the World: Perspectives of Behavior, Ecology, and Conservation: Proceedings of the Portland International Wolf Symposium (1979).* Park Ridge, N.J.: Noyes Publications.

Hoffos, Allan Robin. 1987. *Wolf Management in British Columbia: The Public Controversy.* Victoria, B.C.: Wildlife Branch, Ministry of Environment and Parks.

Jolly, William C. 1976. *North American Wolves and Other Canids: A Perspective on Recent Issues of Congressional Concern.* Library of Congress. Washington, DC: Congressional Research Service, Library of Congress.

Klinghammer, Erich, ed. 1979. *The Behavior and Ecology of Wolves: Proceedings of the Symposium on the Behavior and Ecology of Wolves held on May 23–24, 1975 at the annual meeting of the Animal Behavior Society in Wilmington, N.C.* New York: Garland STPM Press.

Lawrence, R. D. 1986. *In Praise of Wolves.* New York: Henry Holt.

Mech, L. David. 1970. *The Wolf: The Ecology and Behavior of an Endangered Species.* Boston: The Natural History Press.

Morris, Desmond. 1986. *Dogwatching.* New York: Crown Publishers.

Peters, Roger. 1985. *Dance of the Wolves.* New York: McGraw-Hill.

Peterson, Rolf O. 1986. Gray wolf. In *Audubon Wildlife Report, 1986,* 951-967. New York: National Audubon Society.

Pimlott, Douglas Humphreys, J. A. Shannon, and George B. Kolenosky. 1977. *The Ecology of the Timber Wolf in Algonquin Provincial Park*. Ontario: Ministry of Natural Resources.

Savage, Candace. 1988. *Wolves*. Vancouver: Douglas & McIntyre.

Theberge, John B. 1975. *Wolves and Wilderness*. Toronto: Dent.

Turbak, Gary. 1987. *Twilight Hunters: Wolves, Coyotes & Foxes*. Flagstaff, Ariz.: Northland Press.

Zimen, Erik. 1981. *The Wolf: A Species in Danger*. New York: Delacorte Press.

BALD EAGLE

Barrie, Jack A., and Lon E. Lauber. 1990. *Sovereign Wings: The North American Bald Eagle*. Charlottesville, Va.: Thomasson-Grant.

Gerrard, Jonathan M., and Gary R. Bortolotti. 1988. *The Bald Eagle: Haunts and Habits of a Wilderness Monarch*. Washington DC: Smithsonian Institution Press.

Gordon, David G. 1991. *The Audubon Society Field Guide to the Bald Eagle*. Seattle: Sasquatch Books.

Green, Nancy, and R.L. Di Silvestro, eds. 1985. The bald eagle. In *Audubon Wildlife Report, 1985*, 508–31. Washington DC: National Audubon Society.

Hancock, David. 1970. *Adventure with Eagles*. Saanichton, B.C.: Wildlife Conservation Centre.

Lincer, Jeffrey L., William S. Clark Lincer, and Maurice N. LeFranc, Jr. 1979. *Working Bibliography of the Bald Eagle*. Washington, DC: Raptor Information Center, National Wildlife Federation.

Ryden, Hope. 1985. *America's Bald Eagle*. New York: Putnam.

Stalmaster, Mark V. 1987. *The Bald Eagle*. New York: Universe Books.

SANDHILL CRANE

Johnsgard, Paul A. 1981. *Those of the Gray Winds: The Sandhill Cranes*. New York: St. Martin's Press.

————. 1983. *Cranes of the World*. Bloomington: Indiana University Press.

————. 1991. Crane Music: *A Natural History of American Cranes*. Washington, DC: Smithsonian Institution Press.

Tacha, Thomas C. 1988. Social organization of sandhill cranes from midcontinental North America. *Wildlife Monograph* 99:1-37.

Walkinshaw, Lawrence H. 1949. *The Sandhill Cranes*. Bloomfield Hills, Mich.: Cranbrook Institute of Science.

Walkinshaw, Lawrence. 1973. *Cranes of the World*. New York: Winchester Press.

BELUGA WHALE (*see also* Bottlenose Dolphin)

Andersen, Harald T., ed. 1969. *The Biology of Marine Mammals*. New York: Academic Press.

Breton-Provencher, Mimi. 1982. *White Whales of the St. Lawrence River*. Sainte-Foy: Société Linneenne du Quebec.

Evans, Peter G. H. 1987. *The Natural History of Whales & Dolphins*. New York: Facts on File.

Gaskin, D. E. 1982. *The Ecology of Whales and Dolphins*. London: Heinemann.

Gatenby, Gred, ed. 1977. *Whale Sound*. North Vancouver, BC: J. J. Douglas.

Haley, Delphine, ed. 1986. *Marine Mammals of Eastern North Pacific and Arctic Waters*. 2nd ed. Seattle: Pacific Search Press.

Harrison, Sir Richard, M. M. Bryden, and Tony Pyrzakowski, eds. 1988. *Whales, Dolphins, and Porpoises*. New York: Facts on File.

Herman, Louis M., ed. 1980. *Cetacean Behavior: Mechanisms and Functions*. New York: John Wiley & Sons.

International Forum for the Future of the Beluga. 1990. *For the Future of the Beluga: Proceedings of the International Forum for the Future of the Beluga*. Sillery, Quebec: Presses de L'Université du Quebec.

Kleinenberg, Sergei E. and others. 1969. *Beluga* (Delphinapterus leucas): *Investigation of the Species*. Jerusalem: Israel Program for Scientific Translations.

Lockley, Ronald M. 1979. *Whales, Dolphins, and Porpoises*. Newton Abbot, England: David and Charles.

Martin, Anthony R. 1990. *The Illustrated Encyclopedia of Whales and Dolphins*. New York: Portland House.

Payne, Roger, ed. 1983. *Communication and Behavior of Whales*. Boulder, Colo.: Westview Press for the American Association for the Advancement of Science.

Plourde, Suzie., and Elizabeth Rooney. 1990. *The St. Lawrence and Its Belugas*. Sainte-Foy, Quebec: Société Linneenne du Quebec.

Ridgway, Sam H., and Sir Richard Harrison, eds. 1989. *Handbook of Marine Mammals*. Vol. 4, *River Dolphins and the Larger Toothed Whales*. New York: Academic Press.

Slijper, Everhard J., John Drury, trans. 1976. *Whales and Dolphins*. Ann Arbor: University of Michigan Press.

Smith, Thomas G., and Michael O. Hammill. 1990. *A Bibliography of the White Whale,* Delphinapterus leucas. Ste. Anne de Bellevue, Quebec: Dept. of Fisheries and Oceans, Arctic Biological Station.

Smithsonian Books. 1988. *The Magnificent Whales.* New York: Smithsonian Books.

Watson, Lyall. 1985. *Whales of the World: A Complete Guide to the World's Living Whales, Dolphins, and Porpoises.* London: Hutchinson.

POLAR BEAR

Amstrup, Steven C. 1986. Polar bear. In *Audubon Wildlife Report, 1986,* 791–804. New York: National Audubon Society.

Banfield, Alexander W. 1974. *The Mammals of Canada.* Toronto: University of Toronto Press.

Bear Biology Association. 1976. Behavioral aspects of the polar bear. In *Bears: Their Biology and Management: Proceedings of the Third International Conference on Bear Research and Management.* Binghamton, New York and Moscow, Russia. IUCN Publications no. 40. Morges, Switzerland: IUCN.

Davids, Richard C. 1982. *Lords of the Arctic: A Journey Among the Polar Bears.* New York: Macmillan.

Feazel, Charles T. 1990. *White Bear: Encounters with the Master of the Arctic Ice.* New York: Henry Holt.

Jonkel, Charles J. 1970. *Polar Bear Research in Canada: Proceedings of the Conference on Productivity and Conservation in Northern Circumpolar Lands.* IUCN Publication, new series, no. 16: 15-54. Gland, Switzerland: IUCN.

Larsen, Thor. 1978. *The World of the Polar Bear.* Condon, NY: Hamlyn.

Nero, Robert W. 1971. *The Great White Bears.* Winnepeg: Dept. of Mines, Resources and Environmental Management.

Perry, Richard. 1967. *The World of the Polar Bear.* Seattle: University of Washington Press.

Stirling, Ian, Charles Jonkel, P. Smith, P. Robertson, and D. Cross. 1977. *The Ecology of the Polar Bear Along the Western Coast of Hudson Bay.* Canadian Wildlife Service Occasional Paper 33:1-69.

Stirling, Ian. 1990. *Polar Bears.* Ann Arbor: University of Michigan Press.

ADÉLIE PENGUIN

Ainley, David G., Robert E. LeResche, and William J. L. Sladen. 1983. *Breeding Biology of the Adélie Penguin.* Berkeley: University of California Press.

Austin, Oliver Luther, ed. 1968. *Antarctic Bird Studies.* Washington, DC: American Geophysical Union of the National Academy of Sciences–National Research Council.

Davis, Lloyd S., and John T. Darby, eds. 1990. *Penguin Biology.* San Diego: Academic Press.

Ensor, Paul H., and Jennifer A. Bassett. 1987. *The Breeding Status of Adélie Penguins and Other Birds on the Coast of George V Land, Antarctica.* Kingston, Tas.: Antarctic Division, Department of Science.

Gorman, James. 1990. *The Total Penguin.* Englewood Cliffs, N.J.: Prentice Hall.

Jouventin, Pierre. 1982. *Visual and Vocal Signals in Penguins, Their Evolution and Adaptive Characters.* Berlin and Hamburg: P. Parey.

Lloyd S. Davis, and John T. Darby, eds. 1990. *Penguin Biology.* San Diego: Academic Press.

Müller-Schwarze, Dietland. 1984. *The Behavior of Penguins: Adapted to Ice and Tropics.* Albany, N.Y.: State University of New York Press.

Peterson, Roger Tory. 1979. *Penguins.* Boston: Houghton Mifflin.

Simpson, George Gaylord. 1976. *Penguins: Past and Present, Here and There.* New Haven: Yale University Press.

Sparks, John, and Tony Soper. 1987. *Penguins.* New York: Facts on File.

Stonehouse, Bernard, ed. 1975. *The Biology of Penguins.* Baltimore: University Park Press.

Thompson, David Hugh. 1974. *The Penguin: Its Life Cycle.* New York: Sterling Publishing Company.

INDEX

Mutual displays of Adélie penguins, 330, *330*, 331

Narial geysering, 184
National Zoo, 11, 15, 212
Natural selection
    adaptations and, 19–22, 24, 50
    aggressive behavior and, 47–48
    communication and, 40
    genes and, 19–21, 50
Necking contests of giraffes, 147–148, *148*, 149, 152
Nests
    of Adélie penguins, 331
    of bald eagles, 281
    building for young, 58–59, *58*
    danger from predators, 61–62
    of gorillas, 74, *74*
    of greater flamingos, 175
    group, 38, *38*
    of Nile crocodiles, 186–187
    of peacocks, 213
    of sandhill cranes, 293
    sleeping, 74, *74*
    taking shelter in, 31
Nile crocodiles, 178–189
    behaviors of, 189t
    body temperature regulation by, 30, 182, *182*
    breathing of, 181–182
    communal defense of, 188
    conflict behavior of, *178*, 183–184, *183*
    cooperative hunting of, 183
    dominance hierarchy of, 183–184
    feeding behavior of, 180–181, *181*, 183
    friendly behavior of, 183
    humans and, 185
    hunting behavior of, 181, *181*, 183
    locomotion of, 180, *180*
    nests of, 186–187
    parenting behavior of, 63, 179–180, 186–188, *187*
    predators and, 187
    sexual behavior of, 184–186, *186*
    social behaviors of, 37, 179–180, 182–188, 189t
    sounds made by, 184, 186
Nomads, 86
Nonadaptive behavior, 20
Nonbehavioral messages, 41
Nursery building, 58–59. See also Nests; Parenting behavior
Nursing
    of African elephants, *114*
    of beluga whales, 307
    of bottlenose dolphins, 239
    of California sea lions, 252
    of gray wolves, 269
    of lions, *84*
    of polar bears, 313, 317
    See also Parenting behavior

Odors, communication through. *See*

Scent marking
Offspring. *See* Parenting behavior
Opossums, 35, 57
Opportunistic feeders, 25, 64
Opposites, principle of, 42, *42*
Orangutans, 42
Oryxes, *53*
Ostriches, 154–165
    behaviors of, 164t, 165
    conflict behavior of, 155, 159, *159*, *160*, 165
    dominance hierarchy of, 159, *159*, 160, *165*
    feathers of, 158
    feeding behavior of, 156, 162
    flocking of, 158–159
    friendly behavior of, 158–159
    grooming behavior of, 156–157
    humans and, 158
    locomotion of, *154*, 155, 156
    panting of, 157
    parenting behavior of, 63, 162–164, *162*, *163*
    predators and, *154*, 155, 156, 160, 162–163, 164
    rocking display of, *161*, 162
    sexual behavior of, 160–162, *161*
    sleeping behavior of, 157, *157*
    social behaviors of, 155, 158–164, 164t, *165*
    stretching behavior of, 157, *157*
    territoriality of, 160–161
    yawning behavior of, 34, 157–158
Otters, 47, 339
Overcrowding, 50
Owls, 35, *35*
Oxpeckers, 145

Pair bonds, 52
Pandas. *See* Giant pandas
Panthera leo. See Lions
Panting, 30
    of Adélie penguins, 324
    of ostriches, 157
    of polar bears, 312
    See also Body temperature regulation
Parenting behavior, 20–21, 57–63, *57*, *58*, *62*
    abnormal, 65
    of Adélie penguins, 57, 63, 324, 331–332, *331*, *332*
    of African elephants, 57, 59, 102, 106–107, 113–115, *114*
    of bald eagles, 61, 281–282, *281*
    of beluga whales, 59, 62, 306–307, *306*, *307*
    of black rhinoceros, *132*, 134, 140–141
    body temperature regulation, 61
    of bottlenose dolphins, 39, 59, 63, *235*, 238–239, *239*
    built-in shelters, 57–58, *57*
    of California sea lions, 251–252, *252*
    of chimpanzees, 20–21

communal. *See* Communal care of young
    feeding behavior and, 60–61
    gender differences in, 51–52
    of giant pandas, 201–202, *202*
    of giraffes, 59, 63, 150–151, *151*
    giving birth, 59
    of gorillas, *70*, 71, 81
    of gray wolves, 60, 269
    of greater flamingos, 63, 175–176, *176*
    grooming behavior and, 60, 81
    of Komodo monitors, 225, *225*
    of lions, 52, 61, *84*, 86, 95–97, 98
    nest building, 58–59, *58*
    of Nile crocodiles, 63, 179–180, 186–188, *187*
    of ostriches, 162–164, *162*, *163*
    of peacocks, 213, *214*
    of plains zebras, 60, 129–130
    of polar bears, 313, 317, *318*
    protection of young, 61–62
    of sandhill cranes, 293–294, *294*
    schooling, 62–63
    transportation of young, 62.
        See also Egg laying; Nests
Paterson, Francine, 72
Pavo cristatus and *Pavo muticus. See* Peacocks
Paw raising of gray wolves, 262, *262*
Paw swatting of giant pandas, 198, *198*
Payne, Katharine, 108
Peacocks, 204–215
    alarm call of, *204*, 209
    behaviors of, 215t
    body temperature regulation by, 208
    communal defense of, 209
    conflict behavior of, 209–210, *210*
    contact calls of, 209
    drinking behavior of, 207, *207*
    eyespots of, 212
    feathers of, 55, 205–206, 207–208, *208*, 210–211, *211*, 212
    feeding behavior of, 205, 207
    friendly behavior of, 209, *209*
    humans and, 206
    locomotion of, 207
    molting of, 208
    nests of, 213
    parenting behavior of, 213, *214*
    play of, 209
    predators and, 205–206, 209
    preening of, 207, *208*
    roosting of, 209, *209*
    sexual behavior of, 53, 54, 55, 206, 210–212, *211*
    sleeping behavior of, 32, 208
    social behaviors of, *204*, 205, 209–214, 215t
    strutting walk of, 207, 210–211
Pecking order, 48. *See also* Dominance hierarchy
Penguins. *See* Adélie penguins
Perch jostling of bald eagles, 277, 278